Deira

Elmet

York

R. Ouse

Roman
Ridge

Lindsey

Lincoln

ercia

Mercia

R. Trent

Watling St

Leicester

R. Welland

Middle

Anglia

R. Avon

Fosse Way

Great Ouse

Elmham

East
Anglia

Dommoc

Icknield Way

Dorchester

Thames

Icknield Way

Middle
Saxons

London

R. Thames

East
Saxons

East
Saxons

Rochester

Canterbury

Kent

Surrey

axons

Winchester

South Saxons

Icknield Way
Major Roman roads
Dykes
Land over 400 feet
Land over 800 feet
Fens

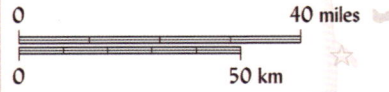

0                40 miles

0           50 km

# THE MERCIAN CHRONICLES

MAX ADAMS is the author of an acclaimed sequence of histories of Early Medieval Britain: *The King in the North*, *In the Land of Giants*, *Ælfred's Britain* and *The First Kingdom*. His other books include *Admiral Collingwood* and the bestselling *The Wisdom of Trees*. He has lived and worked in County Durham, in the North-East of England, since 1993.

ALSO BY MAX ADAMS

*The Museum of the Wood Age*
*The First Kingdom*
*Trees of Life*
*The Little Book of Planting Trees*
*Unquiet Women*
*Ælfred's Britain*
*In the Land of Giants*
*The Wisdom of Trees*
*The King in the North*
*Admiral Collingwood*

# THE MERCIAN CHRONICLES

## KING OFFA AND THE BIRTH OF THE ANGLO-SAXON STATE AD 630~918

## MAX ADAMS

HEAD of ZEUS

An Apollo Book

First published in the UK in 2025 by Head of Zeus Ltd,
part of Bloomsbury Publishing Plc

9 7 5 3 1 2 4 6 8

A catalogue record for this book is available from the British Library.

ISBN (HB): 9781838933258
ISBN (E): 9781838933272

Typeset by Ed Pickford

Printed and bound in Great Britain by
CPI Group (UK) Ltd, Croydon CR0 4YY

MIX
Paper | Supporting
responsible forestry
FSC
www.fsc.org   FSC® C171272

Bloomsbury Publishing Plc
50 Bedford Square, London, WC1B 3DP, UK
Bloomsbury Publishing Ireland Limited,
29 Earlsfort Terrace, Dublin 2, D02 AY28, Ireland

HEAD OF ZEUS LTD
5–8 Hardwick Street
London EC1R 4RG

To find out more about our authors and books
visit www.headofzeus.com

For product safety related questions contact productsafety@bloomsbury.com

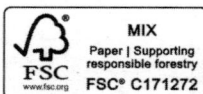

*For Julian 'Jules' Grey,*
*my Mercian brother*
*1970–2023*

# Contents

# List of Maps

For an explanation of the symbols used in these maps see key in endpapers.

# Forespœc

# The Mercian Chronicles

*The Mercian Chronicles*, like my other books about the Early Medieval period, is a portrait of power, how it was exercised at all levels from the humblest household right up to the sceptre-wielding *imperium* of the greatest warrior kings. That power was expressed tangibly in landscape and architecture, in weapons, jewellery, dress fittings, coins and pottery, in roads and harbours and in the written word. Through them we see how relations between people and their lords, and tensions between church and state, propelled post-Roman Britain towards its emergence as a heterogeneous, creative and dynamic force in the Middle Ages and beyond.

Why does Early Medieval power still matter? It matters because the rules by which power was exercised, established by King Oswald and Bishop Aidan, by Saints Wilfrid and Hild, by Archbishop Theodore, King Offa, King Ælfred and his Scandinavian foes, are the same rules by which politicians and corporations play today. If Bede's kings and ecclesiastics were sometimes brutally unsubtle, their behaviours are all the more revealing. Patronage, alliance, feud, othering, the invention of tradition, cultural appropriation and the creation of national identity are not twentieth- or twenty-first-century novelties but hand-me-down tools in an endless struggle for survival, social advantage and a sense of belonging – or not – to community. In

Europe and the Near East those rules were written down in the Old Testament by the scribes of the Peoples of the Book, whose stories have inspired and cautioned politicians of all shades ever since. During and after the profoundly significant Christian conversions of the seventh century, the English were assiduous students of those rules.

Power is also agency, and for those who are curious about where they have come from – and how – trying to piece together how ordinary people in these obscure centuries engaged with and affected their world is the most intriguing and fugitive of the past's many enigmas. Weaving techniques; building foundations; the design of coins; the naming of things, places and routes through the land; artefacts cast aside or deliberately buried; acts of violence and generosity, of spiritual hope and of cruel cynicism are all expressions of human agency. As the inheritors of their landscape, language, material culture and history, modern indigenes cannot escape their legacy.

There is still much to learn from the sharply reflected light cast on Early Medieval peoples by history and archaeology. They kept slaves, sold them; sometimes freed them. They punished offenders brutally. They allowed what seem like arbitrary superstitions to rule their decision making. They ruthlessly exploited political opportunity – careless, it seems, of human consequences. But they were also bound by strict moral codes; by strong ties of kinship and age-old loyalty; by custom, rank, diplomatic courtesy and a strong sense of propriety. Their poetry often speaks in a minor key expressing pain, sorrow, loss, affection and a love of irony. They loved puns and word-play, song and story. Their animalistic, organic but very shiny aesthetic shows a deep affinity with the natural world and with another, fantastical universe of great imaginative exuberance. Above all else they feared exile from family and homeland. Today's Mercians, Londoners, Scots, Welsh and West Saxons live in their shadows.

At first sight the tribal kings of the Dark Ages are no more than reactive thugs, dim shadows against which haloed holy men and women – Cuthbert, Brigit, Juthware, Guðlac, Colmcille

– shine as bright exemplars. Modern politicians and ecclesiastics are more civilised, more subtle, more tightly bound by diplomatic caution and international law. But they can also be unsubtle and ruthlessly thuggish, while their Early Medieval counterparts were likewise capable of deep thought, long perspective and considerable finesse.

These are political books. As an archaeologist I am interested in process – the 'how' more than the 'why'. As a historian the 'why' intrigues me too. It is not the place of either to judge, but to expose, explore and analyse insofar as the very difficult and uneven evidence allows. That does not mean that I am not, myself, political. My politics – perspectives and prejudices – are written in and between the lines of these stories, for those who care to detect them. When I watch the theatrical rhetoric of the contemporary debating chamber and the spin of partisan narratives; the machinations of a thriving patronage system; the great bullying the little, I think of *Beowulf* and of Bede's mead hall; of the momentous council at Gumley; of the punitive tributary status of lesser kings to their more powerful lords and of the finely wrought alliances of the charters. All the world's a stage; and a thousand-mile-long wooden 'O' bounded and connected by its seas and rivers, hills and routeways is a fitting cockpit in which to hear new life breathed into the oldest lines.

Remembering the names of all the kings, queens and bit-part players in these chronicles and retaining in the mind a sense of time, place and the endless ribbon of events is challenging. I have provided what I hope are useful aids to the reader: a Mercian genealogy and table of marriage alliances; brief timelines or chronographies of the most prominent events placed at strategic points in the book; and maps that are essential for grasping the geography. The test of such narratives is passed when – if – the reader casts the light reflected from such exotic forbears back onto themselves and asks, who is the stranger in this foreign country?

# A note on maps

The maps in this book were created using Adobe Illustrator and consist of more than fifty separate layers of information. The sources I have used are many and various – but special mention must be made of David Hill's 1981 *Atlas of Anglo-Saxon England*, now long out of date but still invaluable. The outstanding digital collection in the National Library of Scotland includes Ordnance Survey maps of Roman and Dark Age Britain, plus Land Utilisation Survey maps and much more.

Domesday Woodland was plotted using Wrathmell and Roberts' *Atlas of Rural Settlement in England*.[1] Navigable rivers were plotted using Eljas Oksanen's data.[2] The Droitwich Saltways were drawn using data compiled by Della Hooke.[3] Offa's and Wat's Dykes were plotted using information from Ray and Bapty's 2016 volume.

Any mistakes are of my own making.

# Introduction

# The Kalends of January

I n the winter of 633 King Cadwallon's army, imagined by later Welsh poets as champions of their people, the Cymry, set out from Gwynedd seeking vengeance in battle against a bitter foe, the Northumbrian King Edwin. With his Mercian ally, a heathen warlord named Penda, Cadwallon came to *Hæthfelth* – Hatfield Chase – in the dead flat, watery wilderness between the rivers Trent and Don, where land and sky meet beyond the edges of seemingly limitless horizons. A Welsh chronicle recorded bleakly that 'On the Kalends of January the battle of *Meigen*; and there Edwin was killed with his two sons.'[1] Bede, the Northumbrian monk writing almost a century after the event and assigning it a day in early October, recalled the battle at *Hæthfelth* as a catastrophe, describing how the whole army of the Northumbrians was either slain or scattered. In the seventeenth year of his reign, King Edwin, Christian overlord of all the southern English kingdoms, was cut down and beheaded on the field of battle. One of his half-Mercian sons, Osfrið,* was slain with him; the other, Eadfrið, was taken into Mercia and executed.

---

* The letter 'eth' symbolised with the character ð, capitalised as Ð, is used throughout where sometimes the inconsistent Old English sources use a softer 'th' sound represented by 'thorn' or Þ. The 'eth' sound is like the 'th' in 'then'. In linguistic terms, it is a 'voiced, dental non-sibilant fricative', represented in the International Phonetic Alphabet, appropriately, by the Old English ð'; 'thorn' is unvoiced, represented by Θ.

Those same heathen Mercians, led by Penda, were remembered with loathing in Northumbria, for in that year they also came to the royal township of *Cambodunum*, somewhere in the British-speaking province of Elmet,* and razed it to the ground, together with its church.[2] King Edwin's daughter and another son fled south to Kent and thence across the Channel into the Christian kingdom of the Franks, to their mother's kin. For a year, in Bede's version of events, Cadwallon rampaged through the lands north of the Humber, *be Norðanhymbre*, 'with bestial cruelty', bent on wiping the whole English nation from the face of the earth, while Northumbria lay undefended; ungoverned. Many years later, persistent visions led a priest to seek King Edwin's remains and retrieve them from the field of battle. They were taken to a royal monastery at Whitby on the Yorkshire coast, in Edwin's homeland of Deira, and venerated as the relics of a martyr.

The slaughter at *Hæthfelth* was the first serious check on Northumbrian power in the seventh century. It signalled Mercia's emergence as an independent kingdom and marked the historical debut of a series of Midlands warlords and rulers who dominated, even revolutionised, the Anglo-Saxon political landscape across a century and a half. No other kingdom could match Mercia for power, wealth or sophistication. No kingdom was able to maintain its impressive dynastic stability. Between the twenty-two-year rule of King Penda, beginning as Cadwallon's subordinate ally at *Hæthfelth* in 633, and King Offa's death 163 years later in 796, lie the reigns of just five kings, most of them of Penda's own line, the Iclings.

The use of the word 'king' – Bede's Latin *rex* – is a convention derived in part from Old Testament usage. There is little evidence of ritual anointings or coronation ceremonies among Anglo-Saxon rulers before the end of the eighth century but King Æðelbald† would style himself as both king of the Mercians

---

* Very roughly modern West Yorkshire.
† The Æ or 'ash' ligature in Old English is pronounced like the 'a' in 'hat'.

and, on occasion, *Rex Britanniae*. Penda was the most successful warlord of his generation, portrayed by Bede as a battle-hardened, old-school heathen. Whether or not he was acclaimed as king by the people called Mercians, Bede allows him the title. He seems to have been the first to unite the disparate Midland peoples as a single nation. His sons and successors, Peada (655–658), Wulfhere (658–675) and Æðelred (675–704), responding to new ideas about rational kingship and attracted by the powers of a literate clergy, were converts to the Christian faith, promoting the interests of a newly confident church while expanding Mercia's overlordship, or *imperium*, by force of arms.

Two Mercian kings dominate the history and politics of eighth-century England. King Æðelbald (716–757) was the longest-reigning English monarch before Edward III. All southern Britain lay tributary to him. He promoted London as a major trading centre and began to exploit Mercia's untapped economic potential while ruthlessly suppressing dynastic rivals. During his reign the tensions that drove relations between kings, land, taxes, the church and public works were codified in a set of 'common burdens' that placed the state formally above the individual for the first time. Æðelbald's successor, King Offa (757–796), supposed builder of the eponymous dyke, is the most famous Anglo-Saxon king before Ælfred. He fought against and subdued contemporary rivals in Wales, Kent, Wessex and East Anglia; issued a sophisticated coinage and a law code on which Ælfred would draw; promoted Mercia as an international trading nation and corresponded with the Frankish Emperor Charlemagne. He almost defines the idea of Anglo-Saxon warrior kingship: by turns belligerent, cunning, astute and innovative. Other Mercian rulers were more reticent: two of them abdicated and either retired to a life of monastic contemplation or made the pilgrimage to Rome to end their days in the city of God and St Peter.

These are bald historical facts. But Mercia is, and always has been, an enigma. No independent narrative history of its kings survives – like the West Saxon *Anglo-Saxon Chronicle* or Bede's

*Ecclesiastical History** in Northumbria – to tell of its dynastic rivalries, its political evolutions or its cultural identity. We cannot locate the birthplace of any of its kings and rarely can we place them in the sort of royal residence, like Edwin's Yeavering, or Charlemagne's Aachen, that defined tribal kingship. Their holy men enjoy much better recorded lives; have fuller personalities but are heard, so to speak, as voices from the wings. Even the name Mercians – *Mierce* – the 'borderers' or 'march-dwellers', defies historians who ask: which border; where?

Take a map of England. Trace the courses of a trio of rivers – the Trent, the Severn and the Thames – three sides of a Mercian wedge whose fourth is bounded by the watery margins of the East Anglian Fens. Anyone of these might be the border that defines the peoples calling themselves Mercian. It is often said that the original border lands must have been those that met the British-speaking Welsh, where Offa built his Dyke. But this is too easy a solution, for on close examination one finds Mercians on either side of the Dyke and beyond the Trent and Severn. The Thames Valley, which was certainly of interest to both Mercian kings and their rivals in Wessex, was sometimes a frontier; sometimes not.

If Mercia lay somewhere in the Midlands, it may nevertheless be safer to account for what it was not. The surrounding kingdoms of Northumbria, East Anglia and Essex, Kent and Sussex, Wessex, Gwent, Powys and Gwynedd define it in negative. Within that circling rampart of sometime competitors and periodic allies, lay regions and provinces with exotic-sounding names: the *Wreocensætna*, *Hwicce*[†] and *Magonsætna* along the upper reaches of the River Wye; 'Middle Anglian' peoples called

---

* The *Chronicle* was first compiled in the later ninth century at the court of King Ælfred; several regional versions or recensions survive, known by the letters A to E. Bede's *Historia Ecclesiastica Gentis Anglorum* was completed in 734, a year before his death. The much more disparate material contained in the *Historia Brittonum* was first compiled as a British source in the early ninth century, in a work often credited to a Welsh monk by the name of Nennius.
† Pronounced 'Hweecheh' in Old English.

*Gyrwa* and *Sweodora*, *Gifla* and *Herefinna*; the *Pecsætna* or Peak Dwellers north of the Trent and the *Tomsætna* beside the River Tame between Lichfield and Tamworth – heartland of King Offa's overlordship. Those diverse peoples and their geographies, their inherited histories and fiercely defended identities, many of which survived in recorded documents as late as the Domesday Book, are keys to understanding how Mercia was more than the sum of its parts.

If the Mercian peoples were disparate, they were linked by both the navigable rivers that reach into England's heart from the Channel, the North Sea, the Irish Sea and the Severn estuary, and by a cunningly devised lattice of roads constructed by Britannia's Roman overlords during the first and second centuries of the Common Era. A square, whose corner points are marked by the Roman towns of Chester, Lincoln, London and Bath, is quartered by two diagonals: the major roads of Watling Street, which joins London to Wroxeter and Chester; and the Fosse Way, which joins Lincoln, via Leicester, Cirencester and Bath, to Exeter. They meet at a place known as High Cross, which lies close to the village of Claybrooke Parva in Leicestershire. In the Roman period a small settlement called *Venonis* stood at the crossroads. It is almost the dead centre of England. From here goods, people and armies could travel on metalled roads relatively quickly; smaller roads linked these highways of empire with navigable rivers, with towns and centres of industrial production. The Viking invaders of the ninth century learned the value of Mercia's interconnectedness, its infrastructural wholeness – and took ruthless advantage of that knowledge. Mercia's eighth-century kings learned to exploit it too, for access to continental markets, for military and administrative control, and as a means of unifying its peoples as a single kingdom.

The raw materials of Mercian history are not easily handled. Bede, who could not boast among his enviably well-informed sources a single Mercian, was antipathetic towards Northumbria's principal rivals: we see them through prejudiced eyes. The *Anglo-Saxon Chronicle* records Mercian events almost grudgingly.* More apparently dispassionate are scores of charters, some of them useful forgeries, purporting to record royal Mercian gifts, purchases and occasionally thefts of land – though rarely within the Mercian heartlands themselves. They evoke Mercia's complex human and natural geographies, speak of meadow and woodland, of tolls on ships, of salt, iron and lead concessions and of furiously contested wills. Their lists of witnesses open a window, for the first time, on the careers of secular and ecclesiastical lords and noblewomen – the makers, shakers and bit-part players of Mercian history. Neither Wessex nor Northumbria offers such treasures in this period.

The prospect of writing a Mercian chronicle is daunting, but less so than it would have been even twenty years ago. Coins, in ever-increasing numbers, tell of the movement of goods and traders, of routeways and raw materials, of entrepreneurial bishops and opportunistic royal administrators; and they betray the ambitions and vanities of kings – sometimes queens, too. Excavations at key archaeological sites are beginning to demonstrate how a revolution in farming and production practices, and a pronounced trend towards the secularising of minster estates, underpinned the emergence and flowering of a Mercian economic powerhouse, although whether the process was driven by kings themselves is a much more obscure question. The geography of place names is revealing how Mercian kings pinpointed key strategic sites for the exercise of royal authority, while carefully conceived settlement layouts betray a novel and purposeful architecture, even an idea of efficiency and administrative order: a Mercian cultural imprint. Mercia's most visible legacy, the great earthwork that Offa is said to have built 'from sea to sea', is its own enigma, but archaeologists increasingly understand its nature and purpose as a projection of royal power and reach – a Dark Age Maginot Line.

* Briefly, during the reign of Edward the Elder (899–924), a series of entries in Recension D of the *Chronicle* record Mercian matters in greater detail and is known as the Mercian Register.

The great dyke bearing King Offa's name defines not so much
the limits of his power, but of his ambition.

The achievements of Mercia's kings in forging a state built on
success in trade alongside victory on the battlefield, thoughtful
administration in partnership with coercive alliance and bullying
land-grab, deserve their own chronicle. That chronicle can only,
for the moment, be pieced together in silhouette; nevertheless, the
long shadow cast over English history by a century and a half of
Mercian supremacy is deep and enduring.

# Chronography 1
## 628–716

Unless otherwise stated, narrative source entries are from the *Anglo-Saxon Chronicle*.

### Abbreviations

AC:   *Annales Cambriae*
AU:   Annals of Ulster
EHD:  English Historical Documents 500–1042
HB:   *Historia Brittonum*
HE:   Bede's *Historia Ecclesiastica*
S:    Electronic Sawyer
VB:   Willibald: *Vita Bonifati*
VG:   *Vita Guthlaci*
VW:  Eddius Stephanus: *Vita Wilfridi*

| | |
|---|---|
| 628 | Kings Cynegils and Cwichelm of the West Saxons fight against King Penda of Mercia at Cirencester; they come to 'an agreement'. |
| 633 | King Edwin of Northumbria killed at the Battle of Hatfield/*Meicen* by Penda of Mercia fighting with Cadwallon of Gwynedd. |
| ? 635+ | King Penda defeats kings Egric and Sigeberht of East Anglia in battle at unidentified site. |
| 642 | Campaign against Mercia by King Oswald of Northumbria. Oswald is killed in the Battle of *Maserfelth* (Oswestry) fighting against King Penda of Mercia and possibly Welsh allies.<br>— Cynegils of Wessex dies; succeeded by Cenwalh. He marries a sister of Penda but repudiates her. |
| 643–650 | Periodic raids by King Penda into Bernicia (HE III.8). |
| 645 | King Penda drives Cenwalh of Wessex into a three-year exile among the East Anglians. |

9

649     Possible date of attack on Northumbria by Mercian army, as far north as Bamburgh (HE III.16).

653     Peada of Mercia, King Penda's son, marries Oswiu's daughter Alhflæd and is baptised *Ad Muram*: beginning of Irish mission to Middle Angles (HE III.21, 22). Northumbrian Prince Alhfrið probably marries King Penda's daughter, Cyneburh.

654     King Penda kills King Anna of East Anglia.

655     The 'distribution of *Iudeu*' and campaign against Northumbria by Penda on the ?Firth of Forth; King Oswiu forced to give up a huge ransom (HB 65; HE III.24).

       — Ecgfrið, King Oswiu's son by Eanflæd, held hostage under Queen Cynewise of Mercia.

       — Battle on the River Winwaed: King Penda of Mercia, and thirty chieftains (including King Æðelhere of East Anglia) are killed; Northumbria achieves dominance over Mercia until 658. Peada succeeds Penda as king over southern Mercia.

656     Likely foundation date for first Mercian minster at *Medeshamstede*. King Peada murdered by treachery of his wife, Alhflæd (HE III.24). King Oswiu rules Mercia directly.

658     King Oswiu plunders Mercia (AC).

       — Mercian rebellion by Immin, Eafa and Eadberht expels Northumbrian governors and sets up Christian Wulfhere, son of Penda, whom they have kept concealed, as king of Mercia (HE III.24).

661     King Wulfhere ravages Ashdown in Wessex.

       — New West Saxon see established in Winchester around this time. Wulfhere may also annex Wight after 'ravaging it' and *Meonware* before gifting them to King Æðelwealh of the South Saxons.

664     Synod of Whitby; Northumbria follows orthodox Roman church (HE III.25). A great plague, in which Archbishop Deusdedit dies.

665    King Wulfhere sends Bishop Jaruman to East Saxons to quell upsurge in apostasy after plague (HE III.30).

666    Wilfrid 'carries out episcopal duties' in Mercia and founds minsters at Lichfield and elsewhere (VW XV).

669    Arrival of Archbishop Theodore of Tarsus in England as archbishop of Canterbury (to 690 – HE IV.1); Chad sent to Mercia as its bishop under Wulfhere, succeeding Jaruman. He is granted land at Lichfield to establish see.

672    Church council at *Herutford* (HE IV.5). An early charter records ships berthing at *Lundenwic* as part of a monastic gift at Chertsey (S1165; EHD 479); another describes brine springs at Droitwich.

674    Mercian army under King Wulfhere invades Northumbria and is repulsed (VW 20). Lindsey annexed by Northumbria, later recovered by King Æðelred of Mercia in 678–9.

675    King Wulfhere fights Æscwine of the West Saxons at *Biedanheafod*; Wulfhere dies. Succeeded by his brother Æðelred to 704.
    — Foundation or rededication of Bardney monastery by Æðelred of Mercia and Osðryð, daughter of King Oswiu. Endowed with Oswald's trunk and legs (HE III.11).

676    King Æðelred of Mercia attacks and subdues Kent; ransacks Rochester (HE IV.12).

679    King Æðelred of Mercia and King Ecgfrið of Northumbria fight indecisive battle on the River Trent. Archbishop Theodore oversees peace treaty between Mercia and Northumbria.

681    Wilfrid is released from Northumbrian imprisonment into exile in Mercia under protection of King Æðelred (VW 40).

685    A battle between Northumbrian and Pictish armies at *Nechtansmere*. King Ecgfrið is killed and

Northumbrian overlordship ends (HB 57; AU 686; HE IV.26).

691 Wilfrid expelled from Northumbria again – by King Alhfrið; taken under protection by King Æðelred and made bishop of Middle Anglia at Leicester (VW XLV).

692 Wilfrid is in exile in Mercia, founding monasteries at Oundle, Brixworth, etc. A battle is fought by the Welsh against King Æðelred.

697 Queen Osðryð, wife of King Æðelred, is murdered by Mercians. Guðlac enters the monastery at Repton.

699 Guðlac travels to Crowland to become a hermit (VG XXIII).

704 Æðelred of Mercia abdicates and retires to become abbot of Bardney monastery in Lindsey (HE V.24). His nephew Cœnred (son of Wulfred) becomes king of Mercia.

706 Synod of Nidd in North Yorkshire. Wilfrid's lands and offices are restored to him (VW LX).

709 Bishop Wilfrid dies at Oundle, aged 75. King Cœnred of Mercia abdicates and travels to Rome. Succeeded by Ceolred, son of Æðelred.

714 Death of Guðlac at Crowland.

715 Guðlac's sister, Pega, elevates his incorrupt body and places it in a shrine; visited by Prince Æðelbald while in exile; he enriches the shrine (VG LI).

— King Ceolred invades Wessex; he and King Ine fight at Adam's Grave (Alton Priors, Wiltshire).

716 King Ceolred of Mercia dies, buried at Lichfield. Succeeded first, possibly, by Ceolwald; more certainly the same year by Æðelbald, grandson of Penda's brother Eowa, who died at Maserfelth in 642.

— Boniface departs from *Lundenwic* for Frisia (VB IV).

# Chapter 1

# Penda: War
# 630–658

❖

*Kingship • Penda's rise to power •*
*Where was Mercia? • War • Alliance • The battle*
*on the Winwæd • Succession and legacy*

King Penda was a famous warrior lord, remembered long after his passing as *'viro strenuisimo'*.[1] Such men did not merely prosecute war against their enemies; they personally led their warbands into battle, wielding their own jewel-glinting sword, their 'blood-worm' or 'wound-hoe'. They expected to die in battle and knew that their comrades-in-arms, their *gedryhten* or *comitates*, would light funeral pyres for them, honouring them in grief as they raised an earthen mound over their remains. In turn, kings mourned their fallen comrades, were obliged to be generous with gift, song and feast, with full mead cup and with honour in their own hall. They settled veterans with land on which to farm and raise a family. They tried to ensure that favoured sons succeeded them as kings while the sons of their thegns would, in turn, hope to fight for their heirs.

But kingship was fragile. Rivals watched on from the wings, waiting their turn to bid for the throne. A marriage may not produce heirs. Sickness and the fates might deal any mortal man an unlucky hand. An old English maxim had it that as good fights evil, so youth contends with old age, light with darkness, foe with foe; and that the shape of the future cannot be known.[2]

By the hammer-beaten rules of tribal politics, brother fought brother, brothers fought together against cousins, nephew ousted uncle. Former enemies might unite in expedient alliance against a common enemy. Insult triggered revenge; revenge invited blood feud. Noble women were peace-weavers, prophetesses and matrons in the restive fictive family of the mead hall. They might, on occasion, incite their men to war

– laying a metaphorical sword in their lap just as, in the later raiding centuries of the borderers, wives served their husbands a dish of spurs for their supper. Royal daughters were deployed as diplomatic assets in marriage but they might also marshal their own dynastic and political agency to achieve power and nurture community.

From the seventh century onwards Christianity offered new opportunities for both royal men and women to spend their days in contemplation, prayer and hard work as an alternative to the combative, back-biting benches of the mead hall. The enthusiasm with which sections of the Anglo-Saxon nobility embraced the monastic life, in the process founding a professional cadre of literate thinkers, administrators and entrepreneurs, is the most striking feature of seventh-century Insular* history.

The English-speaking peoples were also poets and philosophers, lovers of wit, rhyme, kenning and word-play. The words and thoughts of those early kings rarely survive – rarely, at least, before Ælfred, the greatest philosopher-king of them all. But when the written word came to the Anglo-Saxon kingdoms in the seventh century, those who recorded the victories, defeats, crimes and virtues of great men and women began to open narrow windows through which posterity might peer into the Early Medieval mind.

King Penda's world view is a mere shadow on a wall; but he was no less real than his more illustrious descendants. Born, perhaps, close to the turn of the seventh century, he was the son of Pybba, son of Crioda, son of Cynewald, son of Cnebba, son of Icel… and so the Mercian royal genealogy reads, all the way back through another seven generations to the god Woden. Anglo-Saxon kings liked to claim ultimate descent from an ancestral pool of otherworldly continental warlords boasting mythic powers of sorcery, arcane wisdom and victory in battle. Woden, or Odin, was the progenitor of choice among warrior kings although the Iclings, as the Mercian royal dynasty came to be known, traced

* That is, of Britain and Ireland.

their immediate ancestry back to a more earthly, tangible man who must have lived sometime in the early sixth century. But none of Icel's descendants before Penda had been kings and nothing of the mythic progenitor Icel's life or deeds is known.

While the narrative histories of several other Anglo-Saxon kingdoms purport to begin in the fifth or sixth centuries with legendary pirate raiders like Hengest and Horsa, Cerdic and Ida carving out coastal territories for Germanic settlers in hostile British territory, only one Mercian king is known before Penda's apparently meteoric rise in the late 620s or early 630s. A Mercian 'king' named Ceorl is recorded by Bede as having given his daughter Cœnburh in marriage to Edwin while the latter was an itinerant prince, an æðeling in exile from his native Northumbria.[3] Ceorl appears in no collection of genealogies and the name is itself distinctly unaristocratic; he earned no entry in the *Anglo-Saxon Chronicle*, nor in the Welsh *Historia Brittonum*. Mercia's first king springs from nowhere, unaccompanied by foundation myth, heroic poetry, a noble line of ancestors or, indeed, any record of how he became king, how long he reigned and what battles he may have fought. The historian must fall back on the age-old rules of tribal politics to draw a thumbnail silhouette of Mercia's first monarch.

Ceorl belonged to the same generation as Edwin's father, the Deiran King Ælle, and that of Edwin's powerful brother-in-law Æðelfrið, the supreme Bernician warlord who wielded *imperium* over northern Britain for more than twenty years until his bloody death at Edwin's hands in 617. Since Anglo-Saxon royal women were diplomatic assets to be deployed in strategic alliance, the implications of the Mercian princess Cœnburh's marriage to the exiled Northumbrian prince are, first, that Ceorl was no ally of Edwin's rival King Æðelfrið, whose Machiavellian attempts to murder Edwin are a dramatic set-piece of Bede's *Historia*. Second, that Ceorl was placing a political bet on Edwin as a theoretically subordinate and pliable future king of Northumbria. It was a bet at long odds: Edwin's prospects in the years before that precocious victory over King Æðelfrið in 617 were not encouraging. Unlike

Edwin, Ceorl disappears from the historical record as suddenly as he emerged, unable to call in the debt; so by the time Edwin was in a position to repay the favour Mercian politics are no more than a blank page again.

When Penda emerged as an ally of the Venedotian* King Cadwallon against Edwin at *Hæthfelth* sixteen years later, the sons born of the marriage between Edwin and Cœnburh were of fighting age. Osfri∂, as Bede says, was killed in the battle; Eadfri∂ deserted his father ignominiously and went over to the enemy alliance; was taken into Mercia and later killed at Penda's behest – perhaps even at Penda's hands – while under his oath-bound protection. This nugget must have been passed to Bede by one of his informants at Whitby, where King Edwin's cult was curated across four generations by abbesses of royal Deiran stock. For the historical detective it is the smoking gun that explains Penda's motives for joining Cadwallon's army. Edwin's sons were dynastic rivals: eligible, through their mother's line, to inherit Ceorl's Mercian kingship, as well as being heirs to the Deiran, and Northumbrian, throne. By the time of the battle at *Hæthfelth* in 632 Ceorl must already have died. If he had not been a victim of Penda's ambition, he may have met his end in battle against Æ∂elfri∂ in about 616 or 617 during a campaign, recorded by Bede, culminating at Chester and whose purpose may have been precisely to eliminate Edwin. In any case, for Penda to assert his own claims to Mercian kingship, Ceorl's grandsons must be eliminated; and he ensured that they were.

How did an aspiring warlord rise from apparent obscurity to be remembered as the founder of a kingdom? As it happens, the single surviving prose account of a contemporary Mercian life, the *Vita* of a saintly hermit named Gu∂lac, records the career of a young freelance warrior from the same milieu. It is a brief tale but it strikes a counterpoint to the poetic epic *Beowulf*, tempering the kaleidoscopic romance of a heroic age and painting the warrior life in altogether more primary colours.

* The adjectival form of Gwynedd.

Guðlac was born sometime in the 670s in the land of the Middle Angles, a conglomerate of small, semi-dependent territories lying between the Trent Valley and the rivers that fed into the Wash: roughly, the historic counties of Rutland and Huntingdonshire, with parts of Northamptonshire, Leicestershire and Cambridgeshire. Guðlac's father, Penwalh, came of 'distinguished Mercian stock'[4] and was possessed of 'abundant goods'. More than that, he was said to be able to trace his line back to that of Icel, the Mercian royal progenitor. He must, therefore, have been a kinsman of Penda. Guðlac's later career as a hermit allows one, tentatively, to locate his father more closely in the territory of *Gyrwe*, where the rivers Nene and Welland flow into the Fens around modern Peterborough. His hagiographer, a priest named Felix embedded at the court of the East Anglian kings, after telling a series of conventional tales of heavenly presentiments and miracles surrounding the saint's birth and childhood, records that the name Guðlac meant 'the reward of war'.[5]

At the age of fifteen the boy took up arms, having been inspired in his dreams by stories of warrior heroes – like Beowulf – passed down through the generations. Guðlac gathered together a band of followers and embarked on a sort of feral rampage that lasted nine years – devastating villages and fortresses and collecting a huge treasury of booty along the way. As his reputation grew, so he attracted 'foreigners' to his warband – eager for their share of blood-lust, glory and the hard currency of looted gold and silver, enveloped by the esprit de corps of the *gedryht*, the warband. Each victory or campaign of violence increased his stock as a warlord, his reputation as a warrior, his portfolio of subject *villae*,* the number of freelance warriors in his retinue and his prospects as a future *dux* fighting at the side of his king; perhaps, ultimately, competing for the kingship itself.

If Guðlac's ancestral home lay among the Middle Angles, what can be said of his kinsman Penda, another Icling? Supposing

---

* The basic unit of land-holding geography, roughly equivalent to a modern township such as survive in the North.

that Penda's career followed a similar course, he eventually felt sufficiently powerful to fight for the greatest prize of all, kingship of the Mercian people, by defeating the incumbent (perhaps Ceorl) or being elected by the king's surviving councillors on Ceorl's death. Even if kings won their crowns by force, they ruled with the consent of, and through the martial and political support of, those nobles whom they were able to reward with land, gifts of treasure and the trappings of rank.

Penda does not have so secure a geographical provenance as Guðlac's family. He emerges from absolute obscurity. An entry in the *Anglo-Saxon Chronicle* under the year 628 records a battle in which Cynegils and Cwichelm, two West Saxon kings, fought Penda at Cirencester, after which the two armies 'came to an agreement'.[6] That may be read as, at the very least, a tactical victory for Penda.* If true, it means that Penda was already an active and capable battle commander in the years before *Hæthfelth*, even if Bede dates his reign from that latter conflict. Setting that account alongside Penda's subsequent fortunes and Bede's *Historia* allows a plausible outline of his early career to be drawn.

At Easter in the year 626 King Edwin was celebrating with his Christian wife Æðelburh, her priest Paulinus and his thegns and followers at a hall that stood by the River Derwent in Deira.† He had made a promise to Paulinus during his years in exile, and to Æðelburh's father,‡ that he would submit himself and his people to baptism in the true faith. At that Easter gathering a West Saxon ambassador named Eomer was present, ostensibly to deliver some diplomatic message or tributary gift; in reality to assassinate the king. His dramatic, violent attempt on Edwin's life was foiled by a loyal thegn named Lilla (so Bede's story

---

* If the *Chronicle* could not 'spin' a West Saxon victory then the encounter was very likely a defeat or stalemate.

† Possibly close to the old legionary fortress at *Derventio*, the later East Riding market town of Malton on the north-west edge of the Yorkshire Wolds. Bede *HE* II.9.

‡ King Æðelberht of Kent, the king converted by St Augustine after 597.

goes), who placed himself between the attacker and his lord and died from the wound inflicted by the assassin's poisoned blade. Edwin was injured but survived. That same night the queen bore a daughter, Eanflæd. King Edwin, in thanks for his deliverance and her delivery, promised Eanflæd to the church. Once healed of his wounds he gathered an army and attacked Wessex in revenge. As it happens, Bede scrupulously recorded the name of the king who sent the assassin: the same Cwichelm who, two years later, is recorded as having fought against Penda at Cirencester.

Now, Penda married a woman called Cynewise and fathered, among his six children, daughters whose names were Cyneburh and Cyneswið. These are names with a distinct West Saxon flavour. One may infer, I think, that the 'terms' drawn up after the battle at Cirencester in 628 included Penda's marriage to a royal West Saxon princess and the marriage of his (unnamed) sister to Prince Cenwealh, who succeeded to the West Saxon kingship in 642.* By the time Penda followed Cadwallon to war against Edwin three or four years later he was already aligned with a pro-Wessex, anti-Northumbrian axis; and, moreover, he had been recognised as king by the ruling West Saxon dynasty. King Cadwallon's own enmity towards Edwin can be traced as far back as the earliest days of Edwin's exile among the British, when they were foster-brothers, and to a time when Edwin had besieged Cadwallon on the island of *Glannauc*, off Anglesey. They were sworn enemies. Such are the politics of the seventh century.

Where was Penda's Mercia? A hundred years ago the distinguished Anglo-Saxon scholar Sir Frank Stenton pointed out that a scatter of place names seems to reflect spheres of influence of various members of the Icling dynasty across a broad swathe of the south Midlands, from Hertfordshire in the east to

---

* Cenwealh was the son of Cynegils. He later repudiated Penda's sister and remarried and Bede says that, as a result, Penda drove Cenwealh out of his kingdom and into exile for three years. Bede *HE* III.7; Sims-Williams 2008, 27.

Worcestershire in the west. Penda's name occurs as an element in Penley (Flintshire), Pinbury (Gloucestershire), Peddimore (Warwickshire) and Pinvin and *Pendiford* (both Worcestershire).[7]

Bede, in whose day Mercian kings were a dominant force among English kingdoms, believed that in former times Mercian territory had lain either side of the River Trent.[8] His information must be set against the testimony of a much-debated document known as Tribal Hidage, which places the 'area first called Mercia' among a list of kingdoms and smaller territories, together with the value of tribute owed by them to an unnamed overlord (possibly King Edwin). The 'area first called Mercia' is not very helpful. Its geography must be constructed in negative. After Mercia in the list comes *Wreocensætna*, a small kingdom based on the former Roman *civitas* capital *Viroconium Cornoviorum* – Wroxeter – close to the east bank of the River Severn in what is now Shropshire. Next comes *Westerna* (literally 'Westerners'), which most historians of the Hidage place on the west, or Welsh, side of the Severn – to be identified with another small kingdom otherwise known as *Magonsæte*, the hinterland of Herefordshire.* Fourth on the list is the land of the *Pecsætna* – the dwellers of the Peak – that is to say, the hilly northern part of Derbyshire, north of the Trent.

A number of smaller entities, which we know to have comprised the territory of the Middle Angles in the East Midlands, follows a more or less clockwise pattern, together with more distant southern kingdoms like Kent, Sussex and *Wihtgara* (the Isle of Wight). Then, listed in a second part of the document, is *Hwinca* [sic], the kingdom of the *Hwicce*. As a sometime independent kingdom and later tributary territory of Mercian kings *Hwicce* is identified, by frequent mentions in charters and by periodic notices in the annals, as the territory flanking the lower Severn Valley – including what are now Worcestershire, Gloucestershire and parts of Warwickshire. Historians have

---

* First historically attested in charters as late as 811. The name is preserved in the place called Maund (Sims-Williams 2008, 40).

made an obvious connection between the apparent coherence of the lands of the *Hwicce* and the boundaries of the medieval diocese of Worcester.[9]

Mercia's core lands ought, therefore, to fill the hole left in the middle: very roughly the modern counties of Staffordshire, West Leicestershire, Derbyshire south of the Peak, South Nottinghamshire and the northern part of Warwickshire. This leaves historians with a conundrum: the place names that may reflect Penda's legacy in the landscape of named settlements seem more likely to be *Hwiccian* than part of the 'original' Mercian core. Did Penda come from the land of the *Hwicce*? If so, his victory against Wessex in 628 may have liberated that kingdom from West Saxon control. It would also imply that, despite Bede's protestation that Penda was an unreconstructed heathen, his background may have been both Christian and British, judging by the continuing presence of an active diocesan church there[10] and, perhaps, by hints that Penda's own name has a Brittonic flavour – *pen* meaning a head, or chief. * Bede regarded the southern edge of the *Hwiccian* lands as being a boundary or frontier zone, to the north of which British bishops could be found in the days of St Augustine's mission to Kent in 597.[11]

Bede's very evident antipathy towards the British church may have coloured his judgement, but even he conceded that King Penda bore no personal or ideological ill-will towards Christians, allowing them to preach in his territory.[12] Penda's early alliance with King Cadwallon of Gwynedd, and apparently close connections with other British kingdoms lying west of the Mercian core lands – the *Magonsæte* and *Wreocansæte* – reinforce the idea of pro-British, and therefore nominally Christian, sympathies. In reality, cultural, linguistic and ethnic identities in these marcher lands are likely to have been complex, nuanced by ambiguous affinities with strong local flavours.

---

* The use of *Pen* as a prefix for many Welsh mountain names derives from the same root. His kinsman Penwalh, father of Guðlac, has a suspiciously similar name.

At any rate, the core Mercian lands either side of the Trent seem to have been Anglicised, at least in material culture, by the beginning of the seventh century. That Penda was able to claim lordship over those peoples and be recognised as overlord by other kingdoms demonstrates his skill as a warrior and canny political operator. Although the details of his career are, to say the least, patchy a series of *Chronicle* entries charts successive victories in battle over West Saxons, East Anglians and Northumbrians. Two years or more after the bloodbath at *Hæthfelth* in 633, Penda led an army into East Anglia. Up until the 620s an East Anglian king, Rædwald – he of the Sutton Hoo ship burial – had been recognised as overlord of all the southern kingdoms.[13] He had hosted Edwin during his exile and led the military campaign during which Edwin reclaimed his kingdom from Æðelfrið in 617 on the River Idle, in what is now South Yorkshire. King Rædwald was succeeded by his son Eorpwald and then by Eorpwald's (probably half-) brother Sigeberht who, during Rædwald's reign, had been forced into exile in Francia. Here Sigeberht had been baptised. When he returned to claim the kingship of the East Anglians he brought in a bishop, Felix,* to minister to his people. He subsequently abdicated, joined a monastic community and handed the reins of power to his kinsman, Egric.

When Penda's army invaded – and nothing is known of the geography of the campaign – Egric called on his predecessor to come to the aid of his countrymen. Sigeberht refused but was dragged from his life of monastic retirement and, so Bede says, took to the battlefield with no more than a staff. In the ensuing battle, both East Anglian kings were killed, their armies slain or scattered.[14]

Virtually no Early Medieval battlefield sites have ever been decisively identified on the ground. What little is known about fighting tactics and military strategy offers scant clues about the nature of warfare in this period. Roman roads were – still are, in some cases – often known by the Old English term *herepað*

---

* Not the Felix who wrote the *Vita* of St Guðlac.

('army road') and Bede makes reference to King Edwin's strategic use of 'highways', along which he had marked springs of clear water where travellers might drink from bronze cups set on posts nearby.[15] Several battles in the seventh century seem to have been fought along the lines of Roman roads where they crossed rivers – including the fight on the River Idle that won Edwin his kingdom in 617. Occasional notices of attacks on defended former Roman fortresses turn up in the annals. The two most obvious routes for an army invading East Anglia from Mercia, or the other way around, are along Roman roads that led through *Durobrivae* (Water Newton) in Middle Anglia and *Durovigutum* (Godmanchester), the former crossing otherwise unforgiving fen land. Even so, in a landscape as uncentralised as that of the British Isles, it was no mean feat to locate an enemy and then bring him to battle.

As late as the Jacobite wars of the eighteenth century, armies had difficulty finding one another in less densely settled parts of the landscape, particularly in hill country. King Ine of Wessex (died 728) is known to have retained a cadre of Welsh horsemen, who may have acted as royal military scouts.[16] When the Viking army of Guðrum so narrowly missed decapitating the West Saxon régime during Christmas 877 at Chippenham, it seems to have been one of King Ælfred's own ealdormen who betrayed his presence there. Perhaps, in an age when kings progressed from one estate to another seasonally, consuming the surplus of each in turn before moving on, their movements were an open secret – one knew where they might be found in any month of the year. Otherwise, one is tempted to suggest that some armies met by prior arrangement – by direct challenge of the sort celebrated in the Irish epic *Táin Bó Cúailnge*: the great cattle raid of Cooley.

Thegns, or *gesiðas*, were oath-bound followers of their lord, from whom they held a life interest in their land holding, or the promise of one. They were obliged to fight in his wars, to die at his side if necessary or be dishonoured. Likewise, tributary kings, once defeated or accepting of their subordinate status, must fight in alliance with their overlord; must sit at his mead bench in correct order of precedence and witness his judgements, receiving in return a suitable share of booty and bathing in the mead-soaked afterglow of glorious victory. In Penda's last battle no fewer than thirty '*duces regii*' were said to have fought – and fallen – alongside him.[17]

It is generally understood that warriors fought with small personal retinues, supported by an unknown number of household followers; that they were equipped with sword, spear and round wooden shield of lime wood, covered with cow hide. The shield might carry a large metal boss to provide extra offensive and defensive punch and be adorned with zoomorphic motifs – battle-charms, as it were. Mailcoats are sometimes portrayed in contemporary images and a few precious fragments of chainmail recovered from excavations show enviable craftsmanship – the preserve of a warrior élite. Princes might enjoy the protection – both physical and psychological, of a helmet likewise decorated with zoomorphic motifs. Six of these have been excavated from English sites over the years – most famously from King Rædwald's funeral ship at Sutton Hoo and from the Coppergate excavations in York in the 1980s. Another, with a splendid wild boar crest, was discovered in 1893 at Benty Grange – in the land of the *Pecsæte* of North Mercia. A second boar-crested Mercian helmet, retrieved from a richly furnished warrior's grave ahead of gravel quarrying at Wollaston in Northamptonshire in 1997, accompanied a bronze hanging bowl and pattern-welded sword.[18] Was this man a Middle Anglian prince; did he fight with King Penda in his wars of conquest?

Those without the means to wield a pattern-welded steel sword made by a specialist swordsmith might inherit such a formidable weapon from their father; others would swing a broad axe. Warriors also carried the short, stabbing sword or knife from which Saxons supposedly took their name: the *scramaseax*. They rode a horse – another élite item in the warrior's equipage – on campaign, but dismounted and then fought on foot. Followers might contribute the equivalent of small arms fire with slingshots. Archers were known in a skirmishing role in Anglo-Saxon England but appear not to have played a significant role in the line of battle, which was traditionally thought of as a closely linked shield wall whose purpose was to prevent the enemy breaking its mental and physical cohesion.[19]

Many battles must have consisted of a lot of shouting, goading and spear throwing, probably quite a lot of shoving but, perhaps, little hand-to-hand combat: the victor was the warlord whose wall held the field at the end of the day. Occasionally, an ambush would be decisive; at other times, the breaking of a weak or ill-disciplined shield wall would be followed by rout; then by slaughter. In a set-piece battle the ultimate aim for both sides would be to kill and dismember the enemy leader – literally to decapitate the régime. A famous story recounted by Bede tells of a Northumbrian king's thegn called Imma who, finding himself wounded and taken prisoner by Mercian soldiers after a battle, tried to convince them that he was not of thegnly rank: not to be executed by the rules of feud but to be ransomed as a poor man – a mere camp follower. His noble bearing and speech gave him away; he was spared, but sold to a Frisian slave merchant in London.[20]

The *Anglo-Saxon Chronicle*, supported by the *Annales Cambriae* or Welsh Annals, Bede's *Historia* and the very rich evidence of the Irish Annals, promotes the idea that kings and kingdoms fought each other in great battles every few years. The sort of lower-level conflict recorded in the early career of Guðlac was probably much more common; much more devastating and ugly in its immediate, local consequences. Many of the battles recorded so tersely in the chronicles must have been one-sided raids for booty, or skirmishes. Such young-blood thuggery, raids, feuds and fisticuffs constituted the martial training for larger, more set-piece conflicts.

In Northumbria, King Edwin's death at the hands of Cadwallon and Penda led to a year (633–4) of apostasy and anarchy in the north – so infamous, according to Bede, that 'those who compute the dates of kings' had effaced it entirely from memory.[21] At the end of that year King Edwin's nephew, the Bernician prince Oswald, came out of exile in the kingdom of Dál Riata (broadly modern Argyll) with a small army consisting of fellow Bernician exiles and contingents from both the retinue of the Dál Riatan kings at Dunadd and island monastic community on Iona. Oswald had been baptised there in a thoroughly Irish milieu and it was in the Ionan abbots' interests that he succeeded in claiming his father Æðelfrið's kingdom.* Oswald's host surprised King Cadwallon's forces somewhere in the vicinity of the old Roman town of Corbridge, just south of Hadrian's Wall on the River Tyne, and won a decisive victory. At *Denisesburn*, on the Devil's Water in hilly country south-west of Corbridge, the rout was completed: Cadwallon was killed; his army shattered. Within a year, King Oswald's Ionan friends had established a new 'Iona in the East', an Irish-flavoured monastery on the semi-tidal island of Lindisfarne within view of the Bernician royal fortress at Bamburgh on the Northumbrian coast.

This stunning victory, the death of Cadwallon and King Oswald's subsequent recognition as Christian overlord of the southern kingdoms, were profound threats to Mercia's security. Oswald's marriage to a West Saxon princess and his sponsorship of King Cynegils's baptism were clear signals of political intent: to isolate Mercia from the club of Christian Anglo-Saxon kingdoms, with their enviable links to Francia, Rome and beyond. King Oswald also established, or re-established, Northumbrian claims over Mercia's immediate neighbour, Lindsey, with the Roman *colonia* of Lincoln at its heart – a once-independent kingdom whose few known kings barely ruffle the pages of the genealogies and who appear nowhere in either Bede's *Historia* or the *Anglo-Saxon Chronicle*.

It is by no means clear how secure Penda's position was in Mercia during the following years, despite his significant victory over East Anglia perhaps in the same year, 635, in which Lindisfarne was founded. The waters of comprehension are

* See *The King in the North*.

muddied by the fact that Penda had at least one brother, Eowa, who may have ruled jointly with him by agreement but who may, alternatively, have been a rival. By the time of his expedition to East Anglia, Penda probably also had a son, Peada. Brothers, uncles and nephews might make successful allies; but the record of such relations in the seventh century shows them more often to have been rivals; sometimes deadly rivals.

Neither the *Anglo-Saxon Chronicle* nor Bede illuminate the dynamics of these tensions in the years leading up to 642. In that year King Oswald led his forces out of the lands beyond the Humber to launch a pre-emptive strike on the western edge of Mercia that might have ended its existence as an independent kingdom. Both sources are silent on the background to, and the progress of, the campaign. Oswald's father, King Æðelfrið, had taken his host to fight in the Marches in about 616, culminating in a great battle at Chester, which he fought against a king of Powys in alliance, inferentially, with Mercia's King Ceorl. Apart from their natural bellicose instincts as warlords, there may have been another motivation for Northumbrian kings risking defeat so far from their homelands. In King Æðelfrið's case he may have hoped to force Edwin's Welsh protectors to give him up; Æðelfrið made strenuous efforts, in the year after Chester when Edwin fled to East Anglia, to force King Rædwald to give up his brother-in-law. King Oswald had no such fear, Edwin's sons having been either killed or taken into childhood exile in Francia, where they died. He may have been prosecuting a war designed to break the alliance between Mercia and the Welsh kingdoms. Equally, he may have hoped to ambush Penda in his heartland; to eliminate the chief threat to his *imperium*.

At any rate, Oswald brought his army to the Marches in the late summer of 642. On 5 August he was defeated and killed by King Penda, likely supported by Welsh allies under Cynddylan of Powys, of whom later poetic tradition recorded that 'when the son of Pyd [i.e. Penda, son of Pybba] requested, he was so ready!'[22] Oswald was cut down and dismembered, his head and limbs impaled on stakes on the battlefield. Bede records the

Old Oswestry: the ancient hillfort, close to where King Oswald was martyred by King Penda in 642. The town is named after the stake on which Oswald's head was impaled.

place of slaughter as *Maserfelth*; the *Historia Brittonum* gives it the name *Maes Cogwy*.[23] Most historians accept Oswestry – in Welsh *Croesoswald* or Oswald's tree – as the location, close to the substantial, and massively fortified, Iron Age hillfort of Old Oswestry. Aside from the Northumbrian king, Penda's brother Eowa (King Offa's great-great-grandfather) was also killed. No-one can say on which side he fought; but his death and that of King Oswald, the principal threat to Penda's rule, allowed the latter, in turn, to claim supreme overlordship over the other southern Anglo-Saxon kingdoms.

One opponent stood in King Penda's way. Oswald was immediately succeeded by his younger brother, Oswiu, a fellow childhood Bernician exile in the kingdom of Dál Riata; likewise a Christian convert and supporter of the church on Lindisfarne. A year after the catastrophe at *Maserfelth* King Oswiu was able to mount a raid against Mercia and succeeded in retrieving the grisly remains of his brother's head and arms from the battlefield.

They were taken into Northumbria and became the focus of a royal cult. The torso, bizarrely, would turn up thirty years later in the hands of a pious Mercian king.*

*Maserfelth* marks a watershed in King Penda's career and in Mercia's emergence as an enduring and cohesive power. He now felt sufficiently emboldened to go after Oswiu, who needed to establish his own web of military and political loyalties in the aftermath of his brother's death. Anglo-Saxon kings might hope to inherit the loyalties of their father's or brother's warriors but, in essence, each new king had to start from scratch.

Over the next decade Penda mounted a number of campaigns in Northumbrian territory but failed to bring Oswiu's forces decisively to battle. In the south he moved to enforce diplomatic initiatives. King Cynegils of Wessex, who had been baptised in the presence of, and who therefore acknowledged the overlordship of, King Oswald, died and was succeeded by his son Cenwalh in about 641.[24] Bede says that, unlike his father, King Cenwalh was not a baptised Christian. He chose as his wife an unnamed sister of King Penda – a clear indication of a change in West Saxon policy and a rejection of Northumbrian influence in the far south.[25] By 645 Cenwalh had repudiated his wife and taken another. King Penda's response to this insult was to drive him out of Wessex and into exile in East Anglia, where he sought protection from King Anna (c. 636–54). Cenwalh was able to return to Wessex three years later in obscure circumstances but having, apparently, been converted to Christianity while in exile. He remained on the West Saxon throne for another twenty years and oversaw the establishment of a second West Saxon bishopric, at Winchester.†

In the ongoing conflict between Mercia and East Anglia, Barbara Yorke, a historian of early Anglo-Saxon kingship, sees

---

* See below, in Chapter 3, p. 91.
† The first, established in about 634 under King Cynegils for Birinus, an envoy of Pope Honorius, was located at Dorchester-on-Thames in Oxfordshire.

attempts by both sides to assert their authority over the peoples
and minor polities comprising Middle Anglia: small territories
ruled by 'kinglets' of the sort termed by Bede *sub-reguli* or
*principes*. Grouped together, their tributary value in Tribal
Hidage totals something like 7,000 hides, equivalent to a second-
tier kingdom like *Westerna*, *Hwicce* or Lindsey and to the size
of Beowulf's legendary kingdom. Minor polities they may have
been, but rights of overlordship were valuable for the tribute
they rendered, while control over the lords of such territories was
a useful asset in war and a buffer against external threats. The
Fens and the rivers that fed into the Wash from the west were
rich in resources – summer grazing, peat, salt, fish and fowl, and
for their access, via navigable channels, to trading and exchange
opportunities up and down the coast and across the North Sea in
Frisia and Scandinavia.

No further pitched battles are recorded between the two
kingdoms before about 654. But one of King Anna's daughters,
Æðelðryð, later a queen of Northumbria and, still later, founder
of an illustrious religious house at Ely, was given in marriage to
a prince of the Middle Anglian *Gyrwa* – clear evidence of East
Anglian influence west of the Fens. A Mercian raid, recorded
only in an obscure Frankish hagiography, the *Life of St Follián*,
seems to have forced Anna into temporary exile in about 650.[26]
Three years later the *Anglo-Saxon Chronicle* records his death.
Bede says, and a detail in the *Historia Brittonum* confirms,
that it was at Penda's hands.[27] It seems very likely that Penda's
renewed campaign against Anna was designed to wrest from him
permanent control of Middle Anglia; if so, it succeeded. By about
653 Penda's oldest son, Peada, had been established as a Mercian
*subregulus* over its peoples.

Penda's periodic attacks on Northumbria were remembered
there long afterwards. One of these became the subject of a Bedan
set-piece miracle story.[28] Since its founding in 635 the monastery
at Lindisfarne had been ruled under a prior/bishop of élite Irish
background named Aidan, a nominee of the Iona community.
One day, sometime between about 645 and 650, during a period

## Mercian and Northumbrian sites and routeways

*Map showing Mercian and Northumbrian sites and routeways, with scale bars of 10 miles and 20 km, and a north arrow.*

Sites and features labelled on the map:

Elmswell, Driffield, Watton, Hornsea, York, Otley, Goodmanham, Beverley, Yeadon, Skipworth, S. Newbold, N. Ferriby, Sherburn-in-Elmet, R. Aire, Barrow, W. Halton, Dewsbury, Flixborough, Riby, Ceceseg, Melton Ross, Doncaster, Kirton, W. Ravendale, Roman Ridge, R. Ancholme, R. Idle, R. Don, Dore, Littleborough, Torksey, Grey Ditch, R. Derwent, Lincoln, Bardney, R. Witham, Wirksworth, Cromwell Bridge, Newark, R. Trent, Sleaford, Garwick, Nottingham, Margidunum, Ancaster, Derby, Threekingham, Hanbury, Repton, Tutbury, Breedon-on-the-Hill, R. Soar, Catholme, Maxey, R. Welland, Seckington, Stamford, Peakirk, Crowland, Tamworth, R. Anker, Leicester, Castor, Medeshamstede, R. Tame, Great Glen, Watling St, Gumley, Oundle, Conington, Fosse Way, Brixworth, R. Nene, Raunds when Aidan was staying in solitary contemplation on the tiny island of Inner Farne, the holy man became aware that the royal fortress at Bamburgh, a few miles away on the mainland across the Inner Sound, was under attack. Bede says that King Penda's Mercian forces had rampaged throughout the kingdom before mounting an assault on the ancestral seat of the Bernician kings – the massive rocky outcrop within whose walls King Oswald's incorrupt right arm was kept as a holy relic of martyrdom. Penda's ambition may have been to capture that relic as a trophy. Failing to take the fortress by siege, he ordered that the buildings in the nearby settlement be torn down and used to raise a 'vast heap' of rafters, beams, wattles and thatch against its walls. When the wind was right he set it alight. Aidan, on Inner Farne, saw the tongues of flame and pall of smoke hanging above Bamburgh and began to pray for its deliverance. God seemingly answering his prayer, the wind veered away from the defenders and into the faces of the attackers so that they fled and gave up their attempt to destroy the fortress. Those with local knowledge of the area's coastal weather might say, pragmatically, that the prevailing south-westerly offshore breezes there frequently reverse their night-time direction and become onshore winds during the day as the land heats up and draws cooler air off the sea. Landlocked Mercians may not have appreciated such meteorological peculiarities. Miracle or no, Bede's informants never forgot the episode and none of his readers would have doubted that God's favour lay with his Christian Northumbrian kings.

Penda raided Bernicia again at the time of Aidan's death, securely dated to 31 August 651.[29] The bishop was, at this time, staying on a royal estate not far from Bamburgh – Bede was probably referring to the great palace complex at Yeavering, *Ad Gefrin* in Glendale, twenty or so miles west of Bamburgh. Bede says that a few years afterwards King Penda came again with an army, destroying everything by fire and sword, burning down the 'village' and the church where Aidan had died. Miraculously, the buttress against which Aidan had been leaning when he drew his last breath survived the fire. That same buttress is locally believed to be preserved in the parish church of St Aidan in Bamburgh.

Raids were an executive branch of tribal rule: a king was obliged to undertake 'hostings' into enemy territory for booty and cattle and to ensure that his warriors were well-trained in the martial arts and in the rigours of campaign – the sort of 'training' undertaken by Guðlac's freebooters in later years. Raids might be accompanied by diplomacy – a combination of carrot and stick; and rival kings became increasingly adept at manipulating, as well as subverting, the customary rules by which such relations were managed. Warfare and alliance were not mutually exclusive diplomatic tools. Penda had offered one of his daughters, Cyneburh, to King Oswiu's son Alhfrið, establishing an alliance whose male progeny would be eligible to rule over either the Northumbrians or the Mercians, or both: an each-way bet. By about 653, around the time when Penda's son, Peada, was established as *subregulus* over the Middle Angles, he in turn sought King Oswiu's daughter Alhflæd in marriage and undertook to be baptised – partly, it seems, at the behest of Alhfrið, 'who was his brother-in-law and friend'.[30] I think one can read this as an indication that the two had spent time in intimate company with each other; and that, in turn, suggests that they may have been foster-brothers in earlier years. Noble fostering was common in Anglo-Saxon, British and Irish cultures. Foster parents like Penda or Oswiu protected and educated their foster-children but, at the same time, effectively held them hostage against their nominal allies and *de facto* rivals. Given that Oswiu and Penda exchanged sons and daughters in marriage, one may infer a relationship of more or less parity. Whose household fostered whose sons cannot be known, although another son of Oswiu's was certainly a hostage at Penda's court.

The baptism, and by implication the marriage, between Peada and Alhflæd, took place, Bede says, at a place called *Ad Muram* 'On the Wall', some twelve miles upstream from the mouth of the River Tyne – possibly at a royal *vill* where Newcastle now stands; or at Newburn, the lowest fording point on the river, where a royal township is known to have existed.[31] Peada now undertook to bring Christian priests into Middle Anglia; and to establish a see there for a bishop.

Now, as Bede stresses in his retelling of these events, Penda was a heathen but he by no means objected to the preaching of the Christian word in his territories. That may be so – and Penda himself may have been baptised or at least affiliated to the British church in *Hwicce* – but King Oswiu's apparent success in making an ally by both faith and marriage of Penda's oldest son, the ruler of the Middle Angles, might also have been seen as an existential threat to Penda's own rule. It might explain why, despite this double marriage alliance and the evident bonds of friendship that existed between their sons – perhaps, also, their daughters – Penda now embarked on a campaign of open war whose aim seems to have been to bring the kingdom of Northumbria to its knees. Might he also have been obliged by the rules of feud to pursue the brother of his brother Eowa's killer?

The ferocity of this campaign was such that Penda's army, including unnamed British allies – perhaps from both Wales and from the British kingdom of Strathclyde, was able in 655 to drive Oswiu as far north as the Forth valley, what is now Scotland's central belt: besieging him, it seems, in the dramatically imposing rocky fortress at Stirling.* The *Historia Brittonum* records that 'Oswiu delivered all the riches that he had in that city into the hand of Penda, and Penda distributed them to the kings of the British, "that is the distribution of *Iudeu*"'.[32] *Iudeu* is identified, at least broadly, by Bede himself in an earlier part of the *Historia*, where he describes *Urbs Giudi* as being half-way along the eastern branch of the sea.† Perhaps drawing on the same source as that cited in the *Historia Brittonum*, he describes how Oswiu was forced to hand over an incalculable store of treasure as the

---

* The identification is broadly accepted, but is not without its problems. Edinburgh, *Din Eidyn*, may be meant. Fraser 2009, 171 and Rannoch Daly, pers. Comm.
† i.e. the River Forth. *HE* I.12.

price of peace, if only Penda would return home.[33] If the *Historia Brittonum* is correct, that huge treasure was handed over; if Bede is right, King Penda refused the offer, being satisfied only with the extermination of the 'whole people'.

Subsequent events suggest that Bede is mistaken here. First, he sets the scene for the culmination of the campaign hundreds of miles further south than Stirling, in what is now West Yorkshire, suggesting that Penda returned satisfied from his northern campaign. Second, Bede says that Oswiu's youngest son Ecgfrið, by his second marriage to Eanflæd,* was at this time a hostage at the Mercian court under the 'protection' of Queen Cynewise. That Ecgfrið survived to become king in Northumbria indicates a satisfactory diplomatic resolution. But whatever the outcome of the *Iudeu* campaign, battle was eventually joined in 655, more or less on the border between the old British kingdom of Elmet and Mercia 'near the River *Winwæd*'. Historians generally identify this as the site where the Roman road (now the A639) between *Danum* (Doncaster) and *Lagentium* (Castleford) crosses a very minor river called the Went, near the village of Thorpe Audlin. Between the road crossing and a medieval crossing of the river at Wentbridge is a confluence with another stream. The road dips down off gentle hills and although the bed of the river is now very narrow, its flood plain is perhaps 500 yards across. Bede says that heavy rains had swollen the river, overflowing its channels and banks.[34] Given the iconic location for such a showdown – along a line on the road between Lincoln and Tadcaster on which several seventh-century battles were fought, one is tempted to suggest that this was less an ambush than a challenge met – a High Noon, death-or-glory shoot-out.

King Penda's army of Mercians and Middle Anglians was reinforced by the hosts of those British kings who had campaigned with him as far as the Forth valley; but overwhelming force, for once, availed him nothing. In total, Bede says, no fewer than

---

* King Edwin's daughter, who had been raised in the Frankish court since her dramatic exile after *Hæthfelth*.

Wentbridge: probably the site of King Penda's catastrophic defeat by King Oswiu in 655.

thirty warlords – *triginta legiones ducibus nobilissimus instructas in bello habuere** – went into battle under the Mercian standard. King Oswiu, by contrast, met them with a 'tiny' army. Bede records that Oswiu's nephew Œðelwald, his *subregulus* in Deira, deserted his uncle and withdrew to await the outcome in a place of safety. Oswiu subsequently had him killed. Nevertheless, by luck or strategy Northumbria's commanders were able to use the flooded river to their advantage, apparently forcing the Mercian army into its swirling, rushing waters so that more were drowned in flight than were killed by the sword. No doubt Bede intended a biblical allusion to Moses. Penda was killed along with many of his thirty commanders and their retinues, including King

* 'thirty legions of soldiers experienced in war'. Bede *Historia Ecclesiastica* Book III, Chapter 24. Colgrave and Mynors 1994, 150.

Æðelhere, the leader of an East Anglian warband and brother of King Anna whom Penda had killed the previous year.[35] The *Historia Brittonum* entry for the battle is bald and telling: 'now was the slaughter at Gaius's Field (*Campo Gai*)', adding that Cadafael, king of Gwynedd, was the only one to escape with his army, by withdrawing in the middle of the night before battle. He was afterwards damned with the epithet *Cadafael Cadomedd* – literally 'Battle-seizer-battle-dodger'. The Mercian overlord was dead after a reign of more than twenty years; Northumbria was triumphantly victorious, its king avenged for the death of his brother Oswald.

The consequences of that day on the field of *Gai* for Northumbrian and Mercian fortunes were profound in the immediate aftermath. King Oswiu was able to assert direct overlordship over the Mercian core lands and over Lindsey while installing Peada, his son-in-law, the *subregulus* of the Middle Angles, as a legitimate native ruler over all southern Mercia – Mercia south of the Trent. He installed his son Alhfrið (whose wife was Penda's daughter, Cyneburh) as *subregulus* of Deira. Mercian military might and reach, its influence over East Anglia and Wessex and its outer core of tributary states; Penda's long-nurtured alliances with British kingdoms: all were shattered. Those territories over which Penda had been able to exercise Mercian authority through proxies – the *Magonsæte*, *Wreocansæte*, *Hwicce* and the lands of the Middle Angles – must now transfer their allegiance to the Northumbrian king or suffer the consequences.

The death of a ruler, especially such a dominant warlord as King Penda, affected all levels of society. Women widowed by war were left without protection, sometimes turned off their land as a new king sought to build his own powerbase by redistributing estates. The fate of Penda's queen Cynewise is unknown, but the survival of her Northumbrian hostage, Ecgfrið, who would succeed his father on the Northumbrian throne, suggests that she was able to assure her own security through shrewd negotiation. King Oswiu's own ealdormen were very likely given large estates in Mercia both as a reward for their valour in battle and so that they might act as Northumbrian proxies there. A general sense of uncertainty, pervasive after such a long reign, induced people to look to portents of ill or good omen – unusual weather phenomena; the appearance of strangers; sickness in their cattle: all invested with new meaning as portents of the future. Even so, farmers continued to sow and harvest their fields, turn their beasts out onto summer pastures and carry on the business of the household: women spinning and weaving, tending fires and children; unfree and poor dependents repairing fences and grain stores, gathering firewood, threshing grain – the seasons unvarying in their obedience to natural laws.

Internal Mercian rivalries, suppressed for so long, bubble invisibly beneath the radar of the historian until they suddenly erupt. At Easter in the year 656, a year after the battle on the *Winwæd*, King Peada was murdered. Bede says that it was by the treachery of his Northumbrian wife, Alhflæd – Oswiu's daughter.[36] In the immediate aftermath, King Oswiu asserted direct rule over Mercian territory.

Penda has traditionally been viewed from the perspective offered by Bede: a brutal pagan warlord, the antagonist of virtuously Christian Northumbrian kings. He is a two-dimensional caricature of what popular history sees as a sort of noble savage. But Penda's known military campaigns between 628 and 655 occupy, at a rough estimate, seven out of twenty-seven years. They define him because they are all we have. Bede's Northumbrian kings, by contrast, are portrayed as politically astute, sometimes self-conscious and psychologically complex. At all times they were instruments of God's plan for the Northumbrian people – a fulfilment of Old Testament providential history. It is, perhaps, time that Penda was afforded some sort of a hinterland. If Oswald – Penda's almost exact contemporary – enjoyed enduring friendships with and intellectual stimulation from his holy men, why should not Penda engage in friendship with his priests? If King Edwin's *mund* – his regal authority – protected women and children from arbitrary violence (as Bede suggests) and if he employed architects to design grandiose expressions of his power and dignity in the fashion of the ancient emperors of Rome, why should not Penda? Bede's kings – Rædwald, who contemplated betraying Edwin to Æðelfrið for hard cash and was taken to task by his queen; Sigeberht, who martyred himself on the battlefield because he had given his life to God – are credibly, even tangibly, human. But no royal township or burial place of Mercia's earliest kings has yet been identified or excavated; their internal spheres of influence are obscure; the geography of their lives invisible.

There is an intriguing footnote to Penda's *curriculum vitae*. His sole possible material legacy is a bagful of scrap weapon fittings and Christian paraphernalia, discovered in a field near the village of Hammerwich in 2009 and now known as the Staffordshire hoard. Sword pommels, buttons, belt buckles, pins, rings, hilt plates and fragments of a prince's helmet were bundled into a sack, buried and somehow forgotten. After nearly 1,400

Dazzlingly fresh and vibrant after 1,400 years in a Staffordshire field.
Battle scrap or part of King Penda's 'distribution of *Iudeu*'?

years interred in a Mercian field they have emerged, dazzlingly fresh and vibrant artefacts, proof that however thuggish the early Anglo-Saxon kings were, they patronised the finest craftsmen, cunning fabricators of a sensational array of martial bling.[37] The burying of the several thousand, mostly tiny, items of gold, silver, cloisonné and garnet cannot be more precisely dated than between the mid-seventh and eighth centuries, but they form very much the sort of collection one might expect from battlefield looting, to be divvied up among comrades, distributed by a king as booty, or to be coveted like a dragon's treasure. No-one knows the circumstances in which their 'owner' – fugitive thief, warlord or king – failed to return to his secret stash to retrieve them. The battle or battles from which they were scavenged might have been one or more of any number during the late seventh century; but historians, with an eye on the find spot in the heart of Mercia, have understandably been tempted to associate the hoard with the 'incalculable treasure' that Penda was able to extort from Oswiu during his northern campaign in 654–5. Is this part of the *Historia Brittonum*'s 'distribution of *Iudeu*'?

After a short period of Northumbrian dominance, Penda's sons Wulfhere and Æðelred revived Mercian fortunes and resurrected their father's policy of Mercian expansion by military and diplomatic means. But they also began to experiment with novel means of expressing and projecting their power. They took tentative steps towards exploiting Mercia's economic resources and explored the new political, spiritual and aesthetic opportunities offered by patronising a generation of ecclesiastic and monastic entrepreneurs. The legacy of Mercia's greatest pagan warlord was thoroughly Christian.

# Chapter 2

# Wulfhere: Conversion
# 658–675

❖

*Wulfhere succeeds* • *Christians in the landscape*
• *A synod and a plague* • *The wīcs* •
*Wilfrid* • *Theodore* • *Battles*

T wo great eighth-century kings, Æðelbald and Offa, used all the tools at their disposal to plot, manipulate and finesse the Mercian succession even if, in the end, their schemes went awry. Charters, coins, marriage alliances and bland *Chronicle* entries recording rivals' deaths are eloquent, if sometimes allusive, witnesses to the ways in which Mercian royal power was exercised in its own interests, investing in the unknowable future and vulnerable to capricious fates.

The machinations of their predecessors defy historical scrutiny, however, and the swirling political and dynastic waters in which Peada's murder was engineered in the spring of 656 are murky. Peada was King Oswiu's son-in-law; his protégé. Alhflæd, Oswiu's daughter, was said by Bede[1] to have plotted her husband's betrayal and, unless some tragic domestic disharmony lay behind the deed, one must suppose that she delivered her husband into the hands of a Northumbrian assassin, an agent of her father's policy. If that is true, one must also suppose that she believed her husband to be plotting rebellion and that her loyalty was primarily to her father, rather than to her spouse. Her subsequent career is unrecorded; if she and Peada had children, their names are unknown.

For two years afterwards Oswiu ruled Mercia directly, while Peada's two younger brothers, Wulfhere and Æðelred, were kept hidden by ealdormen loyal to the old régime. King Oswiu could now claim overlordship and tributary rights over much of central Britain, from the Forth–Clyde isthmus as far south, perhaps, as the River Thames; from the Rivers Severn and Dee in the west to the Fens in the east. With such enormous political and military capital came also a new sort of hegemony under which Christianity was wielded as a tool of Northumbrian overlordship. Northumbria's kings had nailed their colours decisively to the mast of the Ionan church under Oswald, and his brother Oswiu ensured that their investment in rational, literate, Christian kingship – the written codification of the rules of divine lordship – would provide a model for the lands under his rule.

That legacy might have been jeopardised in 658 when three Mercian ealdormen, named by Bede as Immin, Eafa and Eadberht, revolted against Northumbrian rule, expelling the governors imposed on them and installing Penda's second son, Wulfhere – perhaps no more than eighteen years old – as king of Mercia.[2] That same year, unrecorded by Bede but noted in the Welsh Annals, King Oswiu's forces raided Mercia – whether as a cause or as a consequence of the revolt is not certain. Oswiu might have sought to crush the rebellion by force of arms, or he might have hoped for a pliant Mercian leader to recognise his overlordship in return for security; there is no immediate, direct evidence one way or the other. Three years after his assumption of the Mercian throne, Wulfhere is recorded in battle in north-eastern Wessex, ravaging Ashdown; then the Isle of Wight. Kings were obliged to prove their worth in battle; had to provide their followers with opportunities to prove themselves and carry home booty. Wessex was rich campaigning territory: uncentralised; wooded; hilly.

Bede is silent and the *Anglo-Saxon Chronicle* is silent on Wulfhere's relations with Northumbria. As is so often the case in early Mercian history, the political drama is visible only in silhouette. Other, subtler clues must be sought. In the aftermath of the battle on the *Winwæd* in 655, when Peada was installed as proxy ruler over Mercia, the foundations of a church had been laid in the lands of the *Gyrwe* in Middle Anglia:

... a fair spot, and a goodly, because on the one side it is rich in fenland, and in goodly waters, and on the other it has [an] abundance of ploughlands and woodlands, with many fertile meads and pastures.[3]

The *Chronicle* records that King Oswiu and his Mercian client, Peada, had come together in order to establish a religious community here in the name of St Peter. They gave it the name *Medeshamstede*, the chronicler records, because there was a spring there called *Medeswæl*.* In later centuries, after the church of St Peter had been enclosed by a wall, the site became known as the *burh* or fortress of St Peter: Peterborough.

Bede confirms that this new monastic community, the first in Mercia, was entrusted to an Abbot Seaxwulf.† This founding act, deliberately imprinting a physical manifestation of Christian kingship in a new land, is loaded with political as well as spiritual significance far beyond Peada's earlier appointment of a bishop of the Middle Angles, an Irishman named Diuma, in about 653.[4] In the first place, the timing of the *Medeshamstede* foundation sometime between 655 and 658 suggests a parallel to the monastic foundation at Gilling in Yorkshire, a few years previously, which was established as a form of *wergild*, or recompense – guilt money, if you like – by King Oswiu in expiation for his murder of a kinsman, Oswine. Oswiu had also killed Peada's father in the great battle on the *Winwæd*. In supporting his elevation to the kingship in southern Mercia, Oswiu had paid the political debt owed to his son-in-law; *Medeshamstede* seems to have been part-payment for the spiritual debt. But at the same time King Oswiu, as a royal sponsor of the church in Northumbria, was executing a policy of ecclesiastical expansionism, an extension of soft power beyond the Humber.

The new Fen-edge community must have been supported by sufficient and suitable estates to support its material needs, and these could not be conjured up from fresh air. Land must be found; and all land lay under the lordship of either a king or an élite family. Land was held 'of the king' in return for military services and food renders – the *feorm* – delivered to a royal estate

---

* Old English: Medi's well or spring. Watts 2004, 470.
† Later Bishop of Mercia c.676–92. The name has an East Anglian flavour, there being at this time no suitably qualified Mercians.

or *villa regalis*, or consumed by the royal household on the hoof. No *subregulus* or *princeps* of the *Gyrwe* is mentioned in the *Chronicle* entry recording the foundation at *Medeshamstede*. Had its lord been killed in the battle on the *Winwæd*, his lands now available for donation to the church? The earliest land donations to found new churches were often in the order of ten or twenty hides, the hide being a unit of assessment of tax in kind, or render, broadly reflecting the amount of land needed to support a family. Bede used the word *familiae* as a Latin equivalent and in the early charters terms such as *manentes* and *cassati* are also used. Since the form of the original *Medeshamstede* foundation grant is unknown, so the size of the donation – a measure of royal investment, if you will – is also obscure.

Nevertheless, subsequent royal patronage of *Medeshamstede* and its environs is a sign of its importance as a founding house of Mercian Christianity and something of a litmus test of Mercian relations with Northumbria in the first years of King Wulfhere's rule. An immensely long and detailed *Chronicle* entry under the year 656 records how King Wulfhere 'for the love of his brother Peada, and for love of his sworn brother Oswiu', had the building of the new monastery at *Medeshamstede* completed and endowed with huge estates to support its community, 'according to the advice of his brothers Æðelred and Merewalh* and according to the advice of his sisters, Cyneburh (wife of Alhfrið, *subregulus* in Deira) and Cyneswið'.[5] This can be read to mean that the Mercian establishment under Wulfhere wished to cement his predecessor's (if not his father's) co-operative relations with its northern neighbour, and his own support for the wider conversion of the Mercian people. Even if the *Chronicle* entry's minute description of the estate's boundaries and holdings in its first evolution was a retrospective insertion – perhaps even wishful thinking – there

---

* Merewalh was a king of the *Magonsæte*, variously recorded in legends relating to his daughters who were founders of several monasteries. He may have been Penda's son; more likely he was a foster son. See Chapter 3, p. 92.

is no reason to doubt the later Peterborough tradition that Wulfhere was a patron and donor. His endowments must date to a time before 664, when Archbishop Deusdedit, who confirmed the gift as a witness, died.

Mercia became decisively a Christian kingdom under Wulfhere's rule. He implemented a pragmatic policy of rapprochement with his immensely powerful former enemy north of the Humber. The policy seems to have borne fruit: there is no record of any military conflict between Wulfhere and King Oswiu before the latter's death in 670–1, while the sons and grandsons of King Penda remained powerful into the next century. How much the success of Mercian–Northumbrian diplomacy owed to the diplomatic skills of Peada's widow, Alhflæd, or her sister-in-law Cyneburh can only be a matter for speculation – but she came from a family of conspicuously influential and politically active women. If she played the part of peace-weaver, she played it invisibly, but with skill.

Cautious historians have very rightly raised warning flags over the 656 entry in the *Chronicle*.[6] Its length and descriptive detail are out of kilter with other contemporary entries. Two surviving charters, both very close in wording to the *Chronicle* entry, are widely regarded as twelfth-century forgeries.[7] Suspicions are confirmed by the fact that the version of the *Chronicle**in which these entries exclusively appear was itself compiled at Peterborough in the same century – it reflects Peterborough interests. Even so, *Medeshamstede's* later fortunes under Mercian kings show that the traditions recorded in the *Chronicle* are broadly genuine, if retrospectively gilded.

No contemporary English charter survives before the 670s so a written founding record is too much to hope for. But accounts of later donations to the monastery show that a ritual involving the placing of a sod of turf, from the land in question, on a book, sealed the deal.[8] There is something both eternal

* The 'E' version, also known as the *Laud Chronicle*. Garmonsway 1972, XXXIXff.

and elemental about this earthy means of 'booking' land in perpetuity – a religion of 'The Book' planting deep roots in what missionaries believed to be the fertile ground of Anglo-Saxon society. Later compilers of monastic records were keen to show written proof of the traditions handed down to them across the generations and centuries; and they were not above creating records of gifts and privileges that they believed *had* existed or *ought* to have existed.

The monastic community at *Medeshamstede* was established on the north bank of the River Nene – poised between fen and farmland in an area that had been densely populated and exploited during the Roman period. Seasonally flooded water meadows brought the gift of fertility, supporting large herds of cattle and flocks of sheep. Abundant fish and fowl were trapped in the many meres that laced the western Fens and salt could be extracted from the tidal reaches of the Wash, with peat as a seemingly inexhaustible fuel to evaporate it from brine. A famous hoard of Christian silver church plate, indicative of an active late Roman religious community, was found close to the town of *Durobrivae* a few miles upstream from *Medeshamstede* at Water Newton, in 1975.[9] *Durobrivae* had supported a thriving pottery industry manufacturing prestigious colour-coated wares from the second to the fourth century and at Castor, close by, a palatial Romano-British stately home had once stood – a magnificent statement of regional wealth. The Nene gave riverine access upstream, possibly as far as *Hamtun* – Northampton – and downstream to the Wash and beyond.

Given the long-held suspicion that these Fen-edge territories maintained a British-speaking cultural tradition into the centuries after Rome's fall,[10] it is tempting to propose that this first Mercian monastic community was established on receptive ground, not in the Mercian heartlands of the Upper Trent valley, but in the more loosely affiliated lands of the Middle Angles where the memory of active Christian worship may have survived. It is striking, on a map of the region, to see how many Mercian monasteries were founded on the banks of the major rivers that drained into the

Wash from the west and south-west: Witham, Welland, Nene, Great Ouse and Cam. In the eighth century, if not before, rivers brought wealth and opportunity to Mercian minsters.

It is not known when, or in what circumstances, the young King Wulfhere was baptised. His career shows that he was a pragmatic Christian convert. He understood the political and spiritual benefits of a policy of generosity towards the church, a policy entirely compatible with the parallel programme of aggressive territorial expansion expected of a warrior lord, the leader and embodiment of the fortunes of his people. Peada had returned from his own baptism and marriage in Northumbria with four priests: Cedd, Adda, Betti and Diuma, the latter of whom became the first bishop of the Middle Angles.[11] Aside from Diuma, an Irishman, the others were Northumbrians – Cedd one of four celebrated brothers trained by Aidan, the Irish first abbot and bishop of Lindisfarne. Together, they embarked on a programme of evangelism and conversion, first unopposed by King Penda, then actively encouraged by his successors.

It suited Bede's purpose to portray such men as *milites christi*, soldiers of Christ, taking the cross into new, often hostile or unreceptive territory; and there is no reason to doubt that, at times, they faced inertia and opposition. The Old English place names of Mercia preserve a score and more of sites where heathen cults were practised, from Wednesfield and Wednesbury in Staffordshire, containing the name Woden, to Harrowden in Bedfordshire (Old English *hearg*, a heathen shrine) and Weedon in Northamptonshire (Old English *weoh*, also a shrine).[12] But since the western marcher lands of the *Wreocansæte*, *Hwicce* and *Magonsæte* supported British bishops and communities into the seventh century; since it is possible that some or most of the Middle Angles were either British Christians or had been exposed to Christianity; and since Peada and Wulfhere were the mission's active supporters, their task may have been smoother than is often supposed.[13]

That does not mean that men, women, communities and peoples did not actively ponder, doubt, debate, reject and struggle spiritually with conversion. Bede records in some detail the personal and political angst suffered by King Edwin before his dramatic conversion in 627. The hagiographer Eddius Stephanus, writing the life of Bishop Wilfrid, the most controversial churchman of his age, told how once the holy man miraculously restored an infant boy to life in a village, now unidentifiable, called *On Tiddanufri* – in the hinterland of Ripon, North Yorkshire. His price was to demand that the boy be given over to him at the age of seven and taken into the service of God. When that time came, Wilfrid found that the boy's mother had fled the area and was now living with a 'band of Britons' – perhaps in the British lands of Craven, in the western Pennines. Wilfrid sent his reeve or steward, Hocca, to fetch the boy by force and the youth was duly taken into the church. He later died in an outbreak of the plague.[14]

Pope Gregory, at the beginning of that century and in the first birth pangs of the fragile Augustinian mission to Kent, had stressed the value of a hearts and minds approach – bringing the Christian word and its material expressions – icons, relics, books, crosses and murals – into existing places of spiritual value; combining the Christian feasts with their heathen counterparts. In the eighth century some hard-line missionaries showed themselves to be inflexible, as Iona's first evangelist to Northumbria had been. The more successful of their number, like Aidan at Lindisfarne, were wiser; subtler; more compassionate. Many Mercians, Wulfhere among them, must have faced their own dilemmas although, in general, lords followed their king's lead and the ceorls and drengs* of the countryside followed their lords.

The growth of Christianity in Britain says as much about lordship as it does about faith. Monastic endowments gave abbots control over substantial estates – they became *de facto* territorial lords.[15] To benefit from the renders of even ten or twenty hides was to control a substantial set of assets and resources. When

---

* Ceorls – farmers not eligible or obliged to bear arms; drengs – a caste of weapon-bearing minor lords; the Anglo-Saxon squirearchy.

kings donated land to monastic entrepreneurs, alienating those estates permanently from the royal portfolio, they were investing in the loyalty of the church and, very often, securing the careers and compliance of their kinsmen and women: the pioneering abbots and abbesses.

When Wulfhere became king of Mercia in 658, just over sixty years had passed since the arrival of Augustine's embassy from Pope Gregory. His mission to the court of King Æðelberht in Kent had initially been successful but had faltered after the passing of its royal patron. Its greatest success outside Kent had been in Northumbria, where King Edwin, after a long period of hesitation, had decided to follow the religion of his Kentish queen, Æðelburh, and sponsor or coerce the conversion of his nobles to the faith. Paulinus, the first bishop of the Northumbrians, had begun the construction of a church at York and extended his mission into the kingdom of Lindsey; he is said by Bede to have baptised crowds of Northumbrians in the pure waters of its rivers.[16] But on Edwin's death at *Hæthfelth* his immediate successors apostatised; Paulinus fled into exile in Kent, along with Edwin's widow and children.

By a stroke of historical fate, Edwin's nephew, King Oswald, and his younger brother Oswiu, had been baptised and educated during their exile in the Irish-affiliated kingdom of Dál Riata and Oswald's triumphant return to seize the Northumbrian kingdom from King Cadwallon of Gwynedd in 634 brought in its wake a determined mission from Iona, whose first Northumbrian seedlings were transplanted on Lindisfarne. From here, many Irish and Irish-trained priests set out to preach and convert. At the time of Oswiu's decisive victory over Penda at the *Winwæd* in 655, monastic communities had been founded at Whitby under Abbess Hild, at Old Melrose in the Tweed valley under Abbot Eata and in small numbers elsewhere across Northumbria. Older British Christian communities were still in existence at Whithorn in Wigtownshire, at Bangor is-y-Coed on the Dee and probably at many other sites in the west and north. Churches had also been built within the old Roman walls of London (at the site of the later St Paul's Cathedral), at Bradwell-on-Sea in the land of the East Saxons, in East Anglia and at Canterbury and Rochester in Kent. Perhaps the most intriguing possibility is that of a British monastic community still in existence in Penda's day, at the former Roman town of *Letocetum** – Wall, in Staffordshire – a couple of miles south-west of Lichfield, where Mercian bishops – and one archbishop – would have their see. *Letocetum* stood at a strategically vital crossroads where Watling Street intersected with Ryknield Street, controlling the headwaters of the River Trent and giving speedy access south-east towards London, north-west to Chester and the River Dee, and south-west towards the navigable River Severn.

Wall in Staffordshire, Roman *Letocetum*, where Watling Street crosses Ryknield Street.

* Alluded to in a medieval Welsh verse *Marwnad Cynddylan* in which 'book-holding monks' were attacked in a Welsh raid. Kirby 1977, 37; Blair 2005, 30.

In the aftermath of the *Winwæd* victory, Oswiu and Queen Eanflæd – Edwin's daughter, who had spent much of her early life in exile in Christian Francia and in Kent – embarked on a concerted programme of monastic foundation, beginning with twelve sites – six in Bernicia, six in Deira – each founded on a ten-hide estate. This was the first expression of a more organised political movement to plant monasteries as seedlings of royal power and spiritual commitment. Even so, churches must have been a novelty in large parts of the countryside. With one possible exception, at St Paul-in-the-Bail in Lincoln, archaeologists seeking to identify the physical remains of this earliest wave of churches in Mercia have so far drawn a blank. This is partly because such sites have, by and large, continued to be the focus of churches through the medieval period; and partly because the architecture of the early church was probably indistinguishable from that of the modest hall of a thegn or dreng. The first church at Lindisfarne was constructed of oak timbers and thatched with reed 'in the Irish fashion'[17] and it must have been rectangular in shape. Only its fittings, liturgical equipment and the ceremonials enacted inside would have distinguished it from a domestic building; and archaeologists cannot tap into these intangibles at such a remove.

Mercia's first monastic community is likely to have looked much like that at Lindisfarne and, to a modern eye, it would have been both architecturally modest and indistinguishable from a contemporary farmstead, except that it is likely to have had a cross raised in its precinct and to have been enclosed by a ditch and bank or hedge; not so much for physical protection but to separate the exterior, profane world from the spiritually pure, enclosed world of priest, monk, nun and lay brethren. There is also likely to have been a small burial ground, with simple earth-cut graves aligned on an east–west axis – each grave marked by a wooden cross or plain headstone. These first churches stood spiritually remote from the secular world: kings and bishops might visit and be invited to join the brethren in their simple meal but priests and abbots did not often pay court in return. Asceticism had been a virtue in the Irish, perhaps also in the British, church. One of Mercia's earliest bishops, Chad, was admonished by his archbishop for disdaining the use of a horse on his pastoral travels.[18]

The economy that supported such communities consisted in the livestock – sheep and goats, cattle – that they themselves husbanded and in food renders brought in from their estates: ale, grain, bread, honey, firewood, wool and so on. Aside from their evangelical and pastoral roles, they 'behaved' like any other small rural community, of the sort reconstructed for modern visitors on the site of an excavation at West Stow in Suffolk. All the available evidence suggests, however, that the communities of monks and nuns and their abbots and abbesses were socially restrictive, drawn more or less exclusively from the ranks of tribal élites – often collateral members of a ruling family. But, as models of secular households, they also included semi-free and servile members, dependents living on their lands and probably others seeking sanctuary or the relief of debt.

The church as a tool of kingship was in its infancy. King Wulfhere, spreading his political capital and conscious of the need to forge alliances in the south as well as the north, was able in the first instance to enhance his political status by an advantageous marriage to a princess of Kent. Eormenhild was the daughter of King Eorconberht (ruled 640–64) and of Seaxburh, the East Anglian daughter of King Anna. By the rules of such alliances Wulfhere rendered himself tributary to Kent, but was brought within the orbit of the most sophisticated court in the land. Kent enjoyed close links with Francia and Rome and seems to have enjoyed a sort of psychological or moral primacy among the Anglo-Saxon kingdoms. A Kentish queen was a political asset; her offspring would in theory be eligible to sit on the throne of both Mercia and Kent and she delivered diplomatic access to both her father's court and to Canterbury's archbishops. As much as his support of *Medeshamstede*, Wulfhere's marriage was a commitment to membership of the club of Christian kings and queens. It produced a son, Cœnred,* who would become king

---

* The *œ* ligature in Cœnred produces a sound like the short 'e' in fœtid.

long after his father's death; another more dubiously documented son named Berhtwald, who seems to have become a *subregulus* of the *Hwicce*; and at least one daughter, Wærburh. She became a monastic entrepreneur and respected abbess achieving sainthood, and a *Vita* was written to celebrate her life. Her association with a monastery at Hanbury in Staffordshire may be a clue to the location of Wulfhere's own homeland.[19]

\*

The 660s were a momentous decade for the Anglo-Saxon kingdoms. In Northumbria dynastic and religious tensions threatened civil war – one reason, perhaps, why King Oswiu was content to maintain a policy of soft diplomacy with Wulfhere. Similar instability in Essex led to the murder of a king and the threat of apostasy. In East Anglia a succession of short-lived kings – three nephews of King Rædwald – likewise created political uncertainty.

The Northumbrian crisis, which must have been keenly followed in Mercia, reached its catharsis in 664 when King Oswiu, under pressure from both Queen Eanflæd and Alhfrið, his son by a previous marriage and proxy king in Deira, called a council at *Streaneshalch*, almost certainly the monastery at Whitby on the Yorkshire coast. The stakes could hardly have been higher. The Ionan mission to Lindisfarne had infused Northumbrian Christianity, and the psychology of its kings, with a decidedly Irish flavour – of fiercely independent abbots not subordinate to bishops; of ascetic monastic purity over worldly governance; of arcane practices in the monkish tonsure and the dating of Easter – anathema to those, like Eanflæd and Alhfrið, of a Roman persuasion. At Whitby, after heated deliberations recorded in almost painful detail by Bede and by Eddius Stephanus, Rome won the day.[20] Oswiu's cold calculation seems to have been motivated by three considerations that overrode his own spiritual sympathies. A hierarchical church organisation under a single Northumbrian bishop with papal control over monastic governance would sit well with a parallel model of royal administration; second, by aligning himself with Rome, Oswiu might exert some influence on the appointment of Canterbury's powerful archbishops; third, he might hope to diffuse tensions between his own followers in Bernicia and those of his son in Deira. Whitby's outcome was an uneasy truce, with profound consequences: a purge of Irish churchmen opened a rift at Lindisfarne and elsewhere in the northern church that may never have healed. And if Oswiu hoped for a rapprochement with his son he was to be disappointed. Northumbria's discomfort was manna for Wulfhere.

Bede, writing of that same year, 664, when a solar eclipse in May was seen by many as an omen of ill times, knew that soon afterwards a devastating pestilence had spread across the land from the southern kingdoms as far as Northumbria.[21] It reached Ireland at the beginning of August and in the following years became known there as the Great Mortality.[22] Bede had survived an outbreak – perhaps a recurrence – of the plague at Jarrow in 686 when he was a young oblate in the monastery; it wiped out most of the community, leaving just two monks capable of conducting the liturgical offices – one them may have been Bede himself.[23]

It is impossible to know the full extent or impact of the 664 plague on rural communities, or how many of those recorded as having died that year were its victims; but Archbishop Deusdedit and King Eorconberht of Kent (dying on the same day in July),[24] Tuda, bishop of the Northumbrians, Cedd, one of the four priests sent into Mercia by Peada but now bishop of London, Abbot Boisl of Melrose and King Æðelwold of the East Saxons were among those who did not see the year out.[25] The death of Deusdedit, in particular, left a void in ecclesiastical authority exacerbated by the deaths of so many bishops that year. Canterbury's archbishop-elect, Wigheard, sent to Rome for his consecration by King Oswiu and Kent's new king, Ecgberht, died shortly after reaching the holy city.[26] His replacement would not arrive in Kent for nearly five years.

Plagues wreaked havoc on economies reliant on subsistence cultivation and animal husbandry. If fields could not be ploughed, sown or harvested; if winter fodder and fuel could not be gathered, small communities might collapse or take years to recover. So, too, in the aftermath of the devastating first wave of the plague, the fragile shoots of the Christian missions were exposed. Just one bishop, ordained canonically under the Roman rite, survived among all the Anglo-Saxon kingdoms – Wine* in Wessex.[27] Wulfhere's own bishop, the Lindisfarne-trained and Irish-consecrated Jaruman, now became a tool of Mercian diplomacy. Wulfhere had married his niece, Osgyð, to King Sigehere, one of two concurrent rulers among the East Saxons.† Sigehere and his followers had apostatised, restoring derelict temples to their old gods and hoping, Bede says, by reviving the worship of pagan idols, to protect themselves from further ravages of the plague.[28] Wulfhere's legitimate interest in Osgyð's soul might have been one pretext for intervening; but in any case Sigehere's marriage to her was an acknowledgement of Mercian overlordship: Mercia's political interests were at stake. Wulfhere sent Jaruman to Sigehere's court and the Mercian bishop, combining the role of spiritual mentor with hard-headed ambassador, was able to 'correct their error and recall the kingdom to a true belief'.

King Wulfhere's alliance with Kent and his overlordship over Essex also enabled a Mercian king, for the first time, to assert control over the former imperial capital at London. The Roman city had lain substantially uninhabited since the fifth century. A bishop, Mellitus, established his see here in 604 among grandiose but crumbling ruins, in the first optimistic years of the Augustinian mission. But he was exiled by a heathen king of the East Saxons and the see then lay vacant until the 650s. After the plague of 664 the see again fell vacant. By then, a new riverside trading settlement had grown up less than a mile to

* Pronounced 'Weeneh'.
† Osgyð was the daughter of King Friþuwold of Surrey and Wilburh, sister of King Wulfhere, according to a twelfth-century *Vita*.

the west of the old walled city, at a site that became known as Aldwych. *Lundenwic*, as Bede knew it, was 'an emporium for many nations who come to it by land and sea': a bustle of close-set plots where traders weighed, scrutinised and bargained, and craftspeople fashioned and sold their wares.[29] Boats arriving from Frankish ports bringing emissaries and merchants trading exotica such as lava quern stones and spices, pulled up on the sloping beach of the Strand.

The surviving network of Roman roads that radiated from the former capital gave access to the Mercian heartlands via Watling Street and to Kent, while the Thames, the highway of southern England, was navigable to within twenty or so miles of the Severn estuary. Elsewhere, East Anglia enjoyed access to continental trading centres from the substantial settlement of *Gipeswic* (Ipswich) on the River Orwell in what is now Suffolk. Northumbria had a perhaps more modest trading site, or *wīc*, between the rivers Ouse and Fosse close to York (then called *Eoforwic*), and Wessex was traded with Frankish ports out of *Hamwic*, close to modern day Southampton. The East Saxon port at *Lundenwic* was Mercia's only direct trading outlet to the continent: Mercian kings were much exercised by efforts to establish royal authority and trading privileges there.

In the aftermath of the plague an opportunity arose for Wulfhere to effect just such control. He had taken in Bishop Wine, an exile from the court of Wessex after a dispute of some sort with King Cenwealh and, deftly killing two birds with one stone, sold him the see of London 'for a sum of money'.[30] How Mercian kings and bishops were able to exploit *Lundenwic's* enviable trading location for their own profit becomes much clearer in later decades. But Wulfhere and Wine must already have seen the potential for extracting tolls from arriving and departing shipping. London gave direct access to Francia and beyond, from where high-value and high-status goods like precious metals, dyes, furs, wine and oil might be obtained. With control of London, Mercian diplomatic ties with Francia might now be independently and actively promoted.

## The former kingdom of Hwicce

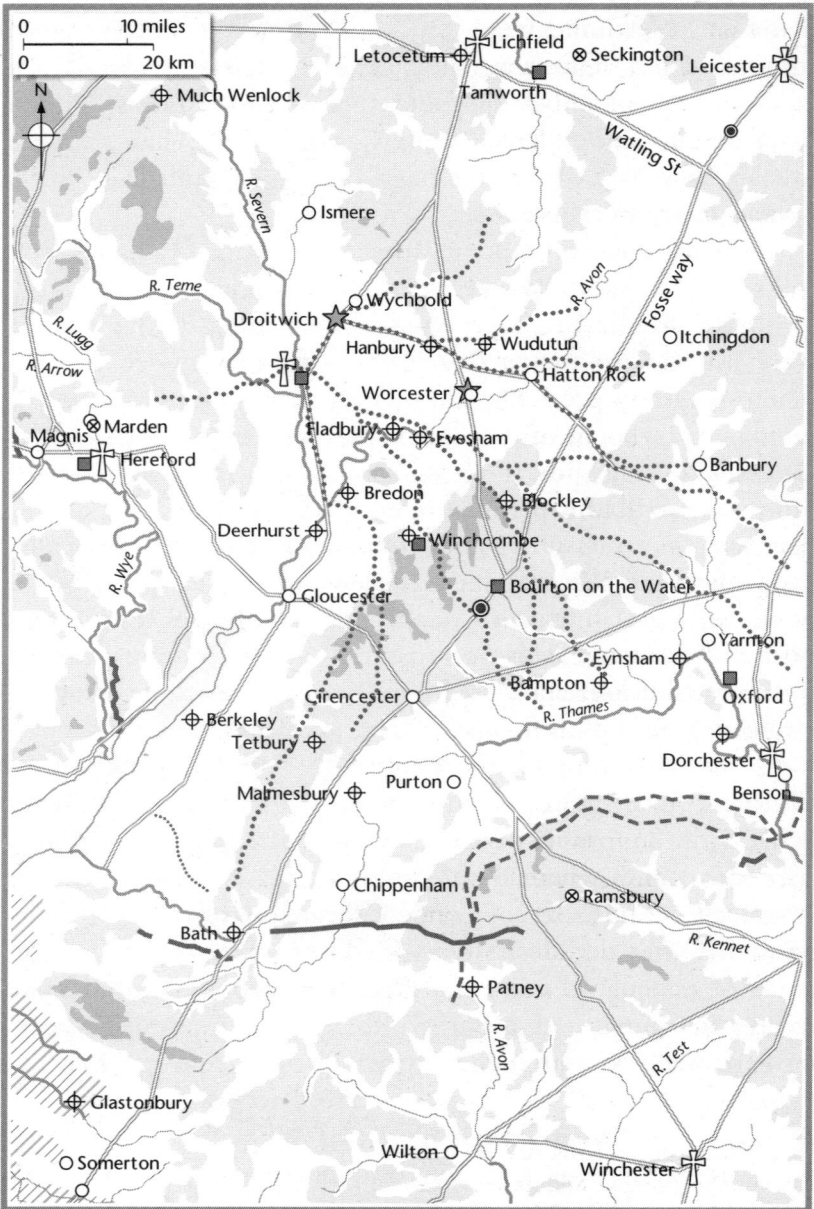

Mercian producers and traders were also able to profit from goods passing down the Thames to London and beyond. Mercia was rich in lead and wool and, in its Staffordshire heartlands, iron. Salt from the celebrated brine springs at Droitwich and in Cheshire was a highly valued commodity used in preserving and enhancing food. Less tangibly, but perhaps more lucratively, the sale and purchase of slaves – prisoners of war, criminals and the human booty of raids – was a thriving business, as Bede himself recognised in a much-quoted anecdote about a Mercian camp follower enslaved in the late 670s.[31]

In the middle of the seventh century Anglo-Saxon farms were not yet producing significant surpluses of textiles, raw materials, dairy products and meat for sale; nor were they yet famous for the beautifully illustrated manuscripts, woollen cloaks and embroidery that made English learning and arts famous across Europe. Anglo-Saxon farmers produced most of what they consumed, and consumed most of what they produced – all else was rendered to a lord.[32] There was little in the way of a functioning coin-based currency.* There were no market towns, aside from the few *wīcs*, to act as a focus for redistribution. Hard cash meant bullion – jewels, recycled coins measured by weight of metal, recyclable scrap and battle loot – or defined quantities of minerals or farm produce, often livestock on the hoof.

Only in monastic communities, where land was held in perpetuity and where investment in enclosures, technology, improved strains of corn and livestock breeding might be sustained over decades, would tax-free surplus allow the first fruits of a medieval market economy to be exploited. Whether kings or bishops – or even foreign traders – were the first to envision and promote the production of a surplus for their own

---

* In Northumbria King Aldfrið (685–705) was the first to issue a coinage carrying his portrait and name. Although only thirty-four of these *sceattas* have so far been found, they represent a significant Northumbrian issue and expanding economy at the turn of the eighth century.

profitable ends is by no means clear; but the busy markets of Francia, Frisia and Scandinavia, and the hustle and bustle of Mediterranean towns increasingly visited by Anglo-Saxon churchmen, offered tempting models.

Relief-carved sculpture from Breedon-on-the-Hill in Leicestershire: the Early Medieval imagination captured for all time.

Profitable concessions could also be deployed as gifts to gain and extend royal influence, prestige and alliances – what is now called networking. Bishop Wine, in purchasing the see of London from the king, became Mercia's man on the spot. The fact that money changed hands demonstrates that Wine was able to profit materially – probably by levying tolls on incoming and outgoing vessels or by owning seagoing and riverine vessels himself – from his position as bishop and *de facto* lord of London. An early charter dating to between 672 and 674, whose confirmatory record survives, tells of a very substantial land grant of 200 hides donated by Friðuwold, the Mercian *subregulus* of Surrey, at Chertsey on the River Thames, to the abbot of the monastery there, along with ten hides by the 'port of London where the ships come to land'.[33] The grant, confirmed by Wulfhere at his royal *vill* of Thame, was also endorsed by Osric, *subregulus* of the *Hwicce*. Friðuwold, the generous donor, seems to have been Wulfhere's brother-in-law and the father of his niece Osgyð,[34] while Eorcenwald, Chertsey's abbot, became bishop of London after Wine.

At the opposite end of the Mercian nexus, a grant by King Wulfhere to the abbot of a monastery at Hanbury in Worcestershire describes the gift of 'fifty hides of land [...]' to Abbot Colman, 'with all appurtenant meadows, woods and salt-pits' (*puteis salinis*).[35] The salt pits in question are likely to have been part of the brine pit complex at the place the Romans knew as *Salinae* – Droitwich in Worcestershire, then at the heart of the kingdom of the *Hwicce*. Rights to extract and trade salt at Droitwich, and to trans-ship it out of London, were jealously protected and exploited throughout the Anglo-Saxon and medieval periods. In partially alienating his rights to precious commodities like salt, Wulfhere was extending his prestige and influence among the *Hwicce* while retaining a majority interest in such perquisites for himself. Rational Christian kingship was also enlightened self-interest.

Early Medieval salt-making, heating brine in lead vats:
a reconstruction from Læsø, Denmark.

The plague of 664, which killed both the king of Kent and the archbishop of Canterbury, plunged the Christian kingdoms into a crisis of political vacuity. King Oswiu and Kent's new king, Ecgberht, had sent their agreed candidate, Wigheard, to Rome to be consecrated by the pope; but he had died there.[36] Without an archbishop, and with just one canonically-consecrated bishop in England, no new Insular bishops could be appointed to fill the gaps left by the plague. King Oswiu's son Alhfrið sent his most favoured priest, Wilfrid – chief protagonist of the Roman cause at Whitby – to Gaul to be consecrated so that he might be appointed to the see of the Northumbrians at York. In Rome, Pope Vitalian was temporarily unable to find a suitable, or willing, candidate to undertake the long journey to Britain and embark on the uncertain challenges of replacing Deusdedit at Canterbury. Wilfrid did not return from Gaul for three years and in his absence Oswiu appointed the Lindisfarne-trained Chad, abbot of a monastery at Lastingham on the north Yorkshire moors, as bishop in Northumbria. Since the see at Canterbury was still vacant, Chad was obliged to be consecrated by Bishop Wine and 'two bishops of the British race'.[37]

Wilfrid returned to England in about 666 to find that his promised see at York had been usurped by the uncanonical Chad.[38] He was forced to 'retire' to the monastery previously given him by Alhfrið at Ripon on the banks of the River Ure, and here he introduced the formal Benedictine rule. Inferentially, his royal sponsor may no longer have been able to support his cause against Chad. Bede hints that Alhfrið had rebelled against his father, and he is heard of no more.[39] While Wilfrid spent much of his time at Ripon, he also made advances to a new potential royal sponsor, King Wulfhere. According to the partisan *Vita Wilfridi*, he...

> ... received frequent invitations from King Wulfhere to carry out episcopal duties in Mercia. Wulfhere had a sincere liking for him. Amongst his [Wulfhere's] good works was the gift, for the good of his soul, of many pieces of land in various places to [Wilfrid, who] used them to found monasteries.[40]

One of these 'pieces of land' lay at Lichfield, which Wilfrid's hagiographer Eddius Stephanus described as 'a place highly suitable for an episcopal see'.[41]

When a new archbishop finally arrived in England in 669, Wilfrid's supporters soon apprised him of Chad's uncanonical consecration and, as a result, Chad was removed from the see at York and Wilfrid installed in his place. Chad, as a consolation, was now canonically consecrated and Wulfhere offered him the see of Mercia, formally located for the first time at Lichfield, as a replacement for Bishop Jaruman, who died that same year.[42]

Wilfrid's choice of location for a Mercian see is part of a pattern that he established over the next decade or so: sites sufficiently close to Roman towns – as at York (*Eboracum*), where he rebuilt Paulinus's church in stone at the heart of the city's former *principium*,[43] Hexham, near Corbridge (*Coria*), Oundle on the River Nene upstream from *Durobrivae*, and Ripon, near Aldborough (*Isurium Brigantium*) – to both quarry them for stone construction materials and absorb something of the reflected *romanitas* that their grand, if crumbling remains inspired. Lichfield lies just a couple of miles north of the Roman town of *Letocetum*, with its important road links and possible former British monastery. While Iona's daughter house on Lindisfarne had been fashioned of hewn oak and thatched in the Irish manner, Wilfrid's churches were built of stone and mortar, with glazed windows and lead roofs – craftsmen being brought in from Gaul to show the English what a proper Roman church ought to look like: eternal; performative; unshakeably monumental.[44]

Chad and Wulfhere established a brief but fruitful relationship at Lichfield. Not far from the church, which may already have been established by Wilfrid, Chad constructed for himself a house or oratory where he could read and pray with a few chosen brethren who had come with him from Lastingham.[45] The king also gave him a generous estate of fifty hides of land in a place called *Ad baruae* in Lindsey, generally accepted as Barrow in north Lincolnshire, close to the southern shore of the Humber estuary. Here he founded a new monastery. Whether he ever visited the place is unclear: Barrow lay in a remote corner of the province and at face value it seems an odd choice of site for a Mercian bishop's foundation. But on closer inspection it makes sense. The grant may have had more to do with Wulfhere's determination to rule over the province (against Northumbrian claims) than with a desire on Chad's part to found a new community there. Perhaps just as significantly, there used to be a small port and ferry crossing close by at Barrow Haven, which took travellers and goods across the Humber. That, and a royal grant of land here, suggests the presence of a beach market or trading harbour, perhaps also a ferry. Later Anglo-Saxon kings were wont to impose on the hospitality of monastic houses on their travels; Barrow would have made a suitable overnight stay for travellers crossing over the Humber into Deira.

Chad died during one of several periodic eruptions of the plague in 672 and was buried at Lichfield. Later, his bones were taken up and placed in a wooden house-shaped shrine and translated into a new church. According to Bede, the shrine incorporated a small aperture into which devotees could reach and take a handful of 'dust', highly regarded for its healing properties in both men and cattle.[46]

Canterbury's new archbishop, Theodore, was a remarkable man – perhaps the most remarkable of all English primates. Bede recounts that he was a native of Tarsus, a Greek settlement in what is now southern Turkey.[47] He found himself in Rome in the 660s among a community of Byzantine monks, at the time when the archiepiscopal see at Canterbury had fallen vacant on Deusdedit's death. Pope Vitalian had first offered the post to a North African scholar named Hadrian. Declining it, Hadrian recommended Theodore, then aged sixty-six, in his place. Theodore accepted the onerous task and was consecrated in Rome in March 668. Hadrian, who had travelled previously in Gaul, was persuaded to accompany him on the daunting journey to the edge of the world.

Bede's assiduous Canterbury correspondents were able to provide him with the sort of detail about Theodore's journey to England that historians crave. He tells us that the party in which Theodore and Hadrian travelled sailed first from Rome's port at Ostia to Marseille; then journeyed overland to Arles. Arles as a staging post suggests passage by boat up the River Rhône to Lyon; thence up the navigable Saône as far as Chalon. The party is next located in Paris, where they overwintered. News of their arrival there reached King Ecgberht in Kent. He sent his reeve, whom Bede names as Rædfrið, to conduct them to the Channel port of *Quentovic*.[*] Theodore spent some time in the port, being too unwell to travel, while Hadrian was detained because of some diplomatic difficulty – bizarre suspicions that he was a spy in the employ of the emperor in Byzantium. Overcoming all

* *Quentovic* lay close to the mouth of the River Canche a little upstream of the modern town of Étaples.

setbacks, Canterbury's new archbishop arrived in Kent in 669. It is extraordinary to think that this already elderly man would remain in office for more than twenty years, until his death in 690 at the age of about eighty-eight.

Assessing the state of the church in the Anglo-Saxon kingdoms, Theodore saw the need for immediate reform of the episcopal structure, in the regularity of worship and in the governance of monastic orders. He and Hadrian embarked on a tour of the English kingdoms, asserting his papal authority and ordaining bishops. King Wulfhere had been excluded from the joint Northumbrian/Kentish initiative to replace Deusdedit. He must now forge his own relations with the new ecclesiastical power in the land and the death, in the year 670 or 671, of King Oswiu, aged fifty-eight, provided him with the opportunity. Oswiu was succeeded as king of Bernicia and Deira by his son, Ecgfrið, whose Deiran pedigree as Edwin's grandson ensured his acceptance among all Northumbrians. Ecgfrið had spent time as a hostage at the Mercian court in his youth: it is hard to believe that he and Wulfhere had not spent time together as foster-brothers – by no means a guarantee of amity. But Ecgfrið, like all Early Medieval kings, needed to establish his own domestic powerbase before attempting to assert any historical overlordship over his southern neighbour. Wulfhere might use those first years to reinforce his own claim to *imperium* over the southern kingdoms.

Against that background Theodore summoned his first church council in 672* at a place called *Herutford*.[48] Of the possible candidates for its location, Hertford, at the crossing of the River Lea close to the Roman Ermine Street, looks the best and most obvious fit. Along with 'many teachers of the church', four bishops attended: from Wessex, East Anglia, Kent and Mercia, while Northumbria was represented by both its new king, Ecgfrið, and by proctors acting for Wilfrid. The see of London was vacant, Bishop Wine having died, and since Hertford lay on the northern

---

* Bede's ascription of the council to the year 673 seems to have been a mistake; and he was followed by the *Anglo-Saxon Chronicle*.

edge of that diocese, Theodore may have chosen the site for its political neutrality.

Bede does not say if Wulfhere was present at *Herutford*, but Ecgfrið's party is likely to have passed through Mercian territory, requiring some form of diplomatic pass; and it is hard to imagine that Wulfhere would allow the new Northumbrian king to attend without insisting on being present himself. Bede, who had access to much Kentish material and who seems to have seen a copy, at least, of the formal record of the proceedings at *Herutford*, records that ten canons were formally agreed, covering marriage and the correct date for celebrating Easter, the independence of monasteries from episcopal interference and provision for future synods. In practical terms, the most significant canon addressed the need for more bishops but, since no firm conclusions on the matter emerged from the council, it is safe to infer that there was opposition from the existing bishops to the idea of dividing their sees.

Theodore's achievement at *Herutford* was to insist on the careful scrutiny and agreement of all parties to the precise wording of the canons, which Theodore dictated himself so that there should be no inaccuracies; no room for misinterpretation. Each member of the council was required to sign them. Among the diverse factions and regional and tribal interests of the Anglo-Saxon kingdoms, Theodore had imposed an idea of uniformity and common consent; of rationality and purpose. In the eighth century Bede, composing his narrative on the origins of a universal church and embedding at its heart an idea of the English nation – the *Gens Anglorum* – as one people, saw Theodore as its architect.

Historians of the Early Medieval period learn to read the signs of the dog that did not bark in the night-time.* Before Whitby and Theodore, mentions of 'uncanonical' British bishops are relatively common. They can rarely be located with anything but guesswork in the south-west, in Wales and in Pictland. Bishop Chad was consecrated by two of them. After Theodore's arrival

---

* The pivotal negative observation by Sherlock Holmes in Conan Doyle's short story 'Silver Blaze'.

and the promulgation of his canons at *Herutford*, the British church is largely excised from the historical record. It suited Bede's purpose to spin the triumph of Roman orthodoxy. But it seems likely that an active diocesan British church survived the first decades after Whitby: in Wales and the south-west; also in the Pennine kingdoms of Elmet and Craven; in Rheged; perhaps, for a while, in *Hwicce* and among the *Magonsæte*. Theodore's episcopal reforms, enacted at the end of the 670s, may finally have sealed their demise or absorption into the Roman milieu.

A year after the council at *Herutford* King Ecgberht died in Kent and, since Wulfhere was the uncle of his two young sons by marriage, one historian has suggested that the grant of land at Chertsey donated by Friðuwold of Surrey and confirmed by Wulfhere may date to this brief period – before Hloðere succeeded his brother in 674 and repudiated any idea of Mercian control in Kent.[49] King Cenwealh of the West Saxons also died within a year of the council at *Herutford*. With Ecgfrið yet to flex his political muscle north of the Humber, Wulfhere, still perhaps in his thirties, might now seek to exert control over much of central and southern Britain.

In 674 he attacked Northumbria directly[*] – the first record of military action by the Mercian king since the early 660s. Wulfhere, being a man of 'proud mind and insatiable will' (more or less a qualification for Early Medieval kingship) assembled a coalition from all the Southumbrian[†] provinces, intent not merely on war but on the enslavement of the Northumbrians – an exaggeration, surely.[50] Ecgfrið, himself one of the most bellicose of Bede's kings, met Wulfhere's forces in battle at an unknown location and comprehensively defeated the Mercian army, if Wilfrid's biographer Eddius Stephanus is to be believed. Bede's more careful phrase is 'put him to flight'.[51] Either way, Mercia

---

[*] The campaign appears in Wilfrid's *Vita* (*VW* 20) but was, oddly, ignored by the *Anglo-Saxon Chronicle*.
[†] Literally south of the River Humber; in effect, all those kingdoms over which Wulfhere had control in 674.

had once more to pay tribute to Northumbria, and it seems that the province of Lindsey, once an independent kingdom, was again brought under Northumbrian control.

Wulfhere's northern ambitions having been decisively thwarted, he turned his attention in the following year to Wessex, where King Cenwealh had been succeeded by his widow Seaxburh, according to the *Chronicle*. Bede's understanding was that Wessex was governed for the next decade under a variety of sub-kings. At any rate, Wulfhere saw an opportunity to invade and extend his overlordship in the south. He had long ago forged an alliance with the South Saxon King Æðelwealh, after his capture of Wight and the lands of the *Meonware* (along the River Meon, in modern Hampshire). Wulfhere had sponsored Æðelwealh's baptism and gifted him those West Saxon territories. Now, in 675, very likely drawing on that alliance for military support, Wulfhere took his army south and fought a West Saxon force under King Æscwine at a place called *Biedanheafod*, whose identity is unknown.*

The *Chronicle* records Wulfhere's death, probably in his late thirties, the same year. Henry of Huntingdon, who wrote a history of England in the twelfth century and who had access to a now-lost version of the *Anglo-Saxon Chronicle*, understood that Wulfhere died of injuries sustained in that battle.[52] The political capital that he had accumulated over his seventeen-year rule was rendered worthless by two catastrophic military failures within a year against apparently weaker opponents. High-stakes warfare brought great rewards; but also great risks.

By the time of Wulfhere's death in 675 only Eadwulf in East Anglia and the two joint rulers of the East Saxons had been in power for a decade. In Kent, Wessex, Mercia and Northumbria young, unproven kings must begin to accumulate the political and military capital necessary to impose themselves on their neighbours, allies and enemies.

---

* *Anglo-Saxon Chronicle. Biedanheafod* may plausibly be identified with Bedwyn in Wiltshire, where a dyke seems to defend the line of the Roman road between Mildenhall and Winchester.

# Chapter 3

# Æðelred: Foundations
# 675–704

❊

*Homestead and township • Mercia's wealth*
*• Council of Hæthfelth • Monastic patrons*
*• Lords spiritual and temporal • Wilfrid*

M ercian artisans went into the woods with their axes to fell timbers for the beams, staves and props with which they would build their homesteads, knowing – like King Ælfred, who would later write eloquently on such matters – that their materials were perishable; that their halls, like their lives on earth, were mere temporary resting places on a longer road.[1]

Farmers are conservative architects and so, across half a millennium, Early Medieval halls were engineered to a consistent form, often on a roughly double-square rectangular plan, with regularly spaced timber uprights, either still in the round or squared off, set either directly in the earth or into trenches or sill beams. Walls might be filled in with hazel wattle panels daubed with a clay and dung mixture, or with planks split and trimmed using adze and axe. Roofs were covered with thatch or reed; either double pitched with gable ends, or hipped. Often, doors were set opposite each other at the centres of the long sides, to allow in as much light as possible and to protect against the wind, whichever way it was blowing. These are expedient technologies, requiring a minimum of conversion from felled 'standard' trees or underwood cut from local coppices. Since no hall survives, archaeologists take their cues from poetry or contemporary description and imagine that, despite such simple engineering, a hall might be much more than strictly utilitarian in appearance. It might be magnificently embellished with painted carvings, dragon-scale roof shingles, and furnished with fine textiles and the martial bling of warrior culture.

Buildings erected in this way could not, unlike the great halls of medieval England, be prefabricated. The most complex joinery takes place where wall meets roof – invisible to archaeologists – where each joint must be fashioned to accommodate natural variations. Problems were encountered and solved expediently: it was a technology of farmers, rather than carpenters. Only the most prestigious buildings – the great halls, like that at Yeavering in Northumberland – benefitted from the work of what we might genuinely call architects, conceived to dominate in grandiose settings, with raking buttresses, porches and lofts for storing rendered produce. When, inevitably, the posts in such earth-fast buildings rotted and the structure was either abandoned or rebuilt, possibly recycling many of the materials, the pattern of the posts – and very occasionally the wood of the post itself – remained in the earth. Those fleeting, unnamed lives leave an enduring record for excavators to interrogate.*

Since the fifth century the typical workshop in such settlements had been constructed over a rectangular pit, perhaps a yard or so deep – for storage or aeration or to provide extra head room when needed – with a post inserted at either end supporting a ridge and tent-like roof that sloped steeply almost to ground level. A plank working floor was often suspended over the pit. These structures, so-called sunken-featured buildings, leave distinctive traces in the earth, even when ancient ground surfaces have been lost through ploughing or erosion. Often the pit, filled in deliberately, or by the sands of time, yields precious, if enigmatic material evidence – loomweights, for example – of the business of running the homestead. If all other material evidence of Mercian lives – weapons, trinkets, tools, skeletons, sculpture, coins and pottery – was lost to us, the outlines of their houses,

---

* These are the dwellings that archaeologists find; but there are not enough of them for the population we suspect must have existed. John Blair has suggested that many more houses must have been built with rubble or earth foundations, traces of which have not survived. Blair 2019, 51.

workshops, granaries and smithies would be eloquent testimony to generations of farmers and craftspeople; to their sense of community and pragmatism; to a lifelong bond with the earth of their homeland.

Close to the village of Alrewas, in south-east Staffordshire, three rivers meet: the Trent, Tame and Mease. The name Alrewas derives, appropriately, from Old English *Alor-wæsse*:* alluvial land with alder trees. The river gravels and sands of the floodplain here were much favoured by prehistoric, Roman and Early Medieval farmers for their light, well-drained and easily cultivated soils. The natural confluence of river valleys – a unifying, unbreakable thread in people's lives – became a landscape of assembly and in its Mercian heyday its people called themselves the *Tomsæte*, dwellers on the River Tame. In the twentieth century those same sands and gravels attracted quarrying companies, to whom archaeology owes its knowledge of those people's daily existence.

This is Mercia's heartland, its 'cultural core': level, fertile land, cleared, de-stoned and cultivated over thousands of years, in which human endeavour and emotional energy were invested not just in agriculture, but also in ceremonial monuments, burials, routeways and, above all, farming settlements.[2] Neolithic henge monuments and a rash of Bronze Age burial mounds still visible to, or stored in the folk-memory of, its inhabitants, reinforce the idea that where rivers meet so too do people. A scatter of Iron Age farms perched on the terrace above the braided channels of the Trent speak of the land's wealth, while a Roman road called Ryknield Street, running dead straight alongside the Trent here, linked two important towns – *Derventio* (Derby) and, via the important brine pits at Droitwich, *Vertis* (Worcester) on the River Severn. Four miles along Ryknield Street to the south-west of Alrewas lies Lichfield and, beyond, the crossroads at *Letocetum* (Wall) where Ryknield Street intersects with Watling Street. Mercia's heartlands were well-connected.

---

* All place-name derivations are from Watts 2004, unless otherwise stated.

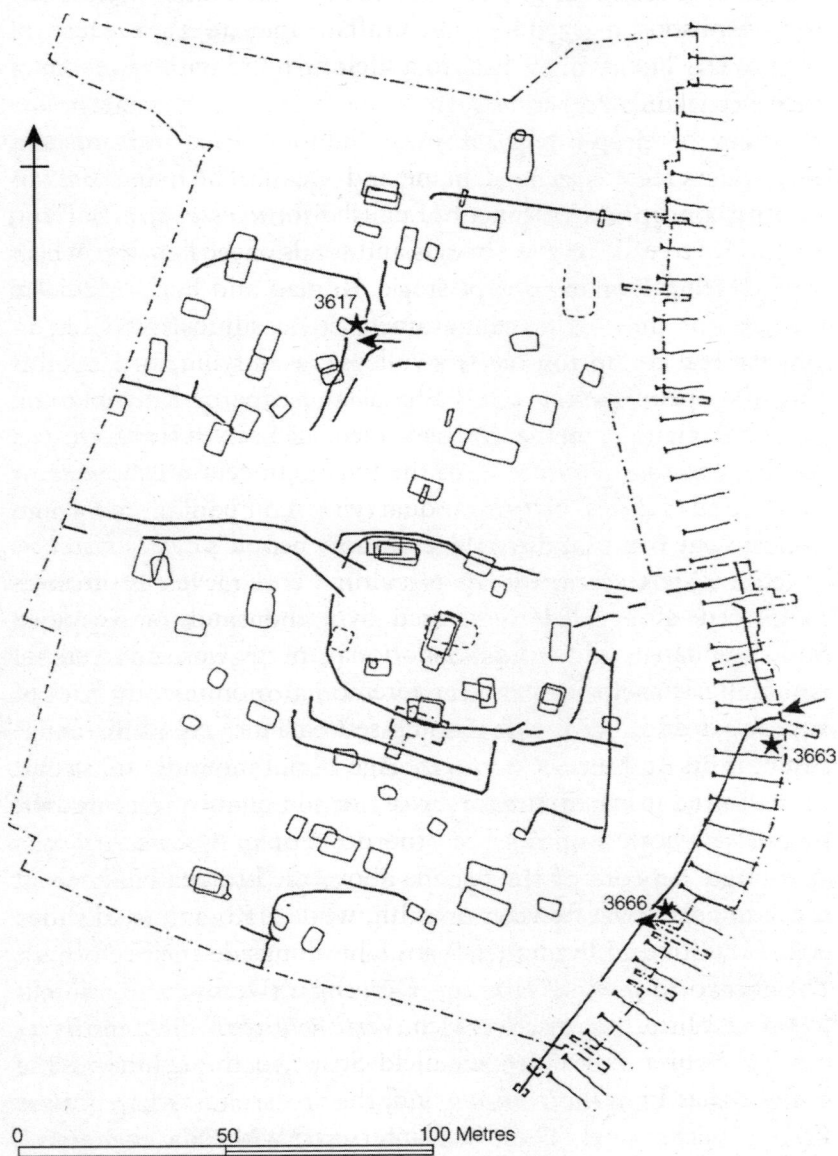

Farmers are conservative: quintessential Middle Anglo-Saxon settlement
forms at Catholme, Staffordshire, in the Mercian heartlands.

North of Alrewas, just downstream of the rivers' confluence and seemingly squeezed between the Roman road and the terrace edge of the Trent Valley, lies Catholme farm. The name credits a Norse speaker, a sometime farmer here – it means Catta's *holmr* or island.[3] A deep-time palimpsest of monuments concentrated along the valley is echoed in other local names that speak of identity, sometimes of function or of distant relations: Walton, on the east bank of the river, is the estate centre of the Welsh – *weala+tūn*; Drakelow is perhaps Old English *draca+hlaw* – a dragon's mound – the haunt of children's fantasies rooted in ancient burial; Burton-on-Trent is *byr+tūn*, a fortified central place; Stretton is *stræt+tūn*, the settlement on the Roman road. Wychnor, a hamlet on the left bank, was Old English *Hwiccenofre*: the slope of the *Hwicce*,[4] a distinctly shaped hill that seems to have marked a key crossing point over the Trent on the route south towards the lands of the *Hwicce*.[5]

In the 1970s a campaign of excavation at Catholme in advance of quarrying revealed the post-hole foundations of sixty-five buildings – halls and workshops – set in an organised and stable landscape of enclosures and trackways dating to the centuries of Mercia's dominance before the Viking Age. This was a substantial community. Evidence of occupation and housing at the south end of the settlement had already been lost to earlier quarrying but, by analogy with a number of other sites of the period, it looks as though the core of the occupied area shifted gradually from south to north over two centuries or more.[6] Halls and workshops ended their useful lives; then were rebuilt on a new site close by. The graveyard serving Catholme's inhabitants has not been found but two cemeteries, discovered during the nineteenth century at nearby Barton-under-Needwood and at Wychnor, hint at the burial practices of the *Tomsæte* in the centuries immediately after Roman withdrawal. These were 'furnished' burials, containing the spears, brooches and pottery that tell of non-Christian ideas about wealth, status and identity in life and the after-life.

Catholme's acid soil did not lend itself to the preservation of bone or cereal grains or other organic materials like leather and textiles – the material evidence of art and craft. Only hints of daily life remained to show where smithing took place or where pottery sherds or loom weights were discarded. Archaeologists must draw on analogies from other contemporary settlement sites – like West Heslerton in East Yorkshire or West Stow in Suffolk – to imagine the richness of material culture and the busy detail of livestock management and food processing, of domestic and industrial craft, that dominated daily and seasonal cycles.

The hearth was the physical and symbolic centre of the household. Preserved meat and fish hung in racks above the fire; grain was stored in granaries close by, raised on stilts to discourage pests. Wool was sorted, carded and spun into yarn on drop spindles even as children were scolded, fires tended, bread baked, floors swept. Chickens pecked on the chaff of threshing floors; pigs gobbled leavings from the table; goats and sheep were milked; cheeses moulded. Women wove cloth and made pots in their workshops, scutched smelly flax on posts in the open air while men ploughed, felled trees, cut underwood. Everyone joined in with the harvest and with seasonal hunting trips. Children and the unfree herded swine, watched sheep, weeded gardens or fished from coracles. Then, as now, householders locked their valuables away or hid them.

The closest contemporary written law code to have survived, that of King Ine of Wessex, dating to between 688 and 694, echoes the proverbial idea that good fences make good neighbours:

A ceorl's homestead [worðig] shall be fenced both winter and summer. If they are not enclosed, and his neighbour's beast gets in through the opening, he shall have no claim on the beast; he is to drive it out and suffer the damage.[7]

Penalties – fines, floggings or worse – were meted out for a range of offences steeped in rural concerns – theft of cattle, or swine feeding on nuts in someone else's wood pasture without consent – while animals carried a tariff of values in case of dispute or compensation: an ewe with her lamb was worth twelve shillings until twelve days after Easter; the horn of an ox was valued at ten pence, while the value of a Welsh rent-payer's life was equal to the fine for fighting in a minster church – worth less, that is, than the life of an Englishman. The fine for 'leaving a lord' without permission – in other words absconding – was sixty shillings, the same as that for burning a tree in a wood: 'for fire is a thief'. Members of the king's own household enjoyed a *wergild* – their head-price – ten times as high as that of a ceorl. Outsiders might be viewed with suspicion: traders appearing with their wares must transact their business before witnesses; and a 'foreigner' found in 'the wood off the track' might be taken up as a thief.[8]

Whatever the prosaic toil of the seasons, whatever the rank or ethnic identity of the people who lived at Catholme – and this is a region of ambiguous affinities if ever there was one – they understood and referenced the past. Between the core of seventh-century compounds and the river stood the remains of a Bronze Age burial mound, and the absence of any intrusive structures here suggests that Catholme's Anglo-Saxon (or Welsh) inhabitants revered or feared it – dragons or no dragons.

Catholme's community was made up of four or more discrete households, judging by the combination of halls, workshops and enclosures rebuilt, time and again, in a consistent configuration across generations. Perhaps each household claimed descent from the same founding family – were kin as well as neighbours. In the busiest enclosure a succession of halls seems to have been served by several ancillary buildings: workshops, barns, perhaps accommodation for dependent members of an extended family or for royal guests. Several of them had been rebuilt more or less in the same place, sometimes on an altered alignment. The last building in the sequence was a more ambitious L-shaped hall associated with a new, perhaps more formal enclosure line. It reflects a general theme: in Catholme's later phases a sense of coherent planning emerges, with buildings constructed at carefully surveyed right angles using a consistent short-perch measure. *

---

* About fifteen feet or five paces: Blair 2018, 70.

If the cultural and economic influences that shaped Catholme's history are obscure, the form and layout of its buildings belong very much to an eastern regional tradition; Catholme is a western outlier, so far as the hall-and-workshop tradition goes. Among the Britons to the west and north, homesteads are much harder to identify archaeologically. A tradition of rubble, turf or mud-wall buildings, like those occasionally excavated in the Pennines and Cumbria, in Cornwall and elsewhere, was much less likely to leave easily detectable traces. Such structures, perfectly sound and practicable in themselves, once abandoned, have weathered away or been ploughed out by later farmers.

A sense of organisation and long-term continuity at Catholme, together with its prime location close to Lichfield, the Roman road and the river confluence, reinforces an idea that its inhabitants enjoyed relative wealth and high status. The larger, L-shaped hall in the central enclosure seems to have become the pre-eminent dwelling and at times it may have hosted Mercia's peripatetic kings as they feasted on the land's surplus. A charter, dating to before 692, which recorded a gift of land by King Æðelred to one of his abbots, mentions the king's own *vill* and chamber at *Tomtun* on the River Tame. *Tomtun* has often been associated with Tamworth, a later 'capital' of Mercian kings; but John Blair suggests that Catholme is a good candidate for this lost royal township, where renders of food and service would be drawn from surrounding communities for periodic consumption by the king's *comitatus*.[9] Even if that is not the case, the charter reinforces the sense that the River Tame and its people, the *Tomsæte*, lay close to the centre of Mercian power, in a paradoxically liminal landscape where the later county boundaries of Derbyshire, Staffordshire, Leicestershire and Warwickshire meet. Was this the original marcher land of a people self-defined as border folk?

Within each household in such settlements a minor model of lordship exercised social control. Slaves and young children stood at the base of the social pyramid; then adolescents and perhaps artisans; then young men of weapon-bearing age and marriageable women, with husband and wife at the top: lord and lady of their modest kingdom, perhaps enjoying the security of inherited wealth and land. Men and women of rank also acted as patrons of their extended families, investing social and political capital in maintaining, extending and displaying their influence through alliance, marriage, gift and canny transaction.

Royal households and monastic communities were models of the same close-knit, custom-bound hierarchies in which ambition, solidarity, love, hate and the tensions of everyday life played out in dispute and resolution, wisdom and folly, violence and compassion. The children of Penda and the brethren of Chad and Wilfrid alike strove to resolve those tensions while deploying their political – and military or spiritual – resources to establish, maintain and extend their influence. Peripatetic kings had to exercise and display their wealth, prestige and power on the move, their households like great caravans of tent-bearing baggage trains, their ox-drawn wagons trundling along patchy roads between residences while reeves acted as their proxies between visits.

Kings were leaders of their warbands and masters of the mead hall; they were arbiters in disputes, judges, brokers and heads of their own households. They spent much time hunting, feasting their dependants and holding counsel with lesser lords or receiving the tribute of *subreguli*. They enjoyed the pick of traders' luxuries and news from afar and indulged their love of song and story by keeping bards or *scops* at their table. By exercising the customary rights and duties of lordship on the move they were able to extend it over huge territories; but such extensive rule rendered them vulnerable to conspiring pretenders.

In the decades after King Wulfhere's death in 675 his younger brother Æðelred began to concentrate royal capital in the Mercian core lands, the hinterlands of the Upper Trent and Tame, while attempting to exploit his enemies' weaknesses and subject them to his *imperium*. He successfully consolidated control over the Mercian satellites of Lindsey, Middle Anglia and the lands of the *Wreocansæte, Hwicce* and *Magonsæte*.

Within a year of his accession a Mercian army marched to Kent and embarked on a campaign of devastation, unmentioned in the *Anglo-Saxon Chronicle* but recorded by Bede, who accused King Æðelred of profaning churches and monasteries and of destroying Rochester, the episcopal seat of Bishop Putta.[10] Bede's informants may have exaggerated painful ecclesiastical memories of what sounds like a raid for booty, to cement the new king's reputation among his warband. This was in no way a war of conquest. So far as one can tell, at no point were Mercian forces engaged by the armies of Kent's King Hloðere; but such military exploits can hardly have endeared the Mercian king to Canterbury's archbishop.

Evidence for Æðelred's domestic policies comes from a disparate and often contradictory set of sources: charters, place names and the minutiae of excavation reports. Cumulatively, they are beginning to tell a coherent story about the origins of the English state. The economy that had propelled Northumbrian kings to lordship over so many other kingdoms was essentially mobile: armies on the move; cattle, horses and sheep rendered as tribute on the hoof; the booty of battle lugged around on wagons in chests full of scrap metal.

Eighth-century economies would begin tentatively to exploit the more marketable wealth of secondary products: wool, salt, high-value textiles, leatherwork, books and metalwork, which opened access to regional and overseas markets and to continental spheres of influence. The seeds of a new economic model were sown in the late seventh century with the emergence of a small number of trading settlements founded in the kingdoms of Wessex, Northumbria, East Anglia, Kent and at the disputed former Roman capital at London – all, it seems, under royal patronage. Some of the impetus for this, as it were, modernising trend came from across the Channel and North Sea. Northumbrian, Kentish and Irish clergy had travelled through Francia, Frisia and the Lombard kingdom of Italy as far as Rome and had witnessed the relative sophistication of continental technology, markets and politics. They took with them the finest Anglo-Saxon goods and, on returning, brought Europe's material treasures and much of its spirit back with them. Canterbury's Archbishop Theodore was steeped in both Roman and Greek culture. Frankish – and exiled native – princesses overtly influenced the policies of their Anglo-Saxon husbands; and the trading settlements at *Hamwic*, *Lundenwic*, *Eoforwic*, *Fordwich* and *Gipeswic* brought Frisian, Scandinavian, Frankish and other even more far-flung traders and commodities to Britain's shores. If Middle Anglo-Saxon kings were not at the forefront of trading initiatives, they were increasingly open to exploiting its opportunities – as King Wulfhere had when he sold the see of London and its attendant perks to Bishop Wine.

Under Æðelred, the Mercian royal family's most conspicuous investments were grants of land for the founding of minsters* – indeed, the years 674 to 704 saw an explosion in the creation of religious houses. A series of charters records donations of land by King Æðelred and his *princeps* Friðuric on which to found a monastery at Breedon-on-the-Hill in Leicestershire as a Middle Anglian daughter house of *Medeshamstede*.[11] The king's niece, Wærburh, seems to have founded houses at Hanbury in the Mercian heartlands eight miles north of Lichfield, at Threekingham in Lindsey and at Weedon in Northamptonshire where her father, King Wulfhere, had an estate. She is said to have been recalled by Æðelred from Ely, where she already held office as abbess, to rule over all the Mercian houses.[12] Other conspicuous grants of land were made to a nunnery at Barking, just east of London; to Abbot Aldhelm at Tetbury in Gloucestershire; of an estate to support a house at Gloucester by two *subreguli* of the *Hwicce*; and by the king himself to found a house at Fladbury in Worcestershire.[13]

---

* Bede uses the term *monasterium* for a religious community and, by association, its church and estates. The term minster, which later came to apply to the great mother churches after tenth-century reformations, is used as a shorthand.

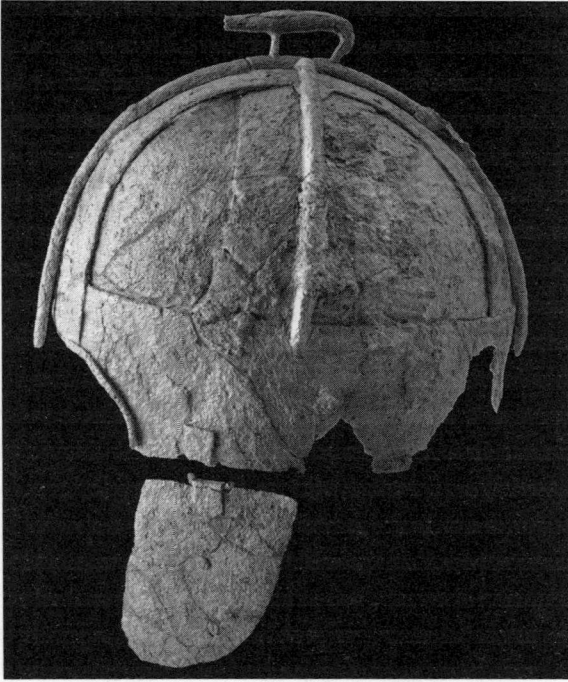

A Mercian Beowulf? The iron boar-crested helmet from a Mercian
warrior burial, found close to the Roman town of Irchester in
Northamptonshire in 1997 and known as the Pioneer.

Archaeologists have long suspected a link between depopulation
in the aftermath of the plague of 664 * and the gift of large tracts
of land – perhaps lying 'waste' and in need of investment – to
religious communities. If this generosity was a calculated punt
on the part of late seventh-century English kings, it paid off
handsomely in the next century. Nevertheless, these land gifts
functioned in complex ways with sometimes unanticipated
consequences, as the heirs of both donor and recipient were to
learn. On the one hand, kings or *subreguli* were acting in a manner
entirely befitting their role as great lords – sharing their wealth by

* See below, Chapter 6, pp. 182–3.

distributing gifts. They were also advertising their personal piety, and one frequently sees in the charters the Latin phrase *pro anima mea* – 'for my soul's sake': investing in the present with the hope of reward in the hereafter. Both the founding gift and subsequent donations of more land or relics, treasures, gospel books and hard cash encouraged prayers to be offered for the king's salvation. Very often those prayers also applied to, and legitimised, the king's chosen heir. Because such land was permanently alienated from the royal property portfolio* it followed that the king's soul would also benefit in perpetuity. Legacy became a psychological driver among kings whose antecedents had expected to die in battle, their earthly kingdom ending with them. Documents recording the gifts of patrons were maintained by many religious houses over centuries, so that prospective donors could see in what illustrious footsteps they were following.

If kings stood to gain spiritually from these gifts, they also enjoyed other benefits. Many, probably most, religious houses were founded under the rule of abbots or abbesses drawn from the king's or queen's collateral family. The estate and its assets were, in a material sense, kept within the family. At the same time, those relatives whose career opportunities had formerly been limited to fighting or child-bearing – both of which might cause kings diplomatic problems – were provided with a comfortable living, relatively free of physical violence or the risks of childbirth. Those royals of an intellectual or contemplative bent might spend their lives in study or in the creation of fine works of art: books, sculpture or embroidery. In turn, they also became patrons to those who sought to live under their rule. It became fashionable for a while, in this golden age of monasticism, for kings to abdicate – rather than waiting to be killed in battle or by the assassin's knife – and retire, either embarking on a pilgrimage to Rome or entering a house of prayer that they themselves had endowed.

---

* It became freehold or *bocland*, as opposed to *folcland*, held for a life interest only, bestowed upon a loyal thegn in return or reward for military service.

The charters in which such donations were recorded – real, forged or advantageously edited in retrospect – often provided for the religious community to be free of such obligations as military service and food renders, foisted upon thegns whose land was held for a life interest only. The first entry in King Wihtred of Kent's contemporary dooms made the terms of this novel social contract explicit:

*Cirice an freolsdome gafola...*
The church shall enjoy immunity from taxation...

*...7 man for cyning gebidde...*
... and the king shall be prayed for.[14]

Within the minster's sacred precinct those obligations, taxes in kind and labour services were regarded as having been 'commuted to prayer'. As Bede would point out in a celebrated letter,* unanticipated consequences followed. To begin with, the available land on which men of rank were raised to fight in the king's warband was, inevitably, diminished as the royal portfolio shrank and some young men of weapon-bearing age and status chose this newly available alternative career path. Saints Cuthbert and Wilfrid both bore arms in their youth, while the Mercian thegn-turned-hermit, St Guðlac, enjoyed a bloody first career as a freebooter. Second, the supply of food and services to the royal court was also diminished, leading to what modern economists would recognise as a tax deficit. As Bede would testify, in his day – he was writing in the 730s – the prospect of an easy life, free of taxation and military service, had seen spurious if not absolutely irreligious houses springing up: a cynical perversion of this age of religiosity.

Seventh-century Northumbrian, Mercian and West Saxon kings may, at first, have been complacent about such things – they could increase the land over which they ruled by conquest or

* See below, Chapter 6, p. 187.

diplomatic subjection and, in any case, the scale of their donations to the church was relatively modest. They either reasoned that the benefits outweighed the risks or they did not consider such matters at all. But by the early decades of the eighth century the tensions stirred by this new tool of kingship forced kings and their religious counterparts – abbots and abbesses, bishops and archbishops – to conceive new ways and means of managing those relationships. One solution was to impose taxation of food and service renders onto religious houses; the other was to bring those alienated lands back into the royal portfolio, by purchase or edict. Both solutions were pursued; both were vigorously contested by the church.

King Æðelred enjoyed the luxury of ignorance: he does not seem to have anticipated such problems. His policy towards the church – at least, the Mercian church – was driven by both a personal sense of piety and by the opportunities it offered to exercise his political capital. In the charters, as in all legal contracts, the devil is in the detail. A fragment from a now-lost charter, preserved among the earliest records of St Paul's Cathedral in London, provides the basic model:

> I Æðelred, king of the Mercians, with the consent and permission of my councillors, give to you, Bishop Wealdhere [of London], a portion of land in the place which is called Ealing, that is, ten hides for the increase of the monastery in the city of London.[15]

Even the apparently still waters of this bald agreement, dating to the decade or so before 704, mask hidden depths. In one sense the king was acting in an absolutely traditional manner as the bishop's lord, favouring his man in London with a gift of land and no doubt ensuring that in the bishop's church, St Paul's, prayers were said for this earthly sponsor. But the right of a Mercian king to donate land at Ealing, in the lands of the Middle Saxons, was also a territorial display – for this region had once been subject to the rule of East Saxon kings. Æðelred's assumption of lordly power in Middlesex was, therefore, also the thumbprint of overlordship, placing the kings of the East Saxons in a subordinate role. The Middle Saxon lands now lay under Mercian lordship and so, too, did the bishop.*

Minsters, landed estates and the careers of ecclesiastics might all be deployed as tools of the state, as well as accumulating royal spiritual capital. It is by such means that historians of the period track the waxing and waning fortunes of kings and kingdoms – for during the previous decade Kentish, East Saxon and West Saxon kings had in turn been able to exert influence over events in London.[16] The decayed but magnificent walled Roman city and its new, bustling commercial *wīc* to the west at Aldwych were prizes worth fighting for.

Increasingly complex charters reflect ever more subtle and ingenious means of exercising influence even over apparently independent religious houses. Sometimes, too, that influence was personal and tangible. Æðelred had taken as his queen a Northumbrian princess, Osðryð, daughter of King Oswiu and Queen Eanflæd, those pioneering monastic patrons.[17] She was also, therefore, the sister of King Ecgfrið, who had once been a hostage at the Mercian court and who succeeded to the Northumbrian throne in 670 or 671 on his father Oswiu's death. The marriage had been brokered either in King Oswiu's day or as a result of some diplomatic initiative between Ecgfrið and the new Mercian king.

In 679, five years into Æðelred's reign, a Mercian army fought against a Northumbrian force somewhere on the River Trent.[18] The encounter ended without a decisive victory for either side but King Ecgfrið's eighteen-year-old brother, Ælfwine, was killed. The death of such a high-ranking member of a royal family necessitated satisfaction through violence or compensation, as Bede recalled in

---

* Wealdhere was consecrated bishop in 693 and died sometime after 705. He is believed to be the author of the earliest known English letter to survive into the present, addressed to Archbishop Berhtwald of Canterbury.

telling the story. Perhaps as a result of an intervention by Osðryð who, as sister of the dead prince and queen of his slayer was in a perfect, if exquisitely delicate, position to act as peace-weaver, the wise counsel of Archbishop Theodore was sought. According to Bede, he 'completely extinguished this great and dangerous fire' and brokered a deal in which the compensation of hard cash – Prince Ælfwine's *wergild* – was paid and peace restored.

It is tempting to suggest that this rapprochement was negotiated at a great church council held at *Hæthfelth* in 680.[19] If this is the place now called Hatfield in South Yorkshire, close to the old course of the River Don, rather than the town in Hertfordshire, the council was held on the ancient border between Mercia and Northumbria. It is also close to the site where King Edwin had met his fate at the hands of Cadwallon and Penda in 633. The historian Ian Wood has suggested that an episode in the *Anonymous Life of Gregory the Great*, in which King Edwin's remains were found on the site of the old battlefield and taken to Whitby to found a royal cult there, may be associated with the council.[20] One might add the possibility that it formed an element of restorative diplomacy after the battle on the Trent: a healing of old wounds. From this time Lindsey became a permanent satellite province of Mercia. The battle on the Trent had, it seems, been fought for control of the province and Æðelred's army had tactically prevailed; but Northumbrian honour had been satisfied.

In these decades, Lindsey was a kingdom of interest to both Mercia and its northern neighbour, being rich in prime arable land and navigable rivers. Mercian kings had already founded religious houses there but had lost control of the province after King Wulfhere's failed invasion of Northumbria in 674. At any rate, in the aftermath of the battle on the Trent five years later, Mercia's king and his Northumbrian queen sought to invest their personal capital in a favoured Lindsey minster at Bardney, on the River Witham some ten miles downstream from Lincoln (formerly the Roman *colonia* of *Lindum*). Perhaps in the hope of pulling off a diplomatic coup in the province, they donated to the community at Bardney the nearly forty-year-old remains, that is to say the torso, wrapped in his purple and gold battle banner, of Osðryð's martyred uncle, King Oswald. No-one knows how, or from where, these relics were acquired, except that King Æðelred's father, Penda, may have kept them as trophies of his greatest victory in 642. Oswald's arms and head, retrieved from the battlefield in a celebrated raid by Oswald's brother, King Oswiu, a year after the battle at *Maserfelth*, already enjoyed a reputation as objects of veneration and for their miraculous healing properties. Now, the minster at Bardney was to be blessed with possession of the rest of Oswald's corporeal remains as sacred relics.

Such gifts were coveted by early minsters. They were likely to lead to further gifts and endowments and ensure a steady and lucrative flow of pilgrims to the site. Unfortunately, as Bede relates, the priests at Bardney took great exception to the idea of housing the relics of this long-dead Northumbrian: a foreigner and former enemy.[21] The fact that one of the patrons was Oswald's niece, another Northumbrian, apparently made the insult worse. Bede, who seems to have heard the story from Abbess Æðelhild, who had witnessed the subsequent miracle herself and who was still alive in the 730s, relates how the carriage bearing the relics was left ignominiously outside the minster precinct, a tent erected over it to protect against the elements. That night, in a classic *deus ex machina*, 'a column of light stretched from the carriage right up to heaven and was visible in almost every part of the kingdom'.[22] The monks of Bardney, duly amazed and humbled, took this as a sign of divine admonition and brought the bones into the minster, washed them, wrapped them in the banner and laid them in a shrine 'with fitting honours'. From then on, miraculous healing powers were attributed not just to the bones, but to the soil on which the water in which they had been washed had been poured. Such were the homeopathic virtues of a martyred king's remains. And so highly prized were those relics that more than a century later, when Lindsey lay firmly under Viking control, another Æðelred, the ealdorman of Mercia, and his wife (King Ælfred's daughter, Æðelflæd) found some means of extracting them and taking them back to their own royal foundation at Gloucester. Few relics were more widely travelled than those of King Oswald.

A more subtle, and consequently invaluable, narrative of monastic patronage survives in a testament found in a thirteenth-century copy of a late eleventh-century *Vita* by Goscelin of St Bertin.[23] Goscelin spent much time in England during the 1080s, researching in the archives of monastic houses. At some point he was resident at the abbey of Much Wenlock in Shropshire, at the far north end of the kingdom of the *Magonsæte* on a spur of Wenlock Edge above the River Severn. Here, Goscelin came across what purported to be the last testament of its seventh-century founder, St Mildburh. She was one of three daughters of a king of the *Magonsæte* named Merewalh (c.625–c.685) – the name means something like 'celebrated Welshman'. Merewalh is mentioned in the dubious *Anglo-Saxon Chronicle* entry for 656, which records the foundation of the Middle Anglian minster at *Medeshamstede*. Here, and in Goscelin's account, he is cited as a brother of both Wulfhere and Æðelred – a son, therefore, of Penda.[24] Historians have cast doubt on this idea, partly because the alliterative 'M' naming policy of the *Magonsætan* royal house suggests that he was a native of that people.[25] It is, I think, likely that he was a native prince of the *Magonsæte*, perhaps a foster-brother to Penda's sons. In any case, he is said by Goscelin to have married a Kentish princess named Domne Eafa, a cousin of Wulfhere's Queen Eormenhild, and to have been buried, around 685, in a minster at Repton on the banks of the River Trent in Derbyshire.[26]

Disentangling the various threads of Mildburh's testimony reveals that the minster at Much Wenlock was founded not by Mercians, nor by Merewalh, but as a daughter house of St Botolph's minster at *Icanhoe* (probably Iken in Suffolk). Botolph purchased the Wenlock site and its estate from King Merewalh – under what circumstances can only be a matter of speculation – and founded a mixed community of nuns and monks under Abbot Æðelheah and Abbess Liobsynde in the decade after about

670. But it may not have been the first religious community on the site. The earliest form of Wenlock is *Wimnicas*, a Brittonic name whose meaning has not been satisfactorily resolved.[27] Intriguingly, excavations at the site of the later medieval abbey uncovered evidence of a Roman-period building, perhaps a former villa, beneath; and elsewhere in England – famously at Lullingstone, in Kent – Roman villas housed private estate churches from the fourth century onwards. Wenlock is a good candidate for a surviving, or a refounded, British church.

Sometime in the 680s, according to her testament, and seemingly on Liobsynde's retirement, Mildburh…

> acquired this place called *Wimnicas* [...] that is, I gave to the worshipful Abbot Æðelheah and to the religious Abbess Liobsynde, in exchange [...] an estate of sixty hides at a place called Hampton.[28]

The confirmation charter drawn up by Æðelheah, witnessed and endorsed by King Æðelred, shows just how substantial these holdings were:

> ... an estate of ninety-seven hides in the place called *Wimnicas*, and in another place, by the River Monnow, an estate of twelve hides, and in another place named *Magna* an estate of five hides, and in the district called *Lydas* an estate of thirty hides...[29]

In later years Mildburh was able to expand the minster's already enviable portfolio of lands through gifts from her brothers Merchelm (the name means 'Helmet of the Mercians') and Mildrið; from King Ceolred (704–9), Wulfhere's son; and through purchase, for 'a large sum of money' from one of the king's ealdormen.[30] These later, smaller acquisitions may have been sought to address particular resource needs: woodland or water meadow, perhaps, or the ironstone for which the nearby Ironbridge Gorge is so famous. As they expanded and consolidated their holdings, the estates of such extensive monastic houses might match for wealth and size some of the smaller territories listed in Tribal Hidage.

Thus invested with both land and the patronage of royal households, Much Wenlock materialised and implanted Mercian power, exercised not on the battlefield but through political investment in its satellite territories and with the support of subordinate royal families. Patronage worked both ways, to control from above and to protect and enhance the reputation of those who owed allegiance to a great lord or lady. Mildburh died in 727, leaving both her personal testament of a career as abbess, and the material legacy of a wealthy and well-endowed minster. She was a successful monastic entrepreneur – able to use her influence and agency as both a holy woman and as a member of the regional élite to attract patrons and, inferentially, enthusiastic nuns, monks and lay brethren to ensure both the spiritual and material functioning of the minster. Such success was also founded on outstanding virtue, natural authority and piety; but it could not be achieved without a sound understanding of territorial lordship.[31]

The rash of new minster foundations during the last quarter of the seventh century took place almost entirely regardless of diocesan authority. When Wenlock was founded, a single bishop – perhaps Wynfrið, or his immediate successor, Seaxwulf – bore episcopal responsibility for the whole of Mercia, notwithstanding Archbishop Theodore's determination, voiced at *Herutford* in 672, to create many more sees. Theodore demonstrated his authority over Mercian bishops when, in the middle of the 670s, he deposed Bishop Wynfrið for 'some act of disobedience', tactfully unspecified by Bede, from which one infers that Wynfrið was resistant to Canterbury's policy of episcopal expansion and division.[32] Mercian kings would rarely, thereafter, exercise uncontested control over the appointment of their bishops; and the archbishop's political influence grew with his intervention after the battle on the Trent in 679. But tensions between abbatial and episcopal authority would rumble on into the next century, and beyond.

By 680 Theodore had accumulated enough political capital to enact his diocesan reforms. In Greater Mercia, over the next decade, he created new sees for the *Hwicce* at *Weogorna* (Worcester), the *Magonsæte* at Hereford, the Middle Anglians at Leicester,[33] the kingdom of Lindsey at Lincoln and, further south, at the former West Saxon see of Dorchester-on-Thames. Theodore was an effective and subtle, if pragmatic politician, ensuring that his new diocesan structure for the English should be based on tribal identity – Bede's *gentes* – aligned with royal authority. What financial provision was made to support these new sees is not clear but, following the examples of Lichfield and Lindisfarne, they must have been endowed with estates, donated from the royal portfolio with the king's acquiescence. Given the location of many new seventh-century sees in former Roman towns, these estates seem likely to have consisted of lands, and the privilege of collecting tribute, within the old walls – providing neatly circumscribed spiritual enceintes – and rights to the resources of their immediate hinterlands. Bishops establishing their sees in a suitably grand manner might quarry old buildings for dressed stone and recycle architectural details, as Wilfrid had done in his crypts at Hexham and Ripon and as the architects of Brixworth later did. Regional episcopal authority was thus steeped in both the material and psychological legacy of Rome and in tribal identity embedded in hereditary kingship.

Those bishops who had enjoyed supreme authority over very large territories – Wynfrið in Mercia and Wilfrid in Northumbria – understandably resisted the fragmentation of their power. But the eventual success of Theodore's reforms, including several new sees in Northumbria, rather speaks for itself: most of his sees survived the chaos of the ninth to eleventh centuries to form the backbone of medieval English ecclesiastical administration.[34] Suitably endowed, cathedrals might become centres of clerical excellence and, crucially, develop *scriptoria* for the copying of charters, bibles, psalm books and liturgical texts. In London, Lincoln, Worcester, Hereford, York and especially at Canterbury, today's incumbents can look back on an almost unbroken written tradition to their earliest predecessors more than 1,300 years ago.

Mercian kings had to find ways to accommodate these new spiritual and territorial lords, owing obedience to another master in Canterbury and ultimately to the pope in Rome; to consider how they might exploit them as political assets. In turn, the newly appointed bishops must fashion relations with the wealthy and independent-minded abbots and abbesses of their dioceses, with fellow bishops and with their kings, while themselves exercising territorial lordship. Bishops could not extend their influence by force of arms; they must explore other means.

The largest centres of population in this period were the great monasteries – 600 'soldiers of Christ' dwelt at Jarrow and Wearmouth in the early eighth century[35] – and the trade-focused riverside settlements at *Gipeswic*, *Hamwic*, *Eoforwic*, *Fordwich* and *Lundenwic*. But the founding of new diocesan capitals in former Roman towns and forts raises the question of how populated they may have been, or may have become under the patronage and wealth of their bishops. The idea of lonely bishops living in and wandering through the shattered ruins of the old empire, almost as squatters, seems at odds with worldly, pragmatic and entrepreneurial ecclesiastics like Wilfrid and Wine, men of great energy and vision and a keen eye for profit. How *did* bishops make a living?

In the hierarchy of Roman towns there was a world of difference between a provincial capital, like London, *coloniae* like York or Lincoln, a large *civitas* capital like Wroxeter or Leicester, and the roadside towns that grew out of early forts. In this context Worcester seems an unlikely choice for the seat of the bishops of· *Hwicce*. It cannot have been chosen for its civic grandeur or its monumental architecture – its Romano-British origins were military and industrial, as a centre for forging and smithing iron and pottery manufacture.[36] Those might be a clue to its value for episcopal income. It seems also to have been a focus of assembly or governance for the people, the *Weogoran*, whose name it shared in the Early Medieval period.* More obviously advantageous was Worcester's position at the southern end of Ryknield Street where it met the River Severn, and its proximity to the valuable Droitwich brine springs, which lay just five miles up the road.[37]

Monastic entrepreneurs were successful in accumulating landed estates as royal gifts, and several of the pioneering abbots also, by virtue of their novelty, held the title of bishop. It is much less clear how Theodore envisaged that his new raft of tribal bishops should fund their apostolic and pastoral missions. Bishop Wine's purchase of the see of London from Wulfhere is a hint that he already enjoyed profitable rights at London's *wīc* on the Strand. Kings granted land personally to their bishops; but we cannot say if those lands were to be held for a life interest or if, in some cases, the see itself retained the rights to them. Either way, the location of Mercian sees at Lichfield, Hereford on the River Wye, Dorchester-on-Thames, Worcester and Leicester – the last four of these lying on major Roman routeways and on or close to navigable rivers[†] – suggests that, from the outset, bishops were choosing, or being offered, sites with profit-making potential.

A now-lost grant from King Æðelred to Oftfor, briefly bishop of Worcester between 691 and 693, describes the gift of 'a shed and two furnaces belonging to the great brine pit at *Wīc*' – that is to say, rights to process salt at Droitwich.[38] If that grant allowed the bishop to sustain the needs of his own household in preserving food, it also encouraged him to profit from the trade in salt from his handily placed riverside holdings at Worcester.

---

* *Wigorna Civitate* S103: Hooke 1985, 80. One cannot exclude the possibility that the status as *civitas* capital followed rather than preceded the placing of the Hwiccian see here.

† Leicester is the most doubtful in terms of navigable rivers but the Soar, on which it lies, may have been a navigable tributary of the Trent as far upstream as Loughborough in the Early Medieval period. Portages (tarbets in the far north) may have connected navigable sections of rivers. Lincoln, uniquely, was connected by canal to the Trent and, via the Witham, to the Wash. Edwards 1987.

London's East Saxon bishops were even better placed. By the end of Æðelred's reign the *wīc* on London's Strand had become a populous settlement: *Lundenwic*. Minsters established on suitably watery islands at Westminster and Bermondsey and downstream at Barking, major churches at St Paul's in the old city and St Martin's, close to the edge of the *wīc*, were able to take advantage of the river, and of London's radiating road network, as trade routes. The Kentish law code of kings Hloðere and Eadric, dating before 685, make explicit provision for trade at *Lundenwic* to be overseen by the king's own *wicgerefa* – his port-reeve – at the king's *sele* (hall) there.[39] *Lundenwic* was no sprawling beach market, but a carefully laid-out network of parallel streets or lanes leading down to the sloping foreshore of the Strand, lying parallel to, but set back from, the river front where boats were drawn up on the beach or sat at quayside wharfs.

Archaeologists, piecing together the fragmentary, but now overwhelming excavated evidence of the *wīc's* origins, credit King Wulfhere with planning and founding the settlement that lay behind the Strand, beneath what are now Drury Lane and Covent Garden.[40] It is equally possible that Bishop Wine, having a personal stake in its success, was its primary instigator, pre-dating an interlude in the 680s when Kentish kings briefly re-established control there and Earconwald was Theodore's bishop in London. London in the late 670s is the setting for a Bedan story of a Mercian prisoner of war being sold as a slave to a Frisian merchant.* King Ine also seems briefly to have imposed West Saxon authority in London;[41] but Mercian control was reasserted by King Æðelred after the death of Bishop Earconwald in 693 when Wealdhere, recipient of the Ealing grant, was appointed to the see in his place. Subsequent Mercian kings vigorously defended their right to control London throughout the eighth

---

* Bede's purpose in telling this story in the *Historia Ecclesiastica*, Book IV, Chapter 22, is that the man, Imma, by means of a heavenly miracle, could not be restrained by earthly bonds. He was eventually ransomed and returned to his homeland.

century; but kings of Kent, Wessex and Essex seem periodically, at least, to have enjoyed rights there too.

Bishop Wynfrið had not been alone in resenting Theodore's episcopal reforms. In about 678 he embarked on a journey to Rome to plead his case before Pope Agatho; at the same time Bishop Wilfrid was also on his way to Rome, having seen his Northumbrian diocese divided too.[42] Eddius Stephanus relates a bizarre episode in which the enmity of Wilfrid's enemies followed him into Francia, with money for bribes intended to ensure that he did not return to Britain. Instead, the thuggish followers of the 'wicked' Duke Ebroin, Mayor of the Frankish Palace, set upon Wynfrið's party – mistaking, as Eddius says, the first syllable of his name – killing some of them, robbing and beating Wynfrið and leaving the luckless former bishop naked and penniless.[43]

It was, perhaps, during these few years of exile (neither the first nor the last, so far as Wilfrid's career went) that Aldhelm, the scholarly and wise abbot of the minster at Malmesbury, wrote to the heads of Wilfrid's many monastic foundations in Northumbria and Mercia. He adjured them to follow their 'father' into exile, drawing an explicit analogy with secular lords – 'worldly men, exiles from divine teaching, [who] desert a devoted lord'. Were they not 'worthy of the scorn of scathing laughter and the noise of mockery from all?'[44] Holy abbots enjoyed, by the cultural codes of Early Medieval society, the status of their secular counterparts: as heads of their household; as patrons; as leaders of the fictive family of the enclosed community or bishop's retinue, which bore to contemporaries such striking equivalence to the *comitatus* or warband.

After many adventures and diversions Wilfrid arrived in Rome in 679. Here, he persuaded Pope Agatho and a large convocation of bishops to support his cause for reinstatement in Northumbria and the removal of his rival bishops there. Returning triumphantly the following year with the pope's written judgement in his favour, Wilfrid presented it to King Ecgfrið 'all stamped with bulls and seals'.* Perhaps unsurprisingly, the king, his councillors and the bishops of the Northumbrian sees rejected Wilfrid's petition. From about 680 he found himself imprisoned in a royal *vill* called *Broninis* – somewhere in Northumbria, but so far unidentified – and later in Dunbar on the coast of East Lothian.

Ecgfrið's and Queen Iurminburh's implacable hostility to Wilfrid, which Eddius attributes to their jealousy of his 'temporal glories',[45] was eventually tempered by the wise counsel of a senior member of the royal family – the Abbess Æbbe, sister of kings Oswald and Oswiu, who must by then have been in her late seventies – and in about 681 they freed Wilfrid, permitting him to go into exile once more with his possessions and companions. Again, Wilfrid sought sanctuary in Mercia, where he had been received with honour before.

King Æðelred had a nephew, it seems, named Berhtwald, whom he had installed as *subregulus* over part, at least, of what is now Gloucestershire at the headwaters of the River Thames.[46] For unknown reasons, Berhtwald gave Wilfrid personal sanctuary and offered him lands on which to found a small monastery. Perhaps they were old friends; Wilfrid's biographer does not say, and nor does he say why Wilfrid did not retire to one of his existing Mercian houses. But the location of Berhtwald's lands in the extreme south-west of Mercian-controlled territory suggests that Wilfrid was not welcome in the Mercian heartlands. As Eddius concedes, Wilfrid's presence in Mercia caused diplomatic problems for the court: Queen Osðryð was King Ecgfrið's sister.

Eddius says that as a result of Northumbrian pressure Æðelred and Osðryð forced Berhtwald to expel Wilfrid and that the troublesome priest now sought refuge with King Centwine, who

---

* An entry inserted incorrectly under 675 in the 'E' or Peterborough version of the *Chronicle* records Wilfrid's return to England in 680 and his presentation of various papal decrees to Theodore at the Council of Hatfield, supposedly called by King Æðelred. This seems hard to reconcile with contradictory accounts of his return, and of the Council, in Bede (*HE* IV.17) and in Wilfrid's *Life* (*VW* 33, 34).

ruled part of Wessex between 676 and 686. But Northumbrian political reach was such that even here Wilfrid would not find safety. Centwine's (unnamed) queen was the sister of Queen Iurminburh of Northumbria, under pressure from whom Wilfrid was moved on once more. Now he sought sanctuary with the heathen King Æðelwalh of the South Saxons and here, having as it were run out of places to hide, he made friends in a land that lay outside the closed shop of his royal Christian enemies. Bede and Eddius agree that Wilfrid now spent some time preaching and converting among the South Saxons.[47] The king gave him a large estate at Selsey,* on whose shores he founded a new religious community in exile. For a few years he does not appear on the Mercian political radar.

The tumblers of historical fate spun again in the year 685, when Northumbria's King Ecgfrið embarked on a disastrous campaign in Pictland and was killed at Nechtansmere† in the Scottish Highlands. Neither of his marriages had produced heirs. The Northumbrian élite – counselled by Ecgfrið's sister, Abbess Ælfflæd and Cuthbert, then bishop of Lindisfarne‡ – now initiated the succession of the dead king's half-brother, Flann Fina mac Ossu: Aldfrið, King Oswiu's son, born to the Irish princess Fina while he was in youthful exile in Dál Riata.

Ecgfrið's death broke Northumbria's diplomatic links with the Mercian court via his sister Osðryð. King Aldfrið's choice of queen was Cuðburh, a West Saxon and a sister of King Ine (688–726). If King Æðelred sensed a threat from this new political axis, he was nevertheless now the first among his peers and continued his policy of expanding Mercian influence beyond its borders. In the year of Ecgfrið's death King Hloðere of Kent also died, fatally wounded in a battle with his nephew and co-ruler, Eadric. Eadric

---

* Old English Seals-ey: Seal's island.
† Now identified with Dunachton in Strathspey. Woolf 2005.
‡ See Bede's Prose Life of Cuthbert, Chapter 24, for Cuthbert's remarkable prophecy made during a meeting with Ælfflæd on Coquet Island. Webb and Farmer 1983, 73–75.

was also a nephew of King Wulfhere's queen, Eormenhild and it is likely that Æðelred supported him in this coup. But Eadric ruled Kent for no more than eighteen months; thereafter, Bede says, 'various usurpers or foreign kings plundered the kingdom'.[48]

King Ecgfrið's death opened a diplomatic window for the now elderly and ailing Archbishop Theodore to restore fractured relations with Wilfrid, exiled among the heathen South Saxons. Eddius recounts how the archbishop summoned Wilfrid to London for a meeting with Bishop Eorcenwold, at which Theodore offered Wilfrid the succession to his archiepiscopal see at Canterbury. At the same time, we are told, he wrote to King Æðelred in Mercia, requesting that Wilfrid's confiscated lands and minsters be restored to him. Eddius quotes from the letter, and records that as a result of Theodore's pleas...

> ... many monasteries were returned to him and lands which he had possessed in his own right. Æðelred treated him with the deepest respect and remained his faithful friend forever.[49]

Historians are frustrated by the difficulty of identifying Wilfrid's foundations in Mercia.* He had regarded Lichfield as a place suitable for the establishment of a Mercian see but had not founded a minster there. Only one minster site can definitely be ascribed to him. Eddius says that when Wilfrid died he was visiting his minster at Oundle, which lies in a bend of the River Nene in Northamptonshire close to the line of a Roman road that once connected it with the town of *Durobrivae* – Water Newton – and within a day's rowing of the Middle Anglian mother house of *Medeshamstede* at Peterborough. The site has not been identified

---

* Outstanding candidates within Northamptonshire include St Andrews at Brigstock, which has an Anglo-Saxon tower and lies on the Roman road between Leicester and a crossing of the River Nene at Thrapston; and St Mary the Virgin and All Saints, Nassington, which also boasts an Anglo-Saxon tower and pre-Conquest sculpture, and which lies some eight miles downstream of Oundle, close to Roman roads and sources of Roman masonry and architecture.

or excavated but Oundle's location bears striking similarities to others of Wilfrid's minsters in Northumbria. The Nene here may, like the Ure at Ripon and the Ouse at York, have been navigable in Wilfrid's day. Like Ripon, Hexham, Lichfield, York and another likely Wilfridan church at Escomb in County Durham, it lay close to both a Roman road and to the remains of substantial Roman towns or forts where *spolia* – monumental masonry like quoins, arches, cornices and the like – could be incorporated into the fabric of his churches: embedding the Roman in the Anglo-Saxon.

A second Northamptonshire minster is sometimes ascribed to Wilfrid on the grounds of its magnificent, basilica-style architecture. All Saints', Brixworth, sits on a prominent hill some six miles north of Northampton. Debates about its date, original form and inspiration are unending; but it boasts the most substantial standing Middle Saxon remains in England. Roman-tile decorated arches, blocked side aisles, ring crypt and gallery baluster shafts are among its most architecturally ambitious features and Wilfrid's surviving crypts at Hexham and Ripon convey the same sense of grandiosity, albeit subterranean. But the twelfth-century Peterborough historian Hugh Candidus believed Brixworth to be a daughter house of *Medeshamstede*.[50] It lacks the most obvious features of Wilfridan sites: a possibly navigable river and close links to the surviving Roman road infrastructure; and it is by no means clear where the Roman material at Brixworth came from. It may be significant that it lies within a day or so's travel of Weedon in the higher reaches of the River Nene close to Watling Street, where King Wulfhere's daughter Wærburh established a minster during this period.

Wilfrid's time in Mercia in the mid-680s lasted two years. Theodore's diplomatic endeavours eventually persuaded King Aldfrið to install him once again as bishop at York and to restore his minsters in Northumbria, including Hexham and Ripon. These he held for five years, during which his relationship with Aldfrið is described by Eddius as 'in and out of friendship' – a result, it seems, of the ongoing feud over the division of the Northumbrian diocese.[51]

In the year 690 Archbishop Theodore, having worn the pallium at Canterbury for twenty-two years, died in his eighty-ninth year and was succeeded, many months later,* not by Wilfrid but by Abbot Berhtwald of Reculver, in Kent. Freed from personal obligation to the primate, King Aldfrið now expelled Wilfrid from Northumbria, as his brother had previously. Once again the troublesome priest sought refuge with his allies in Mercia. King Æðelred offered him the vacant see of Middle Anglia, with his seat probably at Leicester[52] – an indication that Canterbury's new archbishop had yet to impose his own authority on such appointments. We find Wilfrid witnessing a *Hwiccian* charter and consecrating Oftfor as bishop at Worcester in the immediate aftermath of Theodore's death.[53]

The Mercian king's interest in monastic patronage seems to have been consistent throughout his reign. A *princeps* named Friðuric is recorded as having given land at Breedon-on-the-Hill† in Leicestershire to *Medeshamstede*.[54] This may be regarded as the founding charter. Subsequent charters record the gift of additional lands to support the new minster there.[55] Before half of the very prominent hill on which Breedon's priory church stands was quarried away, the minster here must have been an even more imposing sight than it is today, occupying as it does the interior of a great Iron Age fortress lying two miles south of the River Trent, perhaps on the border between Middle Anglia and the Mercian heartlands of the *Tomsæte*.

King Æðelred's sister Cyneburh, who had been married to the ill-fated Northumbrian prince Alhfrið, founded and was abbess of a religious house at Castor, close to *Medeshamstede* and on the site of the former palatial Roman complex north of *Durobrivae*.[56] *Medeshamstede* itself received further endowments and Wilfrid's minster at Oundle, whose founding charter has been lost, is a hint

---

* A disputed succession in Kent, and outside influence from both Wessex and Mercia, prevented an earlier appointment.
† Brittonic *Brigā* – hill – and Old English *dūn*, also for hill: a pleonastic or tautological name in two languages.

Oundle, on the navigable River Nene upstream of Peterborough:
Wilfrid's minster and the place of his death in 709.

that many others across the East Midlands are missing from the monastic map.[57] If a complete list of Mercian foundations from the period survived it would very likely emphasise a pattern that can already be made out on a map of central England showing major rivers and Roman roads: endowments made under Wulfhere and Æðelred across Greater Mercia concentrated royal and religious patronage along the valleys of the River Nene, the Upper Trent and its *Tomsætan* tributaries; along the River Severn from Much Wenlock, via Worcester, down to Gloucester; and along the River Thames where the kings of Essex, Kent, Mercia and Wessex competed for influence and control. A more dispersed group of minsters across Lincolnshire seem also to reference watery routeways and Roman roads.

It may be fanciful to suggest that religious and secular powers in Greater Mercia had set about establishing an explicit network of monastic houses to further their administrative and political ends.\* Kings and their kin, bishops and abbots, found such places convenient – they were, after all, inveterate travellers. The common language and culture of the Roman church, the sharing of literature, liturgy and craft linked these hotspots of intellect, capital and sustained endeavour. But if the 'minsterising' of Middle Anglo-Saxon England was not a matter of forethought and deliberate policy, it at least presaged a century of administrative, economic and cultural evolution – if not a revolution – across the Mercian landscape.

There was no more revolutionary character in late seventh-century England than Wilfrid, sometime bishop of the Northumbrians, indefatigable monastic entrepreneur and arch-Romanist. As bishop of the Middle Angles with his own, not modest, empire of religious houses and estates to rule and enjoy, he nevertheless harboured a strong resentment against his Northumbrian antagonists. Despite periodically enjoying Canterbury's protection, he believed that Theodore's division

---

\* Some of them based on estates whose origins lay in the Roman period. Faith 1997, 25.

of the Northumbrian see had been based on personal prejudice against him; and he still possessed the papal decree, drawn up by Pope Agatho in 679, which had exonerated him and confirmed his right to the see at York. During his years in exile in Middle Anglia in the 690s he seems to have begun to exert pressure on Archbishop Berhtwald to revive that Northumbrian claim.

In 697 the Mercian establishment was shaken by the murder of Queen Osðryð at the hands of an unnamed, but decidedly Mercian assassin.[58] As a daughter of King Oswiu and sister of King Ecgfrið she may always have been vulnerable to anti-Northumbrian sentiment. Now, as Æðelred came into his declining years and thoughts of the succession occupied the Mercian élite, she must have provoked the anger of one or other faction. There seems to have been a long-standing agreement between Wulfhere and his brother Æðelred (or the factions that supported them) that Wulfhere's son Cœnred should succeed his uncle and that Osðryð's and Æðelred's son,* Ceolred, should be king after him. Such sharing arrangements were not uncommon among the Anglo-Saxon kingdoms, where multiple lines of descent from a founding king might compete either constructively or disastrously for the throne. Queen Osðryð may have lobbied, or conspired, to promote her own son ahead of her nephew – what could be more natural? If so, she paid a heavy price. Her widowed husband, now surely in his sixties and politically weakened by this domestic tragedy, took her body to their joint foundation at Bardney, where King Oswald's relics lay, and interred her there.

Such tensions, and the king's inevitable decline, may have prompted Wilfrid to renew his Northumbrian suit while his most powerful ally and sponsor retained a semblance of political

---

* A second son, Ceolwald, mentioned only as briefly succeeding his brother in 716 in a list preserved in Worcester Cathedral, is suggested only by the name. Yorke 2013, p. 111 Kirby (2000: 108–109) cites later tradition that Ceolred was not Osðryð's son. If he was born to a marriage after her death in 697 he can have been no more than ten years old at his accession.

authority. From Eddius's account it seems that he petitioned Archbishop Berhtwald, who agreed to call a council to review Wilfrid's status. Wilfrid and many other bishops, the archbishop and King Aldfrið with his advisers, met at a place identified as Austerfield, probably close to Bawtry on the border between Northumbria and Mercia, in 702–3. The only account of the Council of Austerfield to survive is that of Eddius, who portrays furious arguments among the assembled bishops, an unseemly anti-Wilfridan conspiracy to deprive him of all his lands in both Northumbria and Mercia, and informants sneaking in and out of King Aldfrið's tent; a bitter speech of self-justification from Wilfrid and – the coup de grâce – a definitive expulsion from his see.[59] Eddius claims that Wilfrid was excommunicated. At any rate he returned to Æðelred's court, wounded but fist-shaking. The king, perhaps in view of his own political vulnerability, dispatched Wilfrid, approaching his seventieth year, on yet another journey to seek resolution from the pope – kicking the can all the way down the long road to Rome.[60]

By the time Wilfrid returned to Mercia in 705, once more brandishing a papal judgement in his favour, his royal sponsor had abdicated, retiring to his favourite monastic house at Bardney and leaving his middle-aged nephew Cœnred to succeed him.

# Chapter 4

# Cœnred, Ceolred:
# No country for old men
# 704–716

❖

*Guðlac of Crowland • Exiles and hermits • Synod of*
*Nidd • Abdications and retirements • Æðelbald and*
*Oundle • Ceolred the despoiler*

In the days of King Æðelred there was a nobleman of the Middle Angles named Penwalh. Like Penda's sons, Penwalh traced his descent back to the distant royal Mercian progenitor, Icel. He married a noblewoman, Tette, and they enjoyed the privileged life of Mercian aristocrats, their *mansio* or hall 'furnished with an abundance of goods'.[1] The birth of their first son, Guðlac, in about 673, was attended by miraculous signs that he was marked for greatness. The boy's name meant something like 'Gift of War' or 'War-play' in Old English,[2] suitably propitious for a youth born into an élite culture steeped in the deeds of heroic warriors. Guðlac was baptised and educated in his father's mead hall in the noble ways of the ancients: skills required to lead a *gedryht* or warband onto the field of battle and generously feast and reward his comrades and household with treasure.

At the age of fifteen Guðlac became a *geoguð*, a junior member of his father's household ready to bear arms. Now, inspired by a burning desire to emulate the heroic deeds of the ancients, Guðlac gathered to himself a band of followers and embarked on a bloody, almost decade-long martial career during which he…

> … devastated the towns and residences of his foes, their villages and fortresses, with fire and sword and, gathering together companions from various races and from all directions […] amassed immense booty.[3]

Enjoying success in battle, he overcame his foes and won glorious victories, but conscience moved him to return a third part of his treasures to their owners. Nine years later, at a time when he had earned the rank of a *duguð* or veteran,[4] entitled to enjoy his own land and household and the fruits of marriage, he wearied of the toll of conflict and began to contemplate his own mortality. A sudden, night-time Damascene revelation of the shame and wickedness of his violent past led him to the idea that he must, in future, devote himself to Christ. He told his bewildered companions at arms, his *comitates*, that they must choose a new leader for themselves. Renouncing worldly pomp and the material treasures and booty he had accumulated over a decade of pillage and rapine, he sought entry into the monastery at Repton in Derbyshire, submitted to the tonsure under Abbess Ælfðryð and took on the robes of a monk.

In this way Felix, composer of a *Vita* of St Guðlac sometime between 730 and 740, accounts for the saint's transition from the *comitatus* to a monastic community in about 697 at the age of twenty-four – a story told across twenty short episodes in a work of fifty-three chapters. What follows is, in many respects, entirely conventional hagiography, drawing on exemplars like St Cuthbert and the desert father St Anthony, with whose stories Felix was familiar. But since Felix's *Vita Sancti Guthlaci* contains the sole surviving biographical narrative of any Mercian from the seventh or eighth centuries, it has understandably attracted much historical attention over the decades. It shimmers with tantalising political and social gems.

To begin with, although Guðlac was a man of Middle Anglian stock, and while his career is set against a distinctly Mercian backdrop, Felix's work is dedicated to an East Anglian: his own lord, King Ælfwald. And then, what are we to make of the saint's name and heritage? His father was Penwalh – a name with a Brittonic flavour* – of the line of Icel. It bears suspicious alliterative comparison with the most prestigious branch of that line: Peada, son of Penda son of Pybba. Guðlac's own name breaks the alliterative chain and seems purely martial, even if Felix's spin

* *Pen* = 'high'; '*walh*' perhaps from *wealh* – Welshman or 'foreigner'.

## Middle Anglia

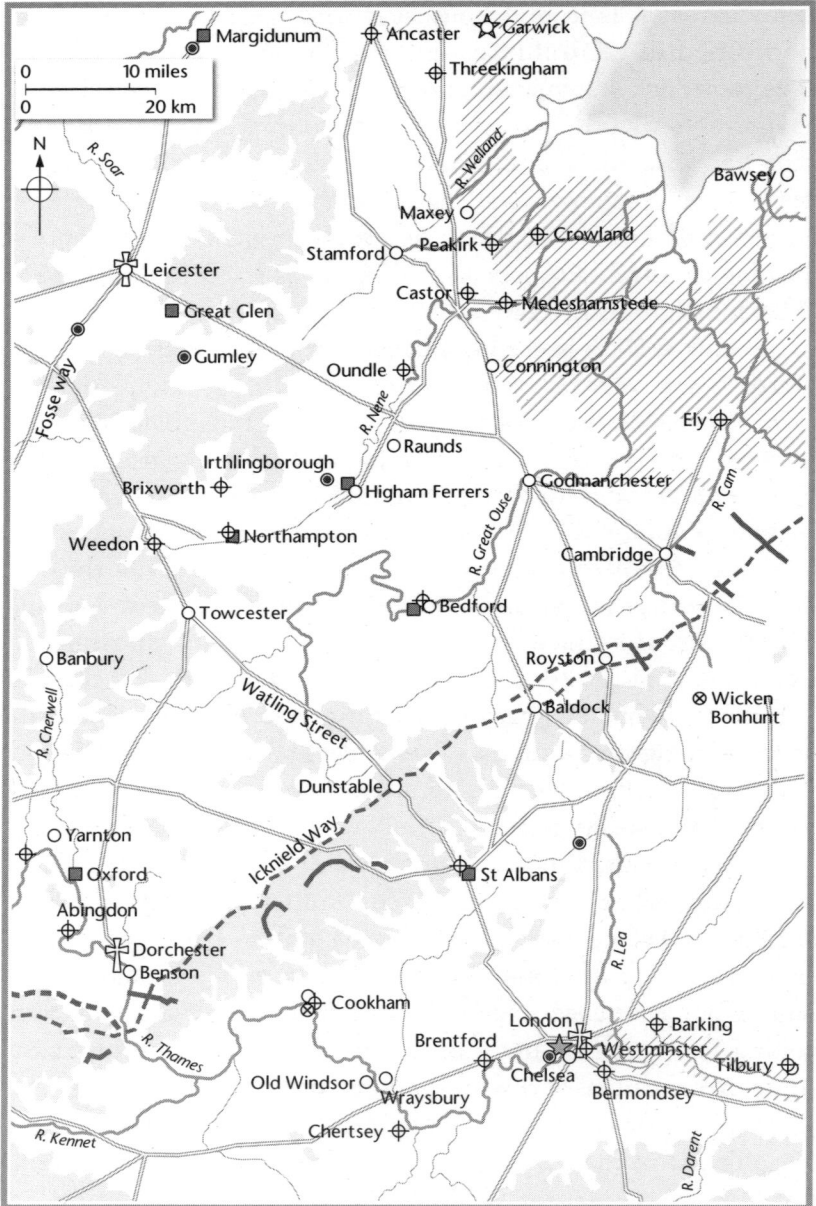

turns it, conveniently, from echoes of a *militia regis* to those a *militia Dei* – a soldier not of the king but of God. Was 'Guðlac' a nickname, conferred by his comrades-in-arms; or an ironic moniker chosen only after he had taken the tonsure?

Scholars have also been bemused by Felix's somewhat circular explanation of how the boy's name derived from that of 'the tribe known as the *Guðlacingas*' – who appear nowhere else in either annal or place name.* Such a people might conceivably be descendants of the youthful warrior's *comitates*, settled on the *folcland* that was their reward for serving their lord; but not the other way round.[5]

If Guðlac were a Middle Anglian by birth it does at least makes some sense that he took orders in the newly founded monastery at Repton. The circumstances of Repton's foundation are obscure, and two charters naming it as a daughter house of *Medeshamstede* are regarded as examples of retrospective wishful thinking on the part of Peterborough's monks.[6] Even so, the tradition of a subordinate relationship between the two minsters is credible and, if the mother house to which Guðlac's family owed allegiance was the principal Middle Anglian minster at *Medeshamstede*, one of its daughter houses might seem a suitable place in which to invest this novel career. King Merewalh of the *Magonsæte* is said, in later tradition, to have been buried at Repton in about 685.[7] If so, Repton must already have been well-established by the time of Guðlac's entry there a dozen years later. Of Abbess Ælfðryð nothing certain is known.†

Excavations at Repton by Martin Biddle and Birthe Kjølbye-Biddle in the 1970s and 1980s, which revealed spectacular evidence of a later Viking fort and mausoleum, also uncovered remains of a seventh-century church beneath those of the later abbey.[8] Repton lay in the core lands of Mercian kingship, on the south bank of the River Trent close to its confluence with the

---

* A single place, Golcar, near Huddersfield (Old English: *Gudlagesarc* – Guðlac's shieling), may preserve the name. Watts 2004, 255.

† But see below in this chapter, p. 126.

The crypt at Repton: Mercian royal mausoleum; perhaps the
most atmospheric surviving building of the age.

River Dove and within a day's journey of Catholme, Lichfield,
Breedon-on-the-Hill and the later royal *vill* at Tamworth. The
Anglo-Saxon crypt at St Wystan's was a site of royal Mercian
burials in the eighth and ninth centuries: so Guðlac's monastic
life began securely within a royal orbit.

What of the bloody fighting career of this precocious *geoguð*,
which preceded his decision to take holy orders? The chronicles
of the period 688–698 are conspicuously lacking in notices of

Mercian warfare, with a single exception – a one-line notice of a 'battle against Penda's son' (King Æðelred) recorded in the Annals of Ulster under the year 692 and in the *Chronicon Scottorum* for the following year.[9] It is not much to go on. Settlements of this period excavated across the Midlands show few signs of the sort of orchestrated violence, or of defences constructed against it, that would offer historians a credible backdrop for such a bloody career. Farms were rarely enclosed by palisades or ditches, and lowland fortified sites of the Iron Age that might be defended rarely provide concrete evidence of reoccupation in this period.* Where to look for a credible theatre of operations lasting almost a decade? The straws at which various historians have clutched in the hope of an answer are sprinkled throughout the later chapters of the *Vita*; but while they add to the intrigue, they do not resolve it. One speculative possibility is that cross-border raiding by freelance warriors was an accepted, if not openly acknowledged, tool of foreign policy – that perhaps Guðlac's warband attacked soft targets across the Fens and into East Anglia with tacit royal approval – like the privateers of the first Elizabethan Age.

After about two years at Repton, during which time he attracted the disdain of his fellows for renouncing intoxicating drink, Guðlac resolved to follow an even more ascetic path, seeking out a desert place in which to pursue the solitary life of the hermit and inspired, it seems, by the heroic exemplars of Sinai and Jordan. He was thus drawn to Mercia's own mysterious wilderness: its peat-dark water margins, which he may have known in his Middle Anglian youth. Felix evokes the liminal, demon-ridden, boggy flatlands of the Fens in a portentously minor key entirely in keeping with the dark, poetic sensibilities that infuse the haunts of Beowulf's deadly adversaries – the monster Grendel and his mother. He describes…

… a very long tract, now consisting of marshes, now of bogs, sometimes of black water overhung by fog, sometimes

---

* But see below in this chapter, for a violent episode at Oundle.

studded with wooded islands and traversed by the windings of tortuous streams.[10]

Setting out from Repton in the late spring of 699, by what Felix calls the most 'direct' route, Guðlac is said to have encountered a man named Tatwine, who may be identified with the abbot of Breedon-on-the-Hill who later became archbishop of Canterbury. * Tatwine told him of a remote and little-known island in a wild place on the west side of the Fens called Crowland – actually a low-lying gravel spit, lying barely two metres above modern sea level and surrounded on three sides by marshes, in the territory of the *Gyrwe* between the Rivers Welland and Nene. In summer, receding waters provided ample and fertile pasture for cattle and sheep and the Fenland historian Susan Oosthuizen has shown that rights to such riches were held in common by communities in neighbouring *vills* lying on higher, drier ground, claiming ancient entitlement to them.[11] As winter approached meadows flooded, reed beds were submerged and the Fens became a waterscape of narrow channels, scrubby alder carrs and broader stretches of open water. The seasonal nature of Crowland's isolation and its perceived privations must have added to its mystique, like the twice-daily tidal inundations of Lindisfarne's ancient causeway.

Looking at a map of the country between Repton and Crowland, it is by no means obvious what the 'direct route' that Guðlac and Tatwine are said to have taken might have been. Neither site lies close to a Roman road. Audrey Meaney, in a study of Guðlac's role as a subject of hagiography, thought it likely that he and Tatwine went first to *Medeshamstede*, Repton's probable mother house; and a recent journal paper argues that Crowland's establishment as a hermitage was *Medeshamstede's* direct initiative.[12] Crowland lies half a day's journey north-east

---

* Breedon lies just a few miles east of Repton; the two houses seem to have been closely linked within the wider *paruchia* of *Medeshamstede*. Dornier 1977. It is not absolutely clear that Guðlac's Tatwine and Archbishop Tatwine are the same man – I think it likely that they are.

Crowland, Guðlac's fenland hermitage and shrine, later the site
of a great medieval abbey.

of Peterborough; the would-be anchorite would have needed the
permission of its prior to establish even so modest a foundation as
a hermitage on *Medeshamstede's* extensive holdings. Perhaps part
of the journey was taken by boat along the River Trent. Monks
and abbots must regularly have travelled between the two; it was
a well-trodden – or paddled – path, if not straightforward.

Tatwine agreed to show Guðlac where this lonely island lay and,
taking a fisherman's skiff, they rowed or punted into the fastness
of the marshes. Finding the place suitably wild and self-denyingly
remote even in the late summer, Guðlac briefly returned to Repton
to say his farewells and then took up residence in the marshes.

Felix claims to draw on the direct testimony of men who
had known Guðlac, when he describes how the future saint
discovered on his little island a mound made of clods of earth
that had been dug into by greedy treasure hunters believing it to
be an ancient tumulus. In the side of the mound Guðlac found

something like a cistern or chamber, over which he built himself a shelter – it looked, perhaps, something like the sunken-floored buildings so characteristic of the region. Here he lived, clothing himself in animal skins and eating the bare minimum of food for survival: every bit the disciple of the desert fathers.[13]

Recent excavation at a site called Anchor Church field, just north-east of the medieval church and town of Crowland, has revealed a deep-time palimpsest of ceremonial and religious structures dating from a late Neolithic or Early Bronze Age henge to a later prehistoric timber circle, numerous barrows and Early Medieval activity, which points to a milieu already steeped in sacred tradition, ceremony and mystery. In the words of the excavators, Guðlac and his contemporaries would have recognised…

… its monumental prehistoric earthworks [as] an obvious choice for hermits to recast into a new form of Christian "holy island."[14]

The attractions and consolations of a state of divine solitude notwithstanding, the new hermit of Crowland suffered despair in his loneliness, restlessly tormented by his past sins and daunted by the enormity of his undertaking. Extreme fasting, instead of bringing him penitent relief, drew him into a delusional state. He suffered terrifying hallucinations – of 'horrible troops of foul spirits'[15] who dragged him through the black pools and thorny scrub of the muddy Fens and then carried him down into the infernal flames of hell. After each of these episodes he was succoured by the appearance of a guiding light, Saint Bartholemew,* on whose feast day in late August he had arrived at Crowland – his faith and fortitude restored.

Early Medieval holy men and women were more or less expected to suffer such trials, temptations and demonic visitations in imitation of scriptural exemplars, as a test of

* An East Anglian, rather than a Mercian cult figure. Higham 2005, 88.

their faith. Exile, privation and extreme mental pressure were necessary companions on a profoundly challenging emotional journey, a metaphysical wandering through a spiritual wilderness towards enlightenment, holy wisdom and purity – virtues that conferred great prestige on those who could stick at it. Reading Felix's *Vita* in an age in which cynicism is sometimes hard to disentangle from critical reasoning, it is tempting to offer various diagnoses for these narrative expressions of spiritual, physical and psychological torture: post-traumatic stress, familiar to many veterans of conflict; the hallucinatory effects of malnutrition and the fevered visions of ergot* poisoning or malarial fever, the latter probably endemic in the Early Medieval Fenlands and a possible source of inspiration for *Beowulf*'s personified monsters and insidious enemies.[16] Whatever the case, Guðlac's spiritual challenges matched, perhaps surpassed, those of his days as a warlord. In victory he rendered the hostile landscape of marsh and endless horizon subject to his lordship; defeated his spiritual enemies and won for himself an eternal glory that transcended the earthly rewards of comradeship and treasure – that, at least, is Felix's purpose in recounting these episodes.

Chapter XXXIV of the *Vita Sancti Guðlaci* tells how, in the days of King Cœnred, Wulfhere's son and Æðelred's nephew and immediate successor on his abdication in 704, the English were at that time 'troubled' by British raids. One very early morning, at cockcrow, Guðlac's dreams were invaded by a tumultuous crowd, a spear-wielding army in whose cries he made out the distinct, 'sibilant' sounds of British speech.† Felix explains that Guðlac had in former times been in exile among that people, and historians have seized on this passage to offer a context for his first career as a warrior. Exile might embrace a number of

---

* A fungus of grain crops whose psychedelic and physiological effects have been likened to those of LSD.

† Sibilant consonants are made by pushing air through the teeth: for example 's', 'sh' and the 'g' in modern English 'Germany'. Perhaps they had pronounced Guðlac's name something like 'Juthlac'.

scenarios for those aristocratic sons whose pedigree made them eligible for the kingship. They might, like the Northumbrian princes Edwin and Oswald, seek a foreign court's protection from their domestic rivals, in which case they would be obliged to fight in the armies of their new lords. Guðlac might credibly have spent time as an exile in one of the Brittonic-speaking kingdoms – in Wales, the lands west of the Pennines, in Strathclyde or even the far south-west – as a potential rival to the sons of Wulfhere and Æðelred.[17] Alternatively, he might, like the future King Ecgfrið of Northumbria, have been sent into fosterage, a peculiarly insecure status in which young men were literally hostages to the fortunes of their hosts.

Some historians have pointed to circumstantial evidence of the survival of British-speaking communities in the Fenlands and the Chiltern hills and sought to embed Guðlac's knowledge of the Brittonic language in an eastern landscape.[18] If he had already been in exile among British communities, bilingualism may have smoothed his passage into the Fenland world.

If Guðlac had spent the later years of King Æðelred's reign as an exile, it is hard to see how he might be perceived as any less of a rival under Æðelred's immediate successors, especially given his record of success as a warrior and the evident loyalty of his warband. Some historians have raised the possibility that the timing of Guðlac's initial retirement to Repton may have been driven by his involvement, actively or otherwise, in the murder of Queen Osðryð in 697.[19] In any case, the fact that King Cœnred's reign (704–9) was marked by British incursions and that there is no record of Mercian reprisals against them indicates the weakness of *his* position. So it is possible that the Fenland margins of Mercian territory, where royal control was always peripheral, might make a suitably discreet internal sanctuary for an exiled prince. There are also examples of kings who took the monastic tonsure as a means of abdication – rendering themselves, as it were, untouchable. The most famous of all Early Medieval exile stories, that of Ireland's Colm Cille or Saint Columba, required him, as a result of too close an involvement with the tribal politics of his day, to decamp to the then remote island of Iona in about 565. Guðlac may have taken the tonsure to save his skin.

If Crowland's new hermit hoped to either discount himself entirely from the politics of his day, or to bide his time and hope for an opportune moment in the future, his opponents seem to have had other ideas. A cleric named Beccel came to Crowland, offering himself as a servant and acolyte – a familiar and ironic problem for the hermit wishing to be left alone. Guðlac's growing reputation is reflected in many passages of the *Vita*, in which he displayed miraculous control over swallows and jackdaws and acquired a reputation for healing – sure signs of God's favour.[20] So it was natural that would-be groupies found themselves attracted to him.

It became Beccel's custom to tonsure the holy man every twenty days, renewing, as it were, the visible symbol of his spiritual commitment; but one day he approached Guðlac with evil in his heart, intending, so Felix says, to murder him and take his place. More credibly, he had been sent as an assassin.[21] At any rate, Guðlac forced a confession from him before he could carry out his scheme, then forgave him and became his friend.[22]

The hermit's self-imposed isolation and resignation from secular affairs did not insulate him from other currents in contemporary politics. Among many visitors to Crowland during Guðlac's fifteen years there, the most frequently named is Æðelbald, son of Alweo; and there is no question that he was a genuine princely exile. Æðelbald's pedigree shows him to have been a grandson of King Penda's brother Eowa and, therefore, a second cousin of kings Cœnred and (from 709 to 716), Ceolred: a direct potential rival for the Mercian throne. Indeed, Felix says that Æðelbald was 'driven hither and thither by King Ceolred' during the period after 709 when he had succeeded his cousin. It seems highly probable, given later friendships and given Æðelbald's prominence and enthusiastic treatment in Felix's account, that the young prince spent at least some of his time in exile at the court of King Ælfwald in East Anglia. From there he was evidently able to visit the holy man of Crowland with impunity and at least two of his personal followers also came to Crowland seeking cures for various ailments.

Bede recorded two episodes in the seventh century in which holy men had been proactive in supporting exiled princes in their ambitions. Paulinus, a member of Augustine's missionary party to Kent, had supported Edwin during his East Anglian exile, in return for a promise to convert if he became king of Northumbria.[23] Bishop Cuthbert had intervened in the debate about who should succeed the childless King Ecgfrið in 685, making the case for Aldfrið, then in exile in Ireland.[24] And Colm Cille's hagiographer Adomnán recalled how the Ionan saint had appeared in a vision to Oswald at Heavenfield, the night before his fateful victory against King Cadwallon in 634, promising him that he would be victorious.[25] Guðlac's – and the East Anglian court's – support for the æðeling Æðelbald is likewise explicit in Felix's account.

Æðelbald and his followers, we are to understand, were suffering a crisis of endurance, of peripatetic exhaustion: constantly on the move, perhaps pursued by men wielding the assassin's knife. One day, he came to Guðlac, as it seems he often did, seeking solace and counsel. His fellow exile, evidently no friend of Mercia's incumbent kings, offered him spiritual comfort. More than that, though: he told the exiled prince that he had prayed to God and God had granted that Æðelbald would rule over his race and become 'chief over the peoples'.[26] Such prophetic endorsement was both a powerful tonic and a token of political capital, to be redeemed at an opportune moment sometime in the unknowable future.

The plight of the exile invoked deep sensibilities in the Early Medieval mind. Among the likely *Fates of Men*, a gnomic verse from the *Exeter Book* – an Old English anthology of poetry and wisdom – the exile suffered this fate…

*Sum sceal on feþe on feorwegas, nyde gongan ond his nest beran…*
*One must needs travel on foot in ways remote*

*And carry his provisions with him and tread the spray-*
    *flung track*
*And the dangerous territory of alien peoples.*
*He has few surviving providers;*
*Everywhere the friendless man is disliked*
*Because of his misfortunes*[27]

Because men needed lords and a sense of belonging to king, tribe and homestead, exile was felt both as physical pain and as metaphorical torture. Weary in spirit, the exile must dip his oar in an icy sea, his mind full of the sorrows of missed hearth-companions and the warmth and good spirits of the mead hall. Equally, the loss of the bonds and rewards of patronage excluded the exile from all society. Voluntary or not, the state of lordlessness broke the fundamental rules of two-way patronage – lord and follower bound in eternal, reciprocal obligation. Aldhelm knew this when he adjured Wilfrid's abbots to follow him into exile. The poet who evoked the travails of *The Wanderer* – another Old English verse from the same anthology – expressed the aching for one's lord as a wound; the pain of waking after restless sleep as like being surrounded by dark waves. Above all, the exile experienced a sense of displacement and loss:

*Hwær cwom mearg? Hwær cwom mago? Hwær cwom*
    *maþþumgyfa...*
*Where has the horse gone; where the man; where the*
    *giver of gold?*
*Where is the feasting place; and where the pleasures of*
    *the hall?*[28]

Holy men and women belonging to a religious house like Repton also experienced and enjoyed fellowship in the fictive family of the cenobitic* household. They served their immediate lord or lady

---

* Cenobitic – communal, as in a monastic household, as opposed to eremitic, the solitary life of the hermit or anchorite.

– the abbot or abbess of the minster to which they were tied by both spiritual and material bonds – and they served a higher lord, too. But there had always been a strain of Christianity that sought and celebrated the idea of self-exile expressed, particularly in Irish Christianity, as a *peregrinatio pro Christo* – a pilgrimage for Christ – which might result, like the journey of Colm Cille to Iona, in the foundation of a new community but which might also lead to extreme asceticism of the sort experienced by Guðlac and by the monks who clung in their tiny beehive cells to the wave-torn, bleak Atlantic rock of Skellig Michael. These eremitic 'ultras' were viewed with suspicion by the orthodox church and with awe and admiration by many. Their friendship, counsel and endorsement were profound and valued signs of heavenly favour. Guðlac and Æðelbald had much in common, besides their regal lineages.

Bede praised the solitary life and admired Cuthbert's withdrawal to the tiny island of Inner Farne – but insisted that the saint had sought the consent of his abbot beforehand; that the arrangement had been steeped in orthodoxy.[29] A more conventional spiritual career path, one probably known by reputation to Felix, was that of the Northumbrian Benedict Biscop, the founder of Bede's own monastery at Jarrow on the River Tyne. Born Biscop Baducing in about 628, he was a *gesið* of King Oswiu who, at the age of twenty-five, was given 'possession of the amount of land due to his rank'. But he...

> ... put behind him the things that perish so that he might gain those that last forever, despising earthly warfare [...] He left country, home and family for the sake of Christ.[30]

The cachet of the self-denying extremist as member of a religious élite, existing as it were on a higher plane, may explain why penitents, the sick, the desperate and secular exiles like Æðelbald sought their company, encouragement and healing skills. It may also explain why they were seen as a threat to secular and religious authority. Bede recalled how, during Augustine's first mission to the English at the turn of the sixth century, he had met with the bishops of the British west to request that they submit to Roman orthodoxy. They, in turn, had consulted a famous hermit who advised that they test Augustine's humility. He failed that diplomatic test disastrously and a long-running and bitter schism was born.[31]

King Cœnred and his cousin and successor King Ceolred cannot have been unaware of the presence of subversives like Guðlac on their borders. Whether they were regarded as a minor nuisance or an active threat is hard to say. The transfer of Mercian power at the end of Æðelred's reign was apparently smooth. He was still alive when the Mercian kingship passed to his nephew Cœnred in 704. His last recorded charter, dated to June 704, authorises an East Saxon grant to Bishop Waldhere of London of thirty hides of land at Twickenham on the River Thames and is confirmed by both his nephew Cœnred and, later, by his own son, Ceolred, both making their mark with a cross.[32] But from the reigns of Æðelred's two immediate successors just three genuine charters survive; and a decade of *Chronicle* entries covering the period is thin on Mercian events. Such still waters hide political undercurrents.

In that year, 704, the now elderly Æðelred had abdicated and retired to the royal monastery at Bardney in Lincolnshire, where his murdered wife Osðryð was buried and where King Oswald's torso also lay. Here he was visited by Wilfrid, returned from his last journey to Rome and once more exonerated by papal authority. The exiled bishop carried a letter from Pope John, from which Eddius quotes directly,* begging that kings Æðelred and Aldfrið in Northumbria should heal their rift with the bishop and hold a synod under the authority of the archbishop of Canterbury to resolve their dispute.[33] According to Eddius, Æðelred summoned his nephew Cœnred and made him swear to uphold the pope's will.

Wilfrid's envoys, sent with the same message into Northumbria, were met with a chillier reception: King Aldfrið dismissed them. But a year later the king was dead; the Northumbrian throne was briefly usurped by the otherwise unknown Eadwulf and Aldfrið's

---

* The original does not survive; Eddius's transcription must be taken on trust.

ten-year-old son, Osred, succeeded. Archbishop Berhtwald was now able to call a synod in accordance with Pope John's wishes and convened it at an unknown site on the River Nidd* in 706. Here, Wilfrid's lands and offices were restored to him, in spite of opposition from many bishops but supported by King Oswiu's elderly daughter Abbess Ælfflæd, acting perhaps as regent for Osred and disinclined to perpetuate Wilfrid's long-running feud with Northumbria.

If King Cœnred attended the synod, his contribution was recorded neither by Bede nor by Eddius. He certainly convened at least one council with his bishops and ealdormen, perhaps in the first year of his reign, to consider the 'reconciliation' of a woman named Ælfðryð.† We know nothing of the detail, nor the location. The only contemporary reference is contained in what Sir Frank Stenton described as the earliest letter between two Englishmen. Bishop Wealdhere of London had been invited to attend the meeting but, without Archbishop Berhtwald's approval and instructions, he refused to attend and had written to explain his position. Ælfðryð's identity is not certain. It is tempting to identify her with the abbess of Repton who had taken Guðlac into the community there in about 697; if so, and if her support for the hermit was a cause of tensions in the Mercian court and among its bishops, those tensions had taken a long time to come to a head.[34]

No minster foundations can be attributed to King Cœnred's five-year reign after 704; no religious house remembered him as its founding patron. What little we know of Mercian politics in these few years, aside from Felix's note of raids by British forces, is that the king was a religious man; that he enjoyed a personal friendship with the East Saxon King Offa (c. 705–9) and that he was concerned with the moral probity of his followers. Bede tells the story of a *comitatus* who, guilty of many unnamed sins, was warned by the king to mend his ways. Even when the man

---

* In the environs of Harrogate in what is now North Yorkshire. Eddius Stephanus: *Vita Wilfridi* 60; Bede, *Historia Ecclesiastica* Book V, Chapter 19.
† It is unclear what spiritual or secular offence she may have committed.

fell gravely ill, struck down by cruel pains, Cœnred's pleas to confess his sins and repent fell on deaf ears. He died, unnamed and unworthy of more than a brief cautionary tale in Bede's *Historia*.[35]

Like his uncle before him, and in keeping with his religiosity, King Cœnred abdicated. The *Anglo-Saxon Chronicle* records that in 709 he undertook the pilgrimage to Rome with his friend Offa, son of King Sigehere of the East Saxons. Bede records that, on their arrival in Rome, both men took the tonsure and lived out their days in holy orders in the city of Saints Peter and Paul.[36] The compiler of official papal annals, the *Liber Pontificalis*, thought the event sufficiently noteworthy to include it in his biographical notes on Pope Constantine, offering a baldly succinct account:

> In his time two kings of the Saxons came with many others to pray to the apostles; just as they were hoping, their lives came quickly to an end.[37]

It would be churlish to doubt, at this distance, the motivations of such men, leaving their homeland and families and renouncing all worldly pleasures to travel dangerous paths on a *peregrinatio pro Christo*. Even so, the ears of the historian, finely tuned to noises offstage, might just detect hints of machinations in the wings. Bede says of Offa that the whole race of the East Saxons had longed for him to take the reins of that kingdom, where he was junior to two second cousins ruling jointly; self-exile was a pragmatic alternative to the risks of an attempted coup.[38] From the year in which Offa departed, Essex was ruled by a distant branch of that line. If similar undercurrents lay behind Cœnred's abdication, a circumstantial clue is the death, that year, of his uncle and sponsor Æðelred in his retreat at Bardney. Cœnred does not seem to have had a son – there is no evidence, even, of a marriage. King Æðelred's own son Ceolred now succeeded, very likely by prior arrangement; but he was not the only possible successor: waiting in the other wing of the Mercian stage was Æðelbald, Guðlac's friend-in-exile and protégé of East Anglian kings.

That other professional exile, Wilfrid, had also retired after his latest triumph at the Synod of Nidd. He spent his last years not in one of his principal churches at Ripon or Hexham, nor even in York, to which see he had been restored, but at Oundle in Middle Anglia, some miles upstream from *Medeshamstede* on a bend in the River Nene. In 709, the year in which Cœnred journeyed to Rome, Wilfrid died at the age of seventy-five, after a life defined by immense spiritual energy, by conflict and tribulation and the accumulation of huge wealth, influence and property. In the ages of Northumbrian and Mercian supremacy there was no more influential or divisive figure than this troublesome priest.

Bede and Eddius record that Wilfrid's brethren carried his body the 150 miles north to his first monastery at Ripon, and there he was buried.[39] The crypt that survives in the cathedral there – claustrophobic, cramped and fuggy as it is – gives visitors and pilgrims alike a direct, tangible link to one of the great figures in Early Medieval English history.

A post-mortem miracle that took place at Oundle after Wilfrid's death offers a clue to volatilities ignored or suppressed by Bede and the Chronicler. Eddius says that…

> … certain exiles of noble birth who were ravaging with an army, because of some wrong done to them, burned the whole of the above-mentioned monastery at Oundle.[40]

The episode is not dated; the miraculous element of the story tells how the bishop's own house would not burn; nor would a cross that stood in the monastic enclosure. Despite long-held suspicions that contemporary sources underplay its effects on both regional politics and the lives of ordinary people, physical evidence for fighting, in weapon-blade injuries, defended settlements and what archaeologists call 'destruction layers' is hard to come by. And yet, Guðlac had spent a decade indulging in violence before turning to the solitude of the hermitage; and Eddius's testimony is also convincing, if short on detail. So this tantalising entry needs some explaining. Who were these exiles of noble birth; what wrong had been done to them; and why Oundle?

It is easy to offer up candidates for noble exiles – those young-bloods of weapon-bearing age with a *comitatus*, eager to forge a power base and needing to accumulate booty with which to reward their followers. Æðelbald was one; Offa of the East Saxons was another; Eadwulf in Northumbria a third contemporary; there must have been many more – men who, in later life, took lands according to their rank, married, became loyal thegns and *principes* of their kings; men whose names pop up as witnesses to charters or those who, like the thegn of Cœnred who would not mend his ways, could not or would not reconcile themselves to maturity. The *Vita Sancti Guthlaci* sets the landscape context for such men – would-be royal dynasts or those whose fathers or grandfathers had been petty kings of minor polities in former days – living in exile for fear of reprisal and assassination, laying up on island fastnesses in the Fenlands where the writ and might of the kings of East Anglia and Mercia was weak. The contemporary history of Wessex is rife with such political and military strife: Northumbria and Mercia, with their strong, long-lasting dynasties, were probably exceptional.

Periodic raids by mounted parties carried out over a few days or weeks, for booty and provisions or to attack the camp of a rival band, evoke romantic ideas of later folk-heroes – not least Hereward the Wake, who was said to have made the Isle of Ely his base for resistance against the forces of William I. He, too, had attacked monasteries – those that had been placed under the rule of Norman abbots.[41] And Hereward had been a rebellious exile even in the days of Edward the Confessor.

Eddius says that some wrong had been done to the marauders. If the exile in question was Æðelbald, it is tempting to contemplate the idea that he had visited Wilfrid in his last years at Oundle, seeking the same consolation, support and legitimacy in his anticipated bid for the Mercian throne as he had from Guðlac. Had Wilfrid, a loyal friend of King Æðelred, refused? If so, had the exiled prince exacted revenge on his monastery after Wilfrid's death?

The feud, that is to say private warfare, was the dark side of Early Medieval ideas of reciprocity in which gift followed gift followed gift with no finality, in an endless cycle of mutual obligation. Likewise, slight was met with vengeance; vengeance with more violence. The efforts of noblewomen and religious leaders to weave peace between feuding parties seem occasionally to have succeeded* and Anglo-Saxon laws made explicit provision, in their head-price or *wergild* coda, for the prevention or termination of feud through cash compensation. A provision in the extensive law code of King Ine of Wessex (688–725) distinguishes between a party of thieves (up to seven men), marauders (up to thirty-five in a party) and an army, here, above thirty-five.[42] Oundle's despoilers were, according to Eddius, an *exercitus* – an army. If Bede and the *Chronicle* are generally silent, or at least tactfully reticent, on the subject of organised violence at the level of the *geoguð*, written contemporary evidence for politics on the near continent suggests that this aspect of Insular history has been under-reported.[43]

Between them, archaeologists and historians may have to accept that the actual incidence of organised violence in the landscapes of eighth-century Mercia – its effects on people's daily lives – is blind ground: a known unknown. But there is little doubt that contemporaries expected that age-old rules would be followed, as one of the so-called Maxims collected in the Exeter Book illustrates:

For a wound, a bandage; for a hard man, vengeance.[44]

Eddius records with satisfaction that those same marauders who had razed Oundle were soon surrounded by their enemies, overthrown and slain. 'And so,' he says, 'the saint [...] avenged his wrongs.'[45] If true, that would rule out Æðelbald from the list of possible suspects; but not necessarily his proxies.

---

* As witness Theodore's intervention after the Battle on the Trent and Eanflæd's insistence on her husband Oswiu's expiation for the murder of a kinsman.

Archaeologists might also have to nuance their understanding of the open nature of settlements in this period: for Eddius says that Oundle was protected by a 'great hedge of thorn, which surrounded the whole monastery'. Religious communities separated themselves from the world without by a physical barrier, sometimes a wall or ditch, which acted also as a symbolic boundary between the sacred and the profane. Eddius's description suggests that a thorny hedge might make just as effective a barrier against lone intruders, predators and petty thieves, albeit insufficient to deter a rapacious warband. For archaeologists, the defensive hedge is another blind spot in the Early Medieval landscape – sometimes suspected; rarely detected.

Oundle's setting is itself intriguing. Bede says that it was the name of a district – *Inundalum* – possibly but uncertainly derived, according to the place-name scholar Victor Watts, from the Old English *un-dāl*, meaning 'without share', 'undivided'.[46] In the seventh and eighth centuries it may have formed part of the territory of the *Westwilla*, rated for tributary purposes in *Tribal Hidage* at 600 hides, like the two provinces of the *Gyrwe*. The *willa* element might be identified with the Willow Brook, which drains the heights of Rockingham Forest between the rivers Nene and Welland and which flows into the Nene some five miles below Oundle.* So far as Wilfrid had been concerned, Oundle was part of the hegemony of Mercian kings; for its indigenes the Mercian affiliation may have been weaker than its regional and local identity. The Nene Valley was to become a highly productive landscape, both spiritually and economically, during the eighth century but, against a background of uncertain Mercian royal politics in the first decades of that century, it may not have been as peacefully contemplative as its holy men and women would have wished.

Ceolred's accession to the Mercian throne in 709, after the abdication of his cousin, was not marked by an immediate

---

* Davies and Vierck (1974, 234) accept a Northamptonshire identification; others have located the *willa* in the area around either Cambridge or Ely in the Fens (Oosthuizen 2017, 60–61).

reassertion of Mercian *imperium* over the southern kingdoms. David Kirby, a historian of early Anglo-Saxon kingship, cites a medieval tradition[47] that Ceolred was not even the son of his father Æðelred's queen, Osðryð. If that is true, it may explain why he did not immediately succeed his father: either he was too young, having been born to a second marriage after Osðryð's murder in 697, or he was the progeny of a less legitimate liaison.

With no genuine charters recorded in his reign, Ceolred's relations with his thegns and with the religious houses in his kingdom and its satellites are obscure, to say the least. Later tradition records that he had the remains of his cousin Wærburh translated and 'elevated' from her original burial place at Hanbury.* He seems also to have given a small parcel of land to the monastery at Much Wenlock. A brief paragraph in the Testament of St Mildburh – another of Ceolred's cousins if, in fact, her father Merewalh was one of Penda's sons – records that Ceolred, 'of high renown for the pre-eminence of his monarchy,'[48] donated to the monastery at Much Wenlock an estate of four hides. More significant than the modest size of the grant was its location: a place called *Peandan Wrye* – probably Wyre Piddle, near Pershore in Worcestershire – which preserves the name of their grandfather, King Penda.†

The amity suggested by the grant may be misleading. A monk from Much Wenlock suffered a profoundly disturbing vision, which he related to the missionary and inveterate man of letters St Boniface (c. 675–754). Boniface, recounting the vision at great length in a letter to Abbess Eadburh of Minster-in-Thanet,‡ recalled how the monk had seen King Ceolred beset by demons charging him with horrible crimes. Abandoned by the angels because of his sins, he was tormented with 'indescribable cruelties'.[49] Thirty

---

* See above, Chapter 3. Thacker 2020, 446.
† The three attestations to the donation, Bishop Cedda, Abbot Elric and Ealdorman Edbrect, are otherwise unrecorded, which casts some suspicion on the legitimacy of the record.
‡ Minster-in-Thanet was founded by Mildburh's sister, Mildrið.

years later, in a letter of admonition to Ceolred's successor, King Æðelbald, Boniface and a group of bishops recalled that Ceolred had attacked the privileges of the church, inflicted violence and extortion on monks, destroyed monasteries and violated nuns. His partner in crime, it seems, was the young King Osred of Northumbria.[50] Ceolred, then, was remembered less as a supporter of the church than as a despoiler, at least in some quarters. Such accusations imply, on one level, a disrespect for the institution and a lack of moral probity. More significantly, they may represent the first signs of a tendency that characterises the eighth century, during which kings and other secular lords attempted to take back control of lands that they or their predecessors had given to the church. If the king kept nuns as his mistresses, it may mean no more than that some minsters were already thoroughly secularised.

Six years into his reign Ceolred felt sufficiently bold to mount a military expedition against Wessex. His army fought that of King Ine close to a Neolithic long barrow known as *Wodnesbeorg* or Adam's Grave.[51] The site of the battle, close to Alton Priors in Wiltshire, in the shadow of the Marlborough Downs and just south of the great Early Medieval earthwork called Wansdyke, lies deep inside West Saxon territory. The outcome is not recorded.

The *Anglo-Saxon Chronicle* blandly noted Ceolred's death and his burial at Lichfield a year later in 716. Boniface, in that same letter to Æðelbald, provided the detail, with relish:

> ... while he sat feasting amidst his companions [he] was suddenly stricken in his sins with madness by an evil spirit [...] Raving mad, talking with devils and cursing the priests of God, he passed on, without doubt, from this life to the torments of hell.[52]

Exit, stage left, King Ceolred. Two recensions of the *Anglo-Saxon Chronicle* and one version of a Worcester regnal list have him succeeded briefly by a Ceolwald who, by alliterative association, might have been a younger cousin or brother.[53] But before the year was out, the exiled Mercian æðeling Æðelbald had seized or been offered the Mercian throne. With Æðelbald, son of Alweo and grandson of Eowa,[54] who had died in the battle with King Oswald at Maserfelth in 642, the line of kings descended directly from Penda was extinguished.

The new king's mentor, the hermit of Crowland, had not lived to see his protégé emerge from the shadows of tortuous exile. That he had been a significant player in Mercian and East Anglian politics cannot be doubted: his later years are characterised, in Felix's account, by the arrival of significant visitors. The first of these, Bishop Headda of Lichfield and, by then, also of Middle Anglia, came with a substantial retinue that included a secretary, Wigfrið, who was said to be able to spot a fake hermit, 'having lived among the Irish'.[55] This was something like an inquisition: a test of authenticity – perhaps also of theological orthodoxy. After detailed discussions both secretary and bishop were, according to Felix, left in no doubt of Guðlac's spiritual credentials. Headda, himself a former priest at *Medeshamstede*, now ordained Guðlac as a priest and consecrated his oratory in Crowland as a church.[56] Behind the obvious prestige of such an endorsement, one might also read a desire that the Crowland hermit should be recognised as Mercia's holy man; and that sense is reinforced by an event that immediately follows in the *Vita*. Abbess Ecgburh, daughter of King Ealdwulf of East Anglia, sent Guðlac a lead coffin, inside which a linen cloth had been folded, requesting that he consent to be buried in it after his death. In return, she wished that Guðlac would reveal to her whom he intended to inherit the hermitage.

In the competitive world of religious patronage, this counts as a shot across Mercian bows – a blatant manoeuvre aimed at appropriating Crowland's holy man for the East Anglian royal house. The combination of lead coffin – lead being a well-known inhibitor of organic decay – and linen, a sign of purity, reflects ideas of anticipated incorruption, echoed in the miraculous discovery of the undecayed remains of both St Cuthbert (died 687) and St Æðelðryð of Ely (died 679) and in the correspondence of St Paul.*

* I Corinthians 15:52.

Guðlac, suitably honoured by what seems, to modern ears, a back-handed gift, replied in a suitably cryptic manner, telling the abbess only that he who would inherit Crowland was, as yet, unbaptised. * Later, when he lay on his deathbed, he instructed his friend Beccel – former would-be assassin – that he wished to accede to Ecgburh's wishes and be buried in the shroud and coffin that she had sent.[57]

Guðlac died in 714, an event sufficiently noteworthy to be recorded in several recensions of the *Anglo-Saxon Chronicle*. He was forty-one years old, prematurely aged by the physical hardship and mental vicissitudes of a life intensely lived. In his last illness he had been overcome by faintness and wracked by internal spasms. On his death Beccel, in accordance with his master's instructions, left Crowland by boat and sought Guðlac's sister Pega, whom later tradition associates with a small foundation close to *Medeshamstede* known as Peakirk (Pega's church).[58] Pega duly came to Crowland to hold a vigil and arrange her brother's obsequies, burying him in his oratory three days later. A year afterwards she returned and exhumed his remains, finding them, in accordance with expectations, incorrupt. His coffin was now raised from the earth and placed in a 'certain monument' around which his friend King Æðelbald later put up 'wonderful structures and ornamentations'.[59]

It is not clear when the site of Guðlac's church became a monastic house – it left no records before the tenth century.[60] A charter purporting to record the foundation of a monastery there by King Æðelbald is a blatant forgery.[61] In the medieval period Crowland was a famous abbey, its massive square tower and oddly stumpy steeple an unmistakeable edifice punching towards the sky out of the flat horizon of the Fens.

At the time of the saint's death Æðelbald was still 'dwelling in distant parts'[62] but when the news reached him he made his way to Crowland and to the grave of his former mentor. Now, answering the exile's prayers, Guðlac appeared to him and told him not to

---

* Cissa, the man in question, was later a primary informant for Felix in composing the *Vita*. Colgrave 1956.

be sad, for the days of his miseries and afflictions would soon end:

> ... for before the sun has passed through its yearly course
> in twelve revolutions you shall be given the sceptre of your
> kingdom.[63]

His prophesy, in the manner of such things, duly came to pass.
Æðelbald became king in 716 and held the reins of Mercian
power for the next four decades. The conventionality of the
*Vita*, its prophecies, miracles and predictable saintly challenges,
does little to detract from its value as a source for historians
of Mercian history. Guðlac was a contemporary exemplar of a
religious career path already trodden by Anthony, Colm Cille and
Cuthbert; recognisable as such to religious men and women, to
kings and would-be kings; and worthy of written testimony. His
*Vita* also affords glimpses into the lives of more or less ordinary
people whose names and careers are otherwise unknowable:
those with ailments and peccadillos and others with more or less
realistic political ambitions. It shows the irresistible attraction of
the shamanic hermit-cum-healer-cum-soothsayer to those seeking
wisdom, solace and legitimacy. Proximity to such people in an
age of peculiar uncertainty brought a sense of vicarious holiness
and virtue and of an animistic connection to the natural world
with whose flying, swimming or crawling denizens saintly types
communed. In the otherwise liminal landscape of the western fen
edge, Mercian and Middle Anglian religious entrepreneurship
was invested across the last decades of the seventh century and the
opening of the eighth, creating a spiritual core land that survived
the next century and laid the foundations for an intensively
managed landscape of material and immaterial productivity.

If the rivalry between Mercia and East Anglia to appropriate
St Guðlac as a royal cult was perpetuated after his death, the
only evidence for its outcome is in the distribution of churches
dedicated to him. They are almost exclusively Mercian and
Middle Anglian.[64]

# Chronography II
## 718–796

Unless otherwise stated, narrative source entries are from the
*Anglo-Saxon Chronicle*.

| | |
|---|---|
| **AC:** | *Annales Cambriae* |
| **AU:** | Annals of Ulster |
| **BC:** | Continuation of Bede |
| **BL:** | *Letters of St Boniface* |
| **EHD:** | English Historical Documents 500–1042 |
| **HB:** | *Historia Brittonum* |
| **HR:** | *Historia Regum* |
| **S:** | Electronic Sawyer |
| **VB:** | Willibald: *Vita Bonifati* |

| | |
|---|---|
| **718** | Boniface's second journey to the continent, via *Quentovic* (VB V). |
| **722** | Queen Æðelburh (wife of King Ine) destroys Taunton 'which Ine had built'; Eadberht the 'exile' fled into Surrey and Sussex. |
| **725** | King Wihtred of Kent dies. |
| **726** | Kin Ine of Wessex abdicates. |
| **729** | King Osric of Northumbria dies. |
| **731** | Bede completes *Ecclesiastical History of the English People*. All southern kingdoms recognise overlordship of King Æðelbald of Mercia. |
| **733** | Mercian army captures Somerton. |
| **736** | Charter: *The Ismere Diploma*. Æðelbald, king of the Mercians and of the South Angles, to Cyneberht, *comes*; grant of ten hides at Ismere by the River Stour |

and land at Brochyl in Morfe forest, Worcestershire, for the construction of a minster (S89; EHD 67). Mercian king styled *Ætdilbalt rex Britanniæ*.

737 'A great drought made the land unfruitful' (BC).
— King Æðelbald of Mercia harries Northumbria.
— Bishop Eadwine of Lichfield dies; Hwitta and Totta (Leicester) consecrated bishops for the Mercians and Middle Angles (HR).

740 A Mercian warband under Æðelbald invades Northumbria while Eadbert is campaigning against the Picts (BC).
— Cuðred succeeds Æðelheard as king in Wessex.

741 Mercian army sacks York; minster is burned (HR).

742 *Council at Clofesho*. King Æðelbald grants privileges to Kentish church (S90).

743 Kings Æðelbald and Cuðred make war against the Welsh.

746 Boniface writes letter of complaint to King Æðelbald of Mercia over church rights (BL LV).

747 Boniface writes to King Æðelbald protesting at his fornication with nuns and violation of monastic assets (BL LVII). Second *Council of Clofesho* convened by Æthelbald. An influential set of canons attempts reform of private minsters and normalisation of episcopal authority and reach.

749 *Council of Gumley* (S92) Mercian king Æðelbald allows church exemption from taxes except bridge and fort building and military service: the Mercian 'common burdens'.

752 King Cuðred fights against Æðelbald at *Beorhford* (Burford, Oxfordshire); puts him to flight. Likely suspension, at least, of Mercian overlordship.

756 King Cuðred of Wessex dies; succeeded by Cynewulf, who rules for thirty years.

757 Æðelbald of Mercia murdered at Seckington near Tamworth; buried at Repton; civil war in Mercia

(HR; BC). He is succeeded briefly by Beornred, deposed by Offa who succeeds to the kingdom of Mercia.

759 'A pestilence occurred, which lasted almost two years, diverse grievous sickness causing havoc, more especially the disease of dysentery' (BC).

760 Outbreak of plague. Offa of Mercia fights Welsh forces in battle near Hereford (AC).

762 King Offa extends Mercian control to Kent after the death of King Æðelberht II.

764 *The great winter*. 'An immense snowfall, hardened into ice, unparalleled in all former ages, oppressed the land from the beginning of winter almost until the middle of spring.' (HR). 'A great scarcity and famine; an abnormally great drought' (AU). *Lundenwic*, *Stretburg*, Winchester, Southampton, York, Doncaster and 'many other places' suffer serious fire (HR).

765 Archbishop Bregowine of Canterbury dies; succeeded by Jænberht (HR).

768 Beginning of Charlemagne's reign in Francia.

769 Possible date of King Offa's marriage to Cyneðryð of ?Wessex.

771 Offa invades kingdom of the South Saxons; defeats Men of Hastings, suppresses native dynasty (HR).

776 Destruction of the South Wales men by King Offa. Mercians and Kentishmen fight at Otford and strange adders are seen in Sussex; probable Kentish victory after which Kent reasserts independence.

778 Devastation of the South Britons by Offa (AC).
— Last sub-king of *Hwicce*, Ealdred, appears in a charter, now referred to as *dux* (S113).

779 King Offa fights King Cynewulf of Wessex at Benson in Oxfordshire; effectively gains control of Berkshire.

c.780 OFFA REX portrait coins appear.

784 The 'devastation of Britain by Offa in the summer' (AC).

785    King Offa now issuing Kentish charters solely;
Ecgberht, last king of Kent, must by now be dead;
Offa takes over and issues coins from existing
moneyers there as *Rex Merciorum*.

786    Pope Hadrian I sends legates to Britain at Offa's
invitation in alliance with King Ælfwald of
Northumbria (HR).

787    *Council of Chelsea*: Offa raises Lichfield to
archdiocese; Offa's son Ecgfrið consecrated as
successor by Hygeberht of Lichfield.

789    First recorded attack by three ships of Norse raiders
on South Coast kills Beauduherd, the king's reeve in
Dorchester.
— King Beorhtric of Wessex marries Offa's daughter
Eadburh; probable date of Ecgberht's flight to
Francia.

790    Alcuin returns to York from Charlemagne's court.
Offa suggests marriage alliance of his son Ecgfrið
to Bertha, daughter of Charlemagne; Charlemagne
embargoes Mercian goods.

792    King Offa gives daughter Ælfflæd in marriage to
Æðelred of Northumbria.
— Offa extends 'common burdens' to Kent,
confirming privileges of churches except for defence
against 'marauding heathens' (S134).
— Offa's 'heavy, late' coin reforms, following those of
Charlemagne, may start in this year.
— Archbishop Jænberht of Canterbury dies; replaced
by Mercian appointee Æðelheard.

793    Lindisfarne attacked and plundered by Vikings.
Famine and evil portents including fiery flashes in the
sky (HR).

794    Offa puts East Anglian King Æðelberht to death.
— *Council at Clofesho*: King Offa confirms privileges
in the Kentish churches (S137).

796    Charlemagne's conciliatory letter to Offa (penned by

Alcuin); rapprochement on trade and pilgrims (EHD 196, 197).
— King Offa of Mercia dies. Succeeded by his son Ecgfrið; he too dies; succeeded by unrelated Coenwulf.
— Battle of Rhuddlan (AC).
— Rebellion in Kent against Mercian authority; elevation of ex-priest Eadberht *Praen* to throne in Kent. Archbishop Æðelheard expelled (or flees) from Canterbury; seeks protection of King Ecgfrið in Mercia.

# Chapter 5

# Æðelbald:
# Rex Britanniae
# 716–735

❖

*The Repton stone • Æðelbald's people •*
*The crown estate • Missionaries and*
*merchants • Concessions • Rex Britanniae*

A weathered sandstone relief carving of a spirited, sword-wielding warrior mounted on a cantering stallion was discovered at Repton church during excavations in 1979 and is now on display in the City Museum in Derby. The composition echoes classical depictions of victorious emperors: the rider dressed in a mail or scale-armour shirt and pleated kilt, with cross-gartered leg coverings. He wears no helmet and his short hair seems to be covered by a braided band or possibly a diadem. His chin is shaven clean but he sports extravagant moustaches. A short fighting knife, the *scramaseax,* is strapped at his belt and he carries a small, round shield or targe, raised at head height, in his left hand. Very faint incisions seem to indicate that the shield originally bore a motif, perhaps a cross or flying bird. The warrior's right hand, now missing, once held a sword – a fragment of the broad blade can be traced above his head. The slab seems to have been one decorative component of a rectangular shaft, perhaps with a cross once attached to its head by a dowel, but now missing. The complete monument is thought to have stood an imperious 10–15 feet tall and its excavators believe the surviving portion to be a 'portrait' of King Æðelbald.[1] If so, it is the first figurative representation of an English king.

The *Anglo-Saxon Chronicle* records under the year 716 that 'Æðelbald succeeded to the kingdom of Mercia and ruled forty-one years. [And...] Æðelbald was the son of Alweo, the son of Eowa, the son of Pybba...' His four-decade reign lies in the shadow of his more famous successor, Offa; but his legacy – the fabric and institutions that underpinned Anglo-Saxon statehood –

The Repton stone: is this damaged stone fragment the earliest
surviving figurative representation of an English king?

was more profound than historians have generally recognised.
He may have envisioned, if not instigated, the construction of a
great earthwork along the Anglo-Welsh frontier; he developed
London's trading settlement into a wealthy, populous international
marketplace; he consolidated Mercian *imperium* over all the
southern kingdoms and brought state and church to a landmark
agreement on land rights and privileges. The circumstances of
Æðelbald's elevation from exiled warlord and saintly protégé to
the Mercian kingship, then overlordship over all the southern

kingdoms, may still be uncertain; but he ushered in a new era of Mercian stability and power, replacing successive, short-reigning, childless kings of whom the first abdicated and the second was said to have died in a frenzied fit. By comparison, their careers are mere footnotes.

Dynastic uncertainty was a recurring feature of Anglo-Saxon kingship and the situation in 716 was particularly precarious for the Mercian state. It is possible to sketch a scenario in which the death in 709 of the old King Æðelred in his self-imposed monastic exile at Bardney precipitated his nephew Cœnred's abdication. The apparent murder of his son and eventual successor, the allegedly dissolute Ceolred, whose excruciating demise during a feast some seven years later sounds as though he was 'poisoned for his sins', seems to indicate complex factional tensions at play.

King Æðelbald belonged to the pool of would-be Mercian kings whose eligibility to rule, by virtue of descent from kings, was unquestionable. Seizing the throne from the short-lived Ceolwald,* Æðelbald took power with the security of ready-made political capital: supported by his contemporary, the East Anglian King Ælfwald, and with the blessing of Mercia's recently deceased but pre-eminent holy man, Guðlac. As a former exile of noble blood, he must already have gathered around him a *comitatus*, the core of a formidable military household. These men must now be rewarded with treasure, with gifts of land and with potential brides; and political wisdom dictates that these would be the daughters of senior Mercian families or those of Mercia's satellites – of *Hwicce*, *Magonsæte* and *Wreocansæte*, Lindsey and Middle Anglia. By right or by main force a new king inherited the estates of his predecessor as well as his own family possessions; was owed by them a *feorm* or food render to support his household, while from the élite who must now swear loyalty to him he drew, or hoped to draw, his political and military strength.

---

* Very likely a brother or cousin of Ceolred; nothing of his very brief reign is known.

For the first time, during Æðelbald's reign, something can be said of the complex structures of geopolitical networking and personal relationships through which a long and successful royal career might be sustained. The names of some of these men have been preserved either in the pages of the Guðlac *Vita* or in the witness lists of Æðelbald's charters. One of them was called Ofa, a 'retainer' during the king's years of exile who found a place in the *Vita* by virtue of having trodden on a thorn, subsequently developing a severe and crippling infection,* of which Crowland's hermit-saint duly cured him. Ofa turns up as a regular and senior charter witness and, as late as 742, he subscribed to one of the most celebrated of all Æðelbald's decrees.[2]

Historians of the period suffer small agonies trying to make sense of the range of Latin terms used to describe members of the king's inner circle – his counsellors, battle commanders and confidants.[3] *Subregulus*, *princeps*, *patricius*, *dux*, *comes* and *minister* are all used, both in charters and by Bede and other secular sources. Each must reflect an Old English notion of rank, nuanced by occasion, political fortune or changing fashion, but not necessarily fixed by title or formal honour. Ofa, veteran of the king's youthful exile and described in the *Vita* as a *comes* – a senior member of the king's *comitatus*, or warband – is at other times a *minister*, *dux*, *princeps* and *patricius*. In Old English those nuances are covered unsatisfactorily by such terms as ealdorman, *geoguð*, *duguð* and *gesið*.† That the king's nobles were differentiated by personal ties to the king, by married status, seniority and military rank seems certain. Hector Chadwick was able to demonstrate that young, unmarried *comitates* – the *geoguð* – were signified by the Latin term *ministri* and that, when married, they bore the rank of *comes*.[4] *Patricius* carries connotations of nobility, of

* Hawthorn (*Cratægus monogyna*) thorns concentrate pathogens and can deliver severe pain and joint swelling. The episode comes in *VSG* XLV. Colgrave 1960, 139–141.
† *Geoguð* – a young, unproven warrior; *duguð* – a senior, proven warrior; *gesið* – a military companion, member of the *gedryht* or *comitatus*.

hereditary rank. *Dux* ought to indicate a senior military role; and *princeps* may have been used to denote the ruler of a large and important territory – it was used of a prince of the *Gyrwe*, for instance.[5] One means of differentiating their standing in the royal household is by the order in which they are cited as witnesses to written documents – precedence in rank being a matter of sometimes extreme sensitivity. Only much later in Æðelbald's long reign do such subtleties emerge into the light.

In 716 a key group of loyal retainers and of the most powerful men in Greater Mercia were the principal political resources available to the new king.[6] He was also able to count on the support of a brother, Heardberht (known only through his consistent appearance as a very senior charter witness), and on that of the now elderly *Hwiccian subregulus*, Æðelric, who had served as a minister under King Æðelred.[7] This is a significant indicator of the new king's political vision. Sudden dynastic change provided opportunities for sub-kings of formerly independent states to reassert their sovereignty – so Æðelric's apparent compliance with the new régime indicates political accommodations on both sides. But nuance is often lost to history. The text of a late ninth-century confirmation charter recalls how King Æðelbald had murdered one Æðelmund, the son of a nephew of two *Hwiccian* kings, and had made a donation to a religious house in expiation. Alliances were often expedient, and fragile.

A man named Beorcol, variously *comes*, *dux* and *patricius*, witnessed the king's charters from the beginning of his reign up to 749 and must be considered, alongside Ofa, as a key member of his original *comitatus*. He attested charters on three occasions in the company of the king's brother. A much later charter, from the reign of King Offa, alludes to an estate that Æðelbald had granted to Offa's grandfather, Eanulf, indicating an otherwise unsuspected alliance between two great Mercian families.[8]

These, then, were the men of the new king's inner circle of trusted followers. What one might term an outer cabinet consisted of men who periodically attested charters and who were recipients of lands in the gift of the king over the next decades. That circle of powerful interests was not confined to secular lords. Bishops of London and Worcester were favoured both for their imprimatur as charter witnesses and as recipients of significant lands and concessions, while the Kentish bishop of Rochester also figured as a recipient of trading concessions. By the early eighth century Mercia was served by no fewer than seven bishops, men for the most part drawn from the same rarefied social ranks as their secular counterparts and more or less fully integrated into the formal political and social structures of the kingdom. All of them appear as charter witnesses and it seems that they provided a regional focus for royal power: bishops belonged to their diocesan seats and these were broadly coterminous with older tribal entities – Middle Anglia, *Hwicce*, *Magonsæte*, Lindsey and Lichfield.

A small but significant number of abbots and abbesses were also recipients of donations from the royal estate portfolio and thus belong to the Mercian establishment. Few charters survive from the Mercian heartlands on either bank of the River Trent to hint at the detail of Æðelbald's dispositions there; but key foundations like Repton, Hanbury and Breedon-on-the-Hill betray their likely frequency in those decades. By such networks of patronage, the power inherent in the lordship of the royal *comitatus* was devolved to a more subtle landscape of belonging in which pastoral care and investment in spiritual-intellectual hubs served to embed royal authority. Monasteries invested in and monumentalised their churches, communities and estates – capitalising them, enjoying the benefits of locations on navigable rivers or major routeways and secure in their everlasting earthly tenancies.

It is a striking feature of Æðelbald's career that, so far as we know, he neither married nor produced eligible heirs. If that was a matter of conscious policy it is difficult to explain; but it echoes the dynastic inertia of his two immediate predecessors, neither of whom produced an heir. It was not a matter of sexual indifference on Æðelbald's part – in later years he was castigated by Boniface for what seems like a commonly known predilection for fornicating with 'religious' women.

The political narrative that historians would like to be able to write for the first decade of Æðelbald's reign, which might help to explain his dynastic thinking, is almost entirely lacking. That is not just a feature of Bede's reticence on Mercian matters; nor merely of West Saxon prejudice in compiling the records of the *Anglo-Saxon Chronicle*. The political histories of Northumbria, Wessex, Kent and East Anglia in those years are equally obscure. At the time of Æðelbald's accession King Ine had been on the West Saxon throne for twenty-eight years and had, the previous year, fought with Æðelbald's predecessor, King Ceolred. But nothing more is known of relations between the two kingdoms before Ine's sudden abdication and departure for Rome a decade later in 726.

The apparent overlordship over Kent, Essex and Sussex which Ine inherited at the start of his reign had weakened during those years. Ine had not chosen to wed a princess of one of the other Anglo-Saxon kingdoms, but instead married a West Saxon noblewoman, Æðelburh. She is said to have led a raid against Taunton in 722, razing a fortress that her husband had constructed in a vain attempt to kill a would-be usurper, Ealdberht, who fled into exile in Sussex.[9] The *Chronicle* entries for Wessex in those decades are notices of civil war and of rival dynasties vying for the kingship; so although Ine might theoretically be able to assert his dominance over Mercia, in reality he was in no such position. There is some evidence that he was able to control the valuable trading settlement at *Lundenwic*, at least in the early part of his reign: his law code describes Earconwald as 'my bishop' of London,[10] even though East Saxon kings still harboured their own proprietary rights there. He was also able to prosecute wars against the British kingdom of *Dyfnas*, to the west, and assert military superiority over the South Saxons.[11] But he and his queen bore no children that we know of and on his abdication she accompanied him on pilgrimage to Rome. Ine's immediate successor, King Æðelheard, may have been her brother; but in the first years after King Ine's death the kingdom was again contested, seemingly by an æðeling named Oswald.[12]

Nevertheless, Ine was the first West Saxon king to promulgate a set of laws, written in Old English, which enshrine the ideologies, perhaps even some of the realities, of royal administration at the beginning of the eighth century. The sorts of crimes that concerned kings – the prevention of blood feuds by the codification of *wergild* or head-price; the protection of church interests; the nature of interpersonal and property offences – frame the ambitions of royal control. Although, disappointingly, there is little mention of trade in Ine's laws, he must have profited from the thriving mercantile settlement at *Hamwic* on Southampton Water; he may even have instigated its formal street pattern and construction. From here West Saxon goods were exported to the Frankish ports of Rouen and *Quentovic*; from here, too, missionaries like St Willibald embarked on journeys that would take them to the Frankish court and then on to Rome. Like his counterparts in Kent, Mercia and East Anglia, King Ine may have found that new sources of revenue from tolls and the export of goods and crafts from his estates and workshops made the risky business of open warfare with powerful neighbours a game less worth the candle.

In Kent, King Wihtred had been on the throne since 694. Like Ine, he promulgated a set of law codes (dated to 695) which have been highly influential in informing historians' understanding of royal administration in the latter part of the seventh century, particularly in defining secular relations with the church. The politics of his reign are equally obscure but there are strong hints that Wihtred and Ine had reached some sort of political entente. Ine had received substantial compensation – said to have amounted to '30,000'* – for the murder of his kinsman, Mul, by Kentishmen. And the more or less contemporary promulgation of their two law codes, which have key features in common, has suggested to some historians a deeper degree of co-operation.[13] During the late seventh century Kentish kings had been able to at least partially control trade out of *Lundenwic*, establishing a

* Probably of silver shillings. *Anglo-Saxon Chronicle* 694.

hall or *sele* there – perhaps a customs house – and appointing a *wicgerefa* or port-reeve to oversee the collection of tolls.[14]

Kent's kings had long been the beneficiaries of close commercial and diplomatic ties with the Frankish kingdoms. A necklace of thriving trading settlements was strung out along the length of the sheltered Wantsum Channel between Reculver, at the mouth of the Thames Estuary, and Sandwich on the Channel. The archbishop of Canterbury – who, under Wihtred, was a native Kentishman named Berhtwald – enjoyed the fruits of his own emporium, called Fordwich. Travellers and merchants passed through Kent on their way to and from other English kingdoms and the ports of Rouen and *Bononia*.* In the decades leading up to Wihtred's rule both Wessex and Mercia had enjoyed periodic dominance over Kent but Wihtred was able to establish a measure of independence and dynastic stability there and on his death in 725 he passed the kingdom to his sons. These were decades of consolidation among the southern kingdoms.

The political landscape of Mercia's western neighbours in Wales at this time can barely be sketched in outline. A celebrated ancient monument – the Pillar of Eliseg – still standing close to the town of Llangollen in what is now Denbighshire, was erected in the middle of the ninth century by the Powysian King Concenn ap Cadell. The inscription which it once carried is now illegible, but it was read and transcribed at the end of the seventeenth century by the antiquarian Edward Lluyd. A fragment of the inscription can be translated from the Latin to read:

> + It was Eliseg who united the inheritance of Powys... however through force... from the power of the English... land with his sword by fire.[15]

Eliseg or Elisedd ap Gwylog was the great-grandfather of Concenn and a contemporary of Æðelbald. The 'English' of the inscription can only mean Mercians during his reign. The

---

* Boulogne, Hauts-de-France.

conflict, if such it was, and the recovery of Powysian lands from Mercian control which it implies, cannot be dated any closer than Eliseg's reign, between about 725 and 755, but certainly took place during Æðelbald's reign.* Powys was the most powerful of the contemporary Welsh kingdoms bordering Anglo-Saxon territories; its formerly close alliance with Penda and his sons seems not to have survived the arrival of the new Mercian dynasty in 716. A conflict between Wessex, Mercia and 'the Welsh' is recorded under the year 743 in the *Anglo-Saxon Chronicle* – but there is no more detail to the entry. Such thin evidence for ongoing conflicts between Mercia and the northern Welsh kingdom provides only a shallow context for the massive investment in frontier infrastructure ascribed to King Offa.

Northumbrian ambitions to extend territorial power south of the Humber had been curtailed after the battle on the Trent in 679, when Lindsey was finally annexed to Mercia. King Aldfrið, the *sapiens* who had been brought from Iona to succeed his brother Ecgfrið in 685 after the latter's death in battle, was succeeded in 704 by a 'usurper', Eadwulf, himself deposed and replaced by Aldfrið's young son Osred a year later. Osred was killed and succeeded by one Cœnred in 716 and during the following twenty years rival branches of the descendants of the Iding dynasty vied for power, their military attentions focused more on Pictland to the north than on Mercia in the south. King Æðelbald was peculiarly lucky in facing so little competition for dominance among his neighbours in the first twenty years of his reign.

Early Medieval kings' ambitions to explore the potential limits of secular power were constrained by a world lacking towns, sophisticated systems of market distribution and currency and the infrastructure of public investment or centralised taxation. They had, nevertheless, to develop a comprehensive understanding of the assets at their disposal and of the tools with which they might

* See Chapter 6 for a possible context in the 740s or 750s. Charles-Edwards 2014, 417.

exploit them. One very conspicuous early investment – or rather, perhaps, repayment – was Æðelbald's construction of 'wonderful structures and ornamentations' around the shrine of St Guðlac.[16] Aside from the political value of such monuments and the support of his loyal followers and counsellors, Æðelbald's primary assets were the food rents and service renders which he was entitled to draw from royal estates. Since surviving Mercian charters* almost exclusively cover lands outside the core territory of the kingdom, it is almost impossible to reconstruct the geography of these landholdings. Royal estate centres – the *villae regalis* – can sometimes be identified, as can the territories of some of the smaller polities swallowed up within Greater Mercia, but often only tentatively.

Only in Bernicia, in north Northumbria, has a credible map of Early Medieval royal estates and their partial histories been reconstructed with some confidence.[17] It is by no means certain that Mercian estates were organised, like those of Bernicia, into 'shires' of six or twelve contiguous 'townships', with a single principal estate centre, the *villa regia*, under the stewardship of a *gerefa* – a reeve – to which food renders and service obligations were drawn. At these royal townships, resources were concentrated so that itinerant kings and their entourages could visit each in turn – probably on an original or theoretical model of ten such shires that would provide for a year's on-the-hoof consumption.†

There is some evidence that such a duodecimal territorial system operated more broadly across southern Britain. Some of these 'shire'-type units might equate to the *regiones* – territorial or tribal units – whose names survived into the era of written charters or were preserved in the time capsule of Tribal Hidage, and whose geography can be tentatively reconstructed. A charter issued in King Æðelbald's name some time before 737 records the

---

* About twenty authentic charters survive from Æðelbald's reign – an average of just five for each decade – which must represent a small percentage of the original.

† For an introduction to this idea, see *The King in the North*.

donation of twenty hides of land by Æðelric, his *comes*, to the monastic church that he had founded in a place called Wootton – *Wudutūn* – Wootton Wawen in Warwickshire on the River Alne.[18] This Æðelric is described as the son of Oshere the king of the *Hwicce* (who died before 717) and is, therefore, the same Æðelric who formed part of the king's inner circle of trusted counsellors and supporters. The charter contains a little gem: the name *Wudutūn* indicates an estate centre where timber was the key resource and it is said to lie in a *regio* that had of old been named from the *Stoppingas*, one of those elusive peoples in the process of being absorbed into larger kingdoms during the seventh century.[19] The later parish of Wootton Wawen, whose surviving late Saxon church may stand on the site of the eighth-century original, appears to mirror the original *regio* of the *Stoppingas*. If this was land in the gift of kings it had very probably been a royal estate of the *Hwicce*. The parish consisted of eleven smaller units, each now a parish in its own right, which seem to have survived almost intact from an original twelve-township territory – that is to say, the *regio* of the *Stoppingas* equates to something like the twelve-*vill* shire familiar from Bernicia.

Another Mercian royal estate, out of which Æðelbald granted a modest ten-hide plot to his ealdorman Cyneberht in 736, lay on the banks of the River Stour.* The land had 'of old' borne the name *Ismere*.[20] A later charter specifically describes *Sture* as belonging *in provincia Usmerorum*: the territory of the *Husmeræ* – another tribal grouping and possible shire-type territory and, like the *Stoppingas*, part of the kingdom of the *Hwicce*. Other apparently very early shire-type territories based on a duodecimal scheme have been identified in Essex[21] and a similar scheme seems to have been fundamental to landholding in Wales.[22] If Catholme, in the Trent Valley near Lichfield, was *Tomtūn*, the *villa regia* of the *Tomsæte*, then archaeologists have the material clues to be able to reconstruct some of the layout of such royal estate centres. Further candidates, characterised by grandiose architecture and by their

---

* In Worcestershire, near modern Kidderminster.

proximity to prime locations such as Roman towns or navigable rivers, have been suggested at Atcham on the River Severn near the Roman town of Wroxeter – perhaps its successor;[23] at Gumley in Leicestershire, where a prominent assembly was held during the reign of Æðelbald; and on the River Salwarpe in Worcestershire at Wychbold, whose name may mean 'fortified enclosure belonging to the *wīc*' and which later saw meetings of royal councils.[24]

As they progressed through their estates the royal household made themselves a visible and accessible presence among their landed élite: reinforcing ties of loyalty or kinship, dispensing summary justice, collecting fines for offences committed against the king's laws and tribute in the form of livestock, horses and bullion. In promulgating or endorsing land gifts and exchanges they confirmed their lordship over subordinates and would-be rivals. At each royal *vill* the mead hall – part theatre and feasting space, part council chamber, part home farm and part hunting lodge – was the focus for consumption of the produce of the land: bread, meat, ale, honey, butter and cheese. The contemporary laws of King Ine describe the annual food rent expected from a ten-hide estate in Wessex, while a specifically Mercian calculation is recorded during the later reign of King Offa in the 790s, for a sixty-hide royal estate in Gloucestershire:

> *...Ðæs gafoles æt Westbyrig twa tunnan fulle*
> *hlutres aloð...*

… The tribute at Westbury: two tuns full of pure ale and a coomb full of mild ale and a coomb full of Welsh ale and seven oxen and six wethers [castrated rams] and 40 cheeses and six long Þero [an unknown commodity, perhaps woollen cloaks] and 30 ambers of unground corn and four ambers of meal.[25]

Mercian kings feasted their companions and their tributary lords; dispensed gifts of land, weapons, decorative metalwork and exotica. They hunted, consulted on matters of state and law, played at martial sports and plotted war; arranged marriages, fosterage and preferments among their chosen supporters and regional supplicants. They were instructed in Christian worship and in the church's idealised view of rational kingship by bishops, abbots and abbesses who were men and women of similar rank – sometimes kinsmen and women.

But food renders were not in themselves sufficient, by the beginning of the eighth century, to sustain royal ambitions or the emerging complexities of expanding kingdoms, military threats and administrative challenges. Mercian kings increasingly enjoyed access to a much wider range of resources, which archaeologists and historians have been able to piece together gradually from fragments and hints. It is now possible to evoke an energetic society poised between conservatism steeped in oral tradition and time-honoured custom and a dynamic sense of a wider world full of marvels and possibilities while hedging against the all-too-real threats posed by violence, disease and capricious fates.

In 716, the year of Æðelbald's accession, a Devonian monk named Winfrið, better known to history as St Boniface, took passage on a boat from the port at *Lundenwic* on the Thames – 'where there was a market for the buying and selling of merchandise' – and sailed across the Channel to the Frisian trading settlement of Dorestad.* This huge site, covering hundreds of acres, was first identified during excavations in the mid-nineteenth century.[26] Dorestad served the fortress of Utrecht (Roman *Traiectum*) at the heart of the Frisian kingdom, linking it to both the North Sea and the great trading highway of the Rhine. As it happened, Boniface's arrival on the continent, as a would-be missionary to work among the unconverted peoples beyond the northern edge of the Frankish empire, coincided with an outbreak of hostilities between the Frisians under King Radbod and the Franks under Charles Martel.† According to Boniface's

---

* Talbot 1954, 35. Now Wijk bij Duurstede on the River Lek, a few miles south of Utrecht in the Netherlands.
† I am simplifying a complex political and military situation. For a summary of the key developments see Fouracre 1995.

hagiographer, Willibald (c.700–787), these years saw a revival of paganism inspired by anti-Frankish sentiment. The timing could not have been worse, and Boniface returned to Britain.

Two years later Boniface made a second journey abroad, again departing from *Lundenwic*. This time he crossed via a more southerly route, landing in the Frankish kingdom and intent on making the long pilgrimage to Rome. His port of arrival was a place called *Cuentwic* by Willibald: the trading settlement of *Quentovic* on the River Canche near the modern town of Étaples.* Boniface, like Wilfrid and Benedict Biscop before him, was taking advantage of an ecclesiastical network linking Rome, Constantinople and Alexandria with Christian communities in Africa, Italy, Francia, Britain and Ireland. They maintained a diplomatic, intellectual and literary web, partly dependent on transport provided by merchant vessels trading in high-value, low bulk goods – gems, dyes, wine, oil, salt and slaves – and partly underwriting that trade. Major trading settlements established in the late seventh century at the English riverine ports on the Itchen, Orwell, Thames and Yorkshire Ouse and at several sites along the Wantsum Channel in East Kent saw an increasing variety and volume of traffic passing across the Channel and the North Sea. Britain was connected not just to its nearest neighbours in Francia and Frisia, but also to Scandinavian ports and, independently, to the emporia of the Mediterranean.†

Landlocked Mercia was blessed with navigable rivers running north-east to the Humber estuary, east into the Wash, south into the Thames and west into the Severn; but *Lundenwic* was the prize that gave the most convenient, most lucrative and prestigious access to continental markets and courts. Mercian kings used all the political tools at their disposal to ensure that London markets and traders operated under their control. Under Æðelbald that control would become absolute.

A decade after Boniface's departure on his second missionary journey, Bede would describe London as an emporium of many

---

* In the Pas de Calais.
† A theme explored in *The First Kingdom* (Adams 2021a).

nations coming to it by land and sea.[27] This has usually been read as Bede's confirmation of the presence of foreign traders in London. It might conceivably also be read as a hint that in his day, control of London's trading profits was shared among several of the English kingdoms.

The Roman provincial capital of *Londinium* on the north bank of the River Thames sat at the centre of a network of radiating roads that ran to all parts of *Britannia*. The tidal Thames brought cargo ships, imperial officials, visitors and news from Gaul, Rome and beyond to wharfs and warehouses lining the embankment. London had boasted fine baths, a grand forum, palatial residences with fashionable mosaics and the temples of an eclectic pantheon of gods. A bridge spanned the river, linking the city to a southern waterfront and suburb at what would later become King Ælfred's *burh* at Southwark.

Both literary and excavated evidence for Roman London's fate indicate that it no longer functioned either as a government, urban or mercantile centre after the fifth century. When Bishop Mellitus founded the church that would become St Paul's Cathedral early in the seventh century it stood within walls that enclosed little more than the residence of a lord of the East Saxons.[28] In the latter part of that century minsters sprang up on prime real estate along this desirable stretch of the river: at Westminster on Thorney Island; at Bermondsey on another island on the marshy south side of the river; and downstream at Barking. Active regional and international trade resumed in the middle of the seventh century; but now merchants drew their shallow-draughted boats and barges onto a gently sloping beach below the Strand, half a mile upstream across the River Fleet and a few hundred metres south of Watling Street – beneath what is now Oxford Street – instead of mooring at the rotting, deep-water hythes of the old city.

Between Aldwych\* and the site of the rail bridge crossing of the Thames at Embankment, and as far inland as Covent Garden where a cemetery lay, *Lundenwic* was an organised cluster of houses, craft

---

\* Literally the 'Old *wīc*' in Old English.

workshops and tenement plots lining formally laid-out lanes.[29] By the end of the seventh century, and very likely at the instigation of King Wulfhere, the riverside had been embanked to allow deeper-draughted vessels to draw alongside. Set back about 100 yards from the shore, the Roman road leading westwards to Silchester and Bath – later to become Fleet Street and the Strand – marked the edge of habitation and industrial activity. As the eighth century progressed open spaces and enclosures used as livestock pens and small garden plots were filled in by new craft and mercantile premises. Somewhere in this D-shaped riverside settlement stood one or more *seles*, the customs houses of Æðelbald's *wīcgerefa* and perhaps those of the Kentish and East Saxon kings. Here too, among the assorted local and regional trading houses and in the sorts of establishments where ships' captains might be found, would-be travellers like Boniface might negotiate their passage across the Channel and the agents handling the affairs of bishops and abbots cut deals on the carrying of goods and letters to and from the Channel ports. The whole settlement enclosed some 150 acres and seems to have supported a population in the region of 6,000–7,000 people during its eighth-century heyday.[30]

From the strikingly fair Deiran slaves who turned up in Rome's markets in the late sixth century,* to Benedict Biscop's purchase of a fine Latin Cosmography in Rome, to the comings and goings of fare-paying missionaries and the dispute between King Offa and Charlemagne at the end of the eighth century over the quality of English cloaks sent to the Frankish court, contemporary writers hint at the sort of commodities travelling across the Channel. These are now supplemented by much more tangible and dispassionate evidence from the efforts of excavators and the metal-detecting records of stray coin finds.

The coin evidence is now compelling. Even if the total number of seventh- and eighth-century coins recovered from England stands only in the low thousands, the true number of coins that

---

* Supposedly the impetus for Pope Gregory's Augustine-led mission to the English in 597. Cf *The King in the North* and *The First Kingdom*.

numismatists can infer from die analysis shows that in these two centuries more coins were in circulation than at any time before the Norman Conquest.[31] Recycled late Roman gold and silver coins and early Frankish issues turn up in small numbers here and there in Anglo-Saxon England, literally worth their weight and used more as bullion than currency – that is to say, coins with a nominal value, regulated for consistent use in transactions. But by the 670s silver pennies, usually known as *sceattas*,* were being minted in large quantities in Northumbria under King Aldfrið.[†] These are very small coins carrying a simple obverse and reverse design – not, initially, bearing a king's head but sometimes inscribed with his name – weighing no more than twenty Troy grains, or 1.3 grams – something like a paperclip or a modern banknote. One would barely feel its weight in the palm of one's hand.

In the decade after 710, in parallel with an expansion of the major trading settlements, the numbers of both English and Frisian silver coins circulating among the Anglo-Saxon kingdoms increased dramatically – many of them seemingly minted at Dorestad, where Boniface had first made landfall in Frisia in 716. Anglo-Saxon coins also turn up in small numbers in Francia, despite the very thin record there of metal-detecting finds.[32] If it is true that the vast majority of coins were coming *into* England through its trading settlements, the implication is that the Anglo-Saxon kingdoms were producing goods of export value to continental merchants.[33]

Although King Æðelbald's name does not appear on any coins during his long reign, numismatists are generally confident that he was responsible for the establishment of mints at London and one other site, perhaps Oxford, within Mercia.[34] During the most productive period of the early eighth century, Mercian coinage copied Kentish designs which themselves owed much to Frankish

---

* Pronounced 'shatters'.

† Only thirty-four coins bearing his name have been found; but they represent something like two million when the dies are analysed. Metcalf 2006.

inspiration. By the 730s and 740s, when economic boom was already turning to bust, distinctive coins were being minted bearing the legend LVNDONIA.[35] By then, a strong geographical correlation is evident between lost coin and the sheep-rearing country of the Gloucestershire Cotswolds and elsewhere in a central triangle bounded by the Rivers Avon, Great Ouse and Thames.[36] Wool, and, perhaps, also finished textiles, was in demand across the Channel.

King Æðelbald was able to profit materially from such trade in several ways. First, he was by far the greatest Mercian landowner: many of the sheep who grazed the Cotswold hills and rich woodland pastures of the Midlands were *his* sheep and the women who processed the wool and turned it into high-value textiles – the famed English cloaks, for example – were *his* women. And then he claimed rights of control over the Mercian ports of trade – principally *Lundenwic* – through his *wicgerefa* – his *wic*-reeve.[37] The king also charged tolls on ships and on commodities, such as salt and wool, entering and leaving the port. A small number of charters recording the king's gift of remissions on such tolls – to his favoured ealdormen, bishops or religious houses – must indicate more substantial opportunities, including judicial rights to the fines and seizure of goods that went with such mercantile activities. Among Æðelbald's appointments, according to the charter evidence, was a *thelonarius* – a tax collector – named Pant.[38] Concentrating the production of finished goods at the *wics*, and asserting control over the minting of coinage, offered further opportunities to rake off profits from trade. In addition, some very high-value exotica such as furs, ivory, gemstones and wild animals from Africa, may have been regarded as royal perquisites. Traders wishing to establish headquarters or warehouses or their own craft workshops at choice locations, and competing for the king's patronage, would consider it worthwhile to offer him gifts of such exotica. The dividing line between gift exchange and profitable sale was much less obvious in the Early Medieval period than it is today.

No such transaction can be directly ascribed to King Æðelbald, but a Kentish contemporary, King Æðelberht (748–54), revealed some of the not-so-subtle mechanisms at play in a fascinating correspondence with Boniface...

… I am sending your grace […] a few gifts: a silver drinking cup lined with gold, weighing three and a half pounds, and two woollen cloaks […] not in the hope of receiving any earthly gift [but for] your prayers. […] There is one other favour I have to ask, namely, to send me a pair of falcons, quick and spirited enough to attack crows without hesitation and bring them back to earth after catching them. […] there are few hawks of this kind in Kent.[39]

To modern readers this must look very much like a commercial transaction in which even prayers had their price. To eighth-century eyes it would be entirely in keeping with the idea of reciprocity – the exchange of gifts or favours which maintained good relations between a lord and his dependents and between friends. Boniface, for his part, sought from another correspondent in Kent – namely, the archbishop – copies of some of the works of Bede. While hardly any manuscripts or books of solidly eighth-century Mercian provenance survive, there can be little doubt that fine illustrated English manuscripts, and the writings of English scholars, were in high demand on the continent.[40] In 716, the year of Æðelbald's succession and of Boniface's abortive first mission to the Frisians, Bede's mentor, Abbot Ceolfrið of Jarrow, set out on a fifth and final pilgrimage to Rome carrying a complete copy of the bible as a gift for Pope Gregory II – one of three enormous, single-volume 'pandects' that he had commissioned from the *scriptorium* at Jarrow on the south bank of the Tyne. Ceolfrið died before he reached Rome; the pandect, now known as the *Codex Amiatinus* after the mountain monastery in Tuscany where it was kept until the eighteenth century, weighs over 75 lbs and contains more than a thousand folios of vellum. The artistic and material investment required for the production of such a work is still astounding: ample proof of the wealth being created in and exported from eighth-century English kingdoms. A hoard of some 300 or so silver coins discovered in Aston Rowant, near Thame in Oxfordshire and dating to about 710, may be the stash of a continental trader making his way up (or down) the Thames on a familiar and profitable journey.[41]

In the organised, tightly clustered streets of the *wīcs*, excavated over decades in often complex and difficult conditions, archaeologists recognise the hand of authority – and they naturally enough make the case that only the kings of Northumbria, East Anglia, Mercia, Kent and Wessex provided the means and drive to establish these trading centres. In their turn, traders were protected by the king's laws and the fruits of his powers of patronage, lubricating diplomatic and commercial transactions across the Channel. But the fingerprint of royal patronage and profit does not mean that all the impetus for the growth in early eighth-century trade came from kings. Charter evidence shows how zealous bishops were able to obtain profitable trading concessions for themselves and their churches. Senior members of the king's household – like Ofa, Beorcol, the *Hwiccian subregulus* Æðelric and the king's brother, Heardberht – may have taken an entrepreneurial lead in trading commodities from their own holdings.

The same goes for the heads of religious houses, such as Mildrið, abbess of Minster-in-Thanet, who obtained from the king remissions on tolls for her abbey's own ship, perhaps operating out of the trading settlement at Sarre on the Wantsum Channel.* But the breadth and productivity of women's agency as ecclesiastics, as proprietors of monastic ateliers, as entrepreneurs in their own right and as courtly consumers of foreign goods is hard to quantify – if for no more obvious reason than that beautifully wrought textiles, the highly valued creations of such

---

* With the fascinating additional detail that the Minster-in-Thanet community had bought the ship from one Leubucus – a name that sounds distinctly un-English. Witnesses and signatories to the charter included the faithful *dux* Ofa and the king's *thelonarius* – his tax collector Pant. Electronic Sawyer S91; S86.

houses, were perishable. The best evidence for such production – and profit – comes from Mildrið's successor at Minster-in-Thanet, the long-serving Abbess Eadburh (known also as Bugga), who enjoyed the benefits of partial remission on tolls from King Æðelbald and who corresponded with Boniface throughout his continental career. In one letter, dated to about 720, she congratulates him on the recent success of his latest mission to Frisia. The postscript reads:

> I am sending you by this same messenger fifty *solidi** and an altar cloth, the best I can possibly do. Little as it is, it is sent with great affection.[42]

More than a decade later Boniface was writing to her with thanks for a gift of 'sacred books' – gospels, or perhaps psalters, an indication that religious women might also be scribes and illustrators of high-quality manuscripts.[43] The transactional mechanism behind such 'gifts' is revealed in a second letter from the same year, 735, in which Boniface, then in Germany, thanks Bugga for the additional gift of 'garments' and then asks if he may commission from her a copy, 'written in gold', of the Epistles of St Peter. He sends, with the priest Eoban whom he trusts to deliver it, the materials that Mildrið will need for the project.[44] Somewhere between the procurement of these 'materials' – inks, dyes, gold leaf, vellum, leather binding, possibly gemstones – and their assembly into the treasured epistles, money was changing hands: at markets, ports and workshops. And the very evident fact that eighth-century Kentish abbesses could dispose of such generous gifts of cash shows how substantial their resources had become. Later in the same century a dispute, whose melodramatic detail survives in a remarkable legal document, shows how closely involved King Offa's Queen, Cyneðryð, became in controlling a key royal asset on the River Thames: the monastery at Cookham.†

---

* A gold coin of some 4.5 grams.
† See below, Chapter 7, pp. 227–30.

If wool and finished textiles were the most important source of export revenue, other commodities allowed Mercian kings to broaden their fiscal portfolios and profit from the rapid expansion of trade. Mercian lands were abundant in lead (from Derbyshire) and salt (from Cheshire and Worcestershire), perhaps also in iron and copper (from Shropshire). Successful military campaigns yielded supplies of slaves for the continental market and bullion that might be recycled as gifts, coined as silver *sceattas* or traded for cash. At the trading settlements themselves craft workshops produced articles made from iron for tools, weapons and armour; from copper alloys for more decorative items like brooches and belt fittings; from flax for fine linen cloth; from leather for shoes, sheaths and belts; from bone for pins, combs, needles and more, and from antler. Such luxuries as otter skins, probably also ermine fur, were in demand from continental consumers. *

In turn, a number of imports can be identified across the Anglo-Saxon trading settlements – not by any means always the same commodities. Imported glass vessels and scrap glass – called cullet – have been found in all the major trading settlements.[45] From *Lundenwic* fragments of a distinctive lava, originating from the Mayen-Niedermendig area of what is now western Germany, are recognised as imports of high-quality quern stones for grinding corn. Hides, oil, wine, ivory and pottery were all being traded across the Channel and North Sea.[46] The excavated evidence of pottery, the almost universal stock-in-trade of the archaeologist, shows that ceramic vessels were being traded to and from the continent both as items of value in themselves and as containers for high-value, low-bulk goods such as salt and dyes, spices and wine. More local wares were traded within the hinterlands of the trading settlements and, alongside coin finds, these patterns of regional distribution show how the economic benefits of trade fanned out

---

* Otter skins were sent by Cuthbert, the abbot of Wearmouth and Jarrow, to a correspondent in Germany (Sawyer 2013, 64). Ermine was a speciality of northern England: the white winter pelt of the stoat, with its distinctive black tail tip, a symbol of luxurious rarity.

from these sites. Occasionally, the remains of exotic plants like figs and grapes turn up in London's waterlogged archaeological deposits – perhaps brought in as dried, luxury items.[47] The vast bulk of material evidence for organic materials – foodstuffs, leather, textiles and wood – has perished. Archaeologists are forever peering myopically through a glass, darkly.

\*

The *Anglo-Saxon Chronicle* is taciturn in the extreme regarding Mercian events during the first decade or so of Æðelbald's rule. If he stamped his mark on his neighbours' territories by raids and open warfare the notices of such exploits do not survive. His energies may have been devoted to Mercia's complex internal politics and to monetising London's enviable trading location. Nevertheless, the evident subordinate status of, for example, the *Hwiccian* royal family and of collateral members of the Mercian nobility, offers hints to otherwise intangible strong-arm tactics: diplomacy backed by the threat of force or by bribes in hard cash. Since warfare was a principal means by which kings acquired treasure to both enrich themselves and reward their followers, it seems surprising that so enterprising and powerful a king as Æðelbald failed to record victories in battle in the early part of his reign. The historian James Campbell suggests that kings increasingly found it more convenient, and it was certainly less hazardous, to raise cash by the sale of land;[48] and, as a corollary, they knew that war hinders trade while peace promotes it.

Commercial land deals are sometimes found in charters. Late in the eighth century, Beonna, the abbot of *Medeshamstede*, leased to the *princeps* Cuthbert ten hides of land in return for 100 shillings and 'annual hospitality' – a one night's feast, valued at thirty shillings.[49] And during the reign of Æðelred (674–704), Abbot Headda of Breedon-on-the-Hill had acquired from the king fifteen hides of land for the monastery of *Medeshamstede*, the value of which was reckoned at 500 *solidi*. The accompanying memorandum specifies the precise merchandise whose value equated to that sum:

A dozen beds, namely feather mattresses and elaborate pillows together with muslin and linen sheets as is customary in Britain; also, a slave, with a slave girl; also a gold brooch and two horses with two wagons.[50]

Just as the productive surplus of the English kingdoms was being commodified, so were political relations. In the first year or so of Æðelbald's reign a flurry of charters shows how he invested in the loyalty of his subordinates and nurtured the shoots of extensive patronage networks through concessions and gifts. Insofar as Mercian kings were the personification of the Mercian people, the *gens Merciorum*, they must also have inherited those aspects of kingship that formed something akin to a 'state'. Æðelbald was able to claim ownership of property over which his predecessors, kin or otherwise, had enjoyed ownership. From their rights to collect tolls exacted at ports and on roads used for trade, he also raised cash. To a large extent their royal estates became his royal estates. From this portfolio he donated lands – that is, the food rents and services due from estates – to members of his *comitatus* or to holy men and women for the establishment of religious houses. Given the lists of witnesses subscribing to these charters, it seems certain that ceremonies, during which such donations were confirmed, formed part of large gatherings that might also have counted as feasts, councils and courts of legal judgement. If so, they are likely to have taken place at *villae regiae* or at traditional tribal assembly sites.* The whole social and political apparatus of the Mercian state was bound in with the land.

Something of the psychology and motivations behind these charters can be reconstructed from their language. One of the earliest substantially authentic examples records that, 'in the name of God, Æðelbald, king of the Mercians, for the redemption of

* Gumley, in Leicestershire, was one; so, probably was Finedon near Irthlingborough on the River Nene; *Clofesho*, wherever it was, probably another.

his soul,* gives three hides to his *comes*, Buca, for the founding of a perpetual dwelling for the servants of God': a tangible as well as spiritual investment. A more complex transaction, involving rights to salt extraction, is prefaced with:

> In the name of the Lord Jesus, I, Æðelbald, by divine dispensation king of the Mercians, having been asked by the holy community of Christ dwelling in the place whose name is Worcester, will concede and grant into their free liberty possession for the redemption of my soul of a certain portion of ground on which salt is wont to be made...[51]

By accepting royal donations, recipients laid themselves under permanent obligation to their lords; accepted (if there was any doubt) tributary status and, if they were religious professionals, contracted to pray for the soul of the king in perpetuity. Secular obligations included service in the king's warband and hosting the royal household – hospitality that might amount to a considerable burden, as the *Medeshamstede* land lease above shows. Increasingly, secular lords attempted to 'book' land from their kings as their spiritual counterparts did, obtaining hereditary rights and privileges.

Much less clear, in these early decades of the eighth century, is the extent to which monastic lands were held to be exempt from food renders and hospitality and 'public' services such as bridge-building, the maintenance of fortifications and military service – the so-called 'common burdens'. Abbots, abbesses and bishops understandably resisted such impositions. And, as a letter written to Bishop Ecgberht of York by Bede in his last years shows, secular lords saw how they might keep land, formerly in the gift of the king and held for a life interest, within their family by founding religious houses of which they themselves were the patrons: so-called proprietary minsters.[52]

Later in the century such matters caused bitter disputes as

---

* *Pro redemptione animæ meæ*. Electronic Sawyer S85.

kings sought new ways to impose their authority and as they began to feel the deleterious effects of large numbers of profitable monasteries (and secular lords) taking more than fair advantage of their generosity. In some cases Mercian kings and their neighbours attempted to place time-limited reversions on gifts of land or to forcibly take them back. Wilfrid's career had shown just how tenaciously ecclesiastical rights might be fought over and how resistant kings might be to compromise.

The combination of charter evidence, the archaeology of the trading settlement at *Lundenwic* and the testimony of silver coin finds in the early part of Æðelbald's reign shows that one of his principal preoccupations was to maintain Mercian control over the most lucrative commercial assets in the kingdom. If *Lundenwic* was the most important outlet and source of royal revenue from those assets, the literal and figurative spring from which much of Mercia's wealth flowed lay some 120 miles away, in the heartlands of the *Hwicce*. Part at least of Æðelbald's early desire to maintain overlordship over formerly independent *Hwiccian* kings was the need to secure for himself the lucrative salt concession at Droitwich.* Salt was a prized commodity because it was essential for preserving meat and fish and for making butter and cheese. It could be transported easily by wagon or pack horse and traded either as a high-value commodity in its own right or as part of the exchange of gifts.

Salt had been won from Droitwich's famous brine springs as far back as the Iron Age; not just any old salt, either: the brine pits there produced it in very high concentrations and of such high quality that it could be stored and transported without being prone to re-liquefying through hygroscopy.[53] The high brine content also meant that less wood was required to fuel the fires beneath huge lead evaporating tanks. That precious lead was itself another valuable Mercian asset, perhaps brought south from mines in the

---

* The 'Droit' element might either mean 'dirty' from Old English *dryht* or noble/princely – from Old English *dryht*. Watts 2004, 195. Campbell (2003, 28) favours the royal over the muddy.

Wirksworth parish church, Derbyshire. The relief-carved figure
known as T'Owd Man is a small monument to the Mercian revival
in lead mining and the wealth it created.

Derbyshire dales around Wirksworth, in whose church a pre-
Norman Conquest sculpted relief figure of a lead miner survives. *
A direct Roman road – the same Ryknield Street that ran past
Lichfield and Catholme – linked the southern Peaks with the Severn
at Worcester, just a few miles downstream from Droitwich on the
banks of the River Salwarpe. A study of Droitwich's hinterland
by Della Hooke has shown that a network of 'saltways', evolving

* Another source has been identified near Stoke Bishop, Somerset (Hooke
1985, 126).

organically over many generations, radiated out from the springs like climbing tendrils seeking light – probing, testing, refining – and she suggests that woodland in the area was specifically managed to produce the prodigious quantities of fuel, either underwood as cut, or as charcoal, voraciously consumed by the furnaces.[54]

Worcester was the diocesan seat of the *Hwicce*. Located on the navigable River Severn and well-connected by Roman roads, its bishops and church community were well placed to take advantage of the salt trade. From the reigns of Wulfhere and Æðelred onwards, grants to concessions in the salt industry had been made to both bishops and to the abbots of favoured houses: for the ownership of brine pits themselves, and for the construction of furnaces.[55] At that time Droitwich, which had been named *Salinae* in the Roman period, was simply *Wīc* – the 'emporium' or market.

Æðelbald's investment in the industry dates from the very beginning of his reign. A charter dated to 716–17 records that the king gave the church of Worcester 'a certain portion of ground on which salt is made' on the south side of the River Salwarpe for the construction of three salt houses and six furnaces, in exchange for six other furnaces in two salt houses on the north side of the river.[56] The net result of that transaction was that the king acquired the rights to the church's existing enterprise while the church built a new facility. That level of investment is a strong indicator of early eighth-century expansion in the industry to suit rising demand and/or prices. The 'convenience' to both parties mentioned in the charter probably reflects the geography of their interests: the king seems to have possessed a royal estate on the north side of the river at Wychbold,[57] while the bishops' interests lay to the south, at Worcester.

Both royal and episcopal households would have consumed much of their own salt output. The king, as well as enjoying the fruits of his near monopoly, profited from taxing the production of salt at source and levying tolls on its transportation. Another charter, of less impeccable pedigree but essentially authentic (and witnessed by, among others, Æðelbald's dependable *comes*, Ofa) records a grant, to the church of St Mary at Evesham, of a portion of a *sele* at *Wīc* – that is, Droitwich – free from all tribute of common tax.[58] As John Maddicott argues in his analysis of the Early Medieval salt industry, *sele* in this context probably means 'brine-share' rather than 'customs house' or hall – hence the use of the term 'portion'. In other words, Æðelbald was granting away part of a royal tax levied on salt production.

Salt, like slaves and textiles, is an intangible asset so far as archaeologists are concerned. Direct material evidence for its destination and consumption as a traded item is not forthcoming. In the prehistoric period a distinctive type of pottery, used for transporting salt, was produced in the Droitwich area and its distribution shows that, by and large, salt was traded to the east and south[59] towards, but not across, the River Thames. Evidence for the medieval reach of the salt trade comes from manorial rights to that trade recorded in the Domesday Survey; before that, clues lie in the activities of the bishops of Worcester, whose close interest in salt production is matched by their acquisition of rights to trade in *Lundenwic*. In about 745 King Æðelbald granted Bishop Milred of Worcester remission from the tolls due on two ships at the port of *Lunduntunes hyðe*.[60] A much later source records a grant made by the king to Bishop Wilfrid of Worcester (c.718–745) of a tenement in the port.[61]

No doubt the bishops' interests were wider than just salt: they may also have possessed estates in the sheep-rich Cotswolds. Wool aside, meat preserved by the bishops' own salt and destined for the London market has a nice commercial logic to it, even if the smoking gun of that trade is elusive. The best evidence comes from finds of a series of *sceattas* concentrated in the south-west Midlands but minted in London.[62] Another mercantile 'hotspot' with links to the salt trade has been identified at Bidford-on-Avon, situated fifteen miles south-east of Droitwich, where a branch of the Roman Ryknield Street crosses the River Avon. Here, a concentration of continental silver coins, high-status metalwork and a large cemetery hint at strong eighth-century commercial activity linking the *Hwiccian* heartlands with the hinterland of the Warwickshire Avon and with points further east and north.[63]

The individual links in the chain that connected Droitwich with *Lundenwic* and the continent beyond are tentative, but suggestive. By mapping the ancient saltways that radiated out from Droitwich, Della Hooke has shown that a key destination for salt was Lechlade, the highest navigable point on the Thames, on what was then the southern boundary of *Hwiccian* territory.[64] Another saltway led to Bampton, now in Oxfordshire, where an important and wealthy minster provided a second, only slightly less direct route to the Thames.* That the Thames was both a commercially vital highway for trade and a focus of productive settlement is indicated by a concentration of place names that indicate crossings, landing places and trading facilities,[65] and by a positive rash of monastic foundations along the length of the river during the late seventh and early eighth centuries.

Although the Thames had been a frontier between Wessex and Mercia in the seventh century, after the time of King Wulfhere it was, as John Blair, the historian of the Anglo-Saxon church, points out, much more a Mercian than a West Saxon river.[66] Mercian grants lay behind the foundation or enrichment of minsters at Cookham, Chertsey, Barking, Bermondsey (a dependency of *Medeshamstede*) and probably others among a string of more than thirty between Cricklade, at the head of the Thames, and Minster-in-Sheppey on its estuary in Kent.[67] Minsters may even have been granted their own rights to exact tolls on commercial traffic and it seems clear that they benefitted directly from riverine trade. In Mercian kings' vision of an emerging state, minsters were key aristocratic components in expanding production, consumption and trade, worthy of patronage and investment for the sake of their souls and their treasure chests. In return, through charter and also, perhaps, hard cash, the Mercian state

---

* Bampton lies more than two kilometres from the modern course of the Thames, but just two metres above it; the two are connected by the Great Brook which, as John Blair hints, might have been artificially enlarged to carry shallow-draughted vessels in the Early Medieval period. Blair 1996, 12.

monumentalised and capitalised the peripheral religious houses of its territories, held political stakes in their assets and took a keen interest in their fortunes.

In 731, during the compilation of his masterwork, *The Ecclesiastical History of the English People*, Bede summed up the present state of the English dioceses. After listing the bishops of Kent, the East Angles, the East, West and South Saxons and those of the provinces of Lindsey, *Magonsæte* and *Hwicce*, he recorded that all those southern kingdoms, right up to the River Humber, were ruled over by King Æðelbald of Mercia.[68] The bishops of Mercia, Lindsey, *Hwicce* and Hereford were already direct Mercian appointments. Ingwald in London was also a client of Æðelbald's, the recipient of generous trading concessions at *Lundenwic*.[69] In that year, 731, Tatwine, former priest at Breedon-on-the-Hill and confidante of Guðlac, succeeded Archbishop Berhtwald at Canterbury, confirming Mercian power to intervene in Kentish affairs – or, at least, reflecting the desire of Canterbury's religious community to elect archbishops acceptable to Mercian interests. Archbishop Tatwine's two immediate successors were also Mercian appointees. Æðelbald's patronage of bishops of Rochester and of the wealthy nunnery at Minster-in-Thanet strongly supports Bede's testimony of *imperium*.

Confirmation that Æðelbald's assumption of overlordship among the southern kingdoms was manifestly self-conscious comes from three charters, dating between about 736 and the last year of his reign in 757, which variously describe him as king not only of the Mercians, but of all the southern Angles. In one of these, the so-called Ismere diploma, he styled himself *Rex Britanniae*.[70]

Reading the entries in the *Anglo-Saxon Chronicle* for the first part of Æðelbald's reign, thin as they are, leads to the supposition that his assumption of overlordship cannot date earlier than the middle of the 720s. Literally nothing is known of his political and military fortunes before then. In 725 King Wihtred died and the Kentish kingdom split into its historical east and west components under his sons. A year later, King Ine of Wessex abdicated. In 729 King Osric died in Northumbria. Of the kings

The so-called 'Ismere diploma', an original Mercian charter of 736: in which King Æðelbald styled himself *Rex Britanniae*.

who ruled at Æðelbald's succession, only Ælfwald in East Anglia, whose father had sponsored him in exile, remained. And only he, and King Eliseg of Powys, might have taken exception to the idea that they were subordinate to Mercia.

Kentish and West Saxon royal power might already have been in decline before the mid-720s. In any case, by the early 730s King Æðelbald felt able to flex his muscles over former rivals. His imposition of a Mercian archbishop in Canterbury is matched by his capture of the West Saxon royal estate at Somerton in 733, his first recorded act of territorial aggression – to which Wessex seems to have been unable to respond. Up to that point our evidence for Æðelbald's political development is allusive; from then on the sources suggest that he played an overtly active role in consolidating Mercian power between the Humber and Thames, and then beyond.

What little narrative history survives to tell of the rise of the Mercian kings from Penda to Æðelbald relies very heavily on Bede. If the motherlode of his story was the relationship between church and state and the providential fortunes of Northumbrian kings and their priests, he had also been, as a consummate historian, a recorder of the intimate detail of people's lives, of the political motivations of kings and queens, holy men and women; of their pride and glory, weaknesses, triumphs and failures. He mapped out for future generations not only the exhilarating birth of a new European order, but also a vision of a single people, the *Gens Anglorum*, united under one church. Even those events and machinations on which he is infuriatingly taciturn can be inferred from his subtle, sometimes contradictory but always revealing asides or lacunae. But in 735 the great chronicler himself died. From this point, the framework of Mercian, and English, history must fall back on much less confident, if no less suggestive, sources.

# Chapter 6

# Æðelbald:
# England's anvil
# 736–757

❖

*Peace dividend • War again • Praying, fighting,
working men • 'Productive' sites • Church
privileges • The Common Burdens • Networking*

The year is 733. Seventeen years into his reign, King Æðelbald's first recorded act of military aggression is an attack on Wessex, during which he captures the territory around the key royal *vill* of what will later become Somerset's county town, Somerton. It lies at the east end of a hilly lozenge of land between the rivers Cary and Parrett, with the marshy levels to the west and south. In winter it must often have been cut off by floods; but the lower reaches of the Parrett gave access to the Bristol Channel and a few miles to the south-east lay the former Roman town of *Lindinis* (Ilchester), strategically sited on the Fosse Way, the route to the Mercian heartlands. This victory over King Æðelheard, five years into his own reign, seems to confirm Bede's acknowledgement that Æðelbald now held sway over the southern kingdoms. More than that, it shows a new energy in Mercian ambitions to extend control deep into the territories of its most powerful southern rival.

But it rather begs the question: why, in the previous four decades, is the record of warfare between the major Anglo-Saxon kingdoms so sparse? King Ceolred had fought against Wessex the year before his death in 716. King Ine had met the South Saxons and the 'Britons' in battle during those years but had otherwise been preoccupied with dynastic rivalries; Northumbrian kings had periodically campaigned in Pictland. But the litany of bloody encounters recorded before about 690 was followed by a period of apparent stasis, during which political tools seem to have been the weaponry of royal choice. The violent episodes recorded in the *vitae* of Guðlac and Wilfrid suggest that raids for booty and martial exercise wreaked local mayhem and that minor campaigns and skirmishes were fought beneath the radars of Bede, the Chronicler or the Welsh Annalists. Even so, major offensives, of the sort in which kings got themselves killed, speak loudly by their absence.

That same forty-year period is also conspicuous for a dramatic flowering of monastic foundations and for a rapid expansion of economic activity, which brought the kingdoms of southern Britain back into the mainstream of European trade and diplomacy. The establishment of the trading settlements alongside a plethora of minster foundations shows that royal power was invested in both religious institutions and ambitious mercantile projects. An idea of statesmanship seems to have taken root. Kings stopped trying to kill each other and became rich and pious.

Archaeologists have been aware for some decades now that in this period the nature and location of rural settlements also underwent significant, if undramatic, changes. They see evidence of expansion from the easily tilled and drained alluvial soils of river basins, the old cultural core lands, towards higher, more intractable but more fertile clay lands. They see increasing evidence of formal layout in settlements: of architecture and design; of specialisation in agriculture and its manufactured products. Even more significantly, perhaps, many of the place names that underlie the medieval and modern landscape were in the process of being formed during the eighth century and many of them are suggestive of the conscious hand of royal administration. Excavators have been finding unexpected signs of commercial production at sites known to be early minster foundations, while other contemporary settlements seem to lie somewhere along a spectrum embracing secular and ecclesiastical extremes. Abbots and abbesses seem to have been increasingly adept at commodifying their generously endowed assets. Archaeologists look, with increasing confidence, to confirm those observations, in one hand brandishing the written words of charters while with the other they point to the ever-increasing and reassuringly dispassionate testimony of coin, metalwork and pottery finds.

In the first half of the eighth century something is afoot in the English landscape; and the rules of a new game seem to have been enthusiastically embraced, if not invented, in Mercia.

Students of the Early Medieval period are taught never to commit the sinful fallacy of evoking causality: 'A' happened; then 'B' happened; therefore A caused B. We inhabit a universe of floating probabilities with occasional glimpses of a sort of demonstrable reality. But suppose one were to throw the following litter of historical confetti into the air: the plague that visited Britain with deadly consequences in the year 664 and which returned periodically at least until the middle of the 680s; * the installation and twenty-one-year tenure of the remarkable Archbishop Theodore at Canterbury from 668; the foundation of scores of minster churches and attendant monastic communities across all the kingdoms of southern Britain during the following five or so decades; the establishment of trading sites at Fordwich, *Lundenwic*, *Hamwic*, *Gipeswic* and *Eoforwic*; the activities of missionaries and monastic entrepreneurs like Wilfred, Boniface and Benedict Biscop travelling to and from continental Europe; the establishment and expansion of silver coinage; the rise of Anglo-Saxon export trade in textiles and other high-value goods like salt, slaves and furs; evident changes in settlement location and layout; and increasing tensions between church and state. Seeing coherent political and social trends in this blizzard of trends and innovations is a challenge.

It is impossible to calculate with any certainty the effect of the plague of 664 – probably caused by the same pathogen that lay behind the infamous Justinianic plague of the sixth century and which would cause the Black Death almost 700 years later.[1] It seems to have accounted for several kings and a number of senior church figures and, by Bede's testimony, it had devastating effects

---

* Bede is the eloquent chronicler of those events, of which he was a lucky survivor (see above Chapter 2 and Maddicott 1997). Other notices are contained in the works of the Ionan Abbot Adomnán (*Vita Colombae* II.46) and Wilfred's biographer Eddius (*Vita Wilfridi* XVIII).

on many local communities – 'it laid low a great multitude', he recalled.[2] Carried by infected rats, *Yersinia pestis* must have arrived on the back of cross-Channel travel at a time before the *wīcs* had been formally established; it is intangible proof, in fact, of the existence of such traffic in that decade. By analogy with the fourteenth century it is possible that a quarter, even a third of the population of England died in those plague decades.

If, in the aftermath, kings found themselves unable to muster the armies needed to fight their wars, or farmers to provide their food rents, it would not be a surprise. For ecclesiastics, brought up on an astringent diet of Old Testament prophecy and in the sure knowledge that the end of days might come in their time, there was a strong element of divine providence in such visitations.* The *Gens Anglorum* had, through lack of faith, lost the Lord's protection – and when the Anglo-Saxon translator of the Vulgate Bible used the term *Dryhten* for the Lord God, it was meant in a very real, contemporary and relevant sense of spiritual and temporal lordship. When the kings of Mercia and other rulers gave their lands away with such apparent enthusiasm in the following decades, they did so '*pro animabus suis*' – for their souls' sake. They may have believed that such piety would, as they had been promised, give them everlasting life at God's side; that perhaps it would protect them from the capricious fates of disease, battle and coup d'état. They may also have seen the economic value of sponsoring the intensive capitalisation of monastic lands through perpetual tax-free grants: to invest in a protected labour force and in agricultural innovation so that farmer-monks and their well-connected abbots and abbesses were incentivised to increase the land's wealth – especially if that land lay unkempt and unploughed because of the plague.

Like the plague and like mercantile goods, the transmission of ideas followed routes taken by churchmen and women or

---

* Bede himself took a more dispassionate, even scientific view. Maddicott 1997, 19.

their correspondence – along coastal shipping lanes to safe and accessible harbours and shores and along navigable rivers and former Roman roads linking centres of ecclesiastical administration or communities of monks and nuns. Their ghostly shadow, so evocatively caught in the narrative webs of hagiographers and in the pages of Bede, is also mapped by the accidental loss of silver coins – the tell-tale footprint of movement, knowledge exchange and commercial trade.

If monastic entrepreneurs chose their advantageous riverside or port sites *because* of their economic potential, or if the latter *emerged* from the properties of the former, the result was the same: monastic estates, held in perpetuity, focused investment in labour, agriculture, architecture, sculpture, literature and technology and, against the background of a population shrunk by plague, drove a new economic impetus. The heavy, expensive but labour-saving mouldboard plough, for which there is now seventh-century evidence from Lyminge in Kent, is likely to have been a monastic import, allowing a shift in agriculture and settlement away from alluvial soils onto clay lands and at the same time, perhaps, shedding grief-stricken associations with plague-ridden sites.[3] In turn, agricultural innovation fed both a new trend towards farming specialisation and a tradeable, taxable surplus. If kings and their ministers saw into the future and invested in their monastic protégés with profit in mind, they may have been more cunning than historians give them credit for. If the drivers were, on the other hand, canny ecclesiastics and their merchant friends, it would be something less of a surprise.

The lacuna in inter-kingdom warfare briefly pricked by the capture of Somerton in 733 was more decisively ended in the years after 737, when the Northumbrian King Ceolwulf was forcibly 'retired' to Lindisfarne and replaced by his kinsman Eadberht, at which time King Æðelbald took the opportunity to 'harry' Northumbria with an army – coincidentally or otherwise during an extended period of drought.[4] Three years later he mounted a full-scale invasion of Northumbria while King Eadberht was away campaigning in Pictland.[5] In that same year, 740, King

Æðelheard died in Wessex and the *Chronicle* entry that records his death also records that his successor, Cuðred, 'resolutely made war against King Æðelbald'. *

The following year, King Æðelbald again took his army into Northumbria, sacked York and burned down the minster.[6] This grossly impious deed may have had more to do with high politics than with anti-clerical sentiment on Æðelbald's part: the Northumbrian king's brother, Bede's correspondent Bishop Ecgberht, had received the pallium of a metropolitan archbishop[†] in 735 and might almost be seen as the *subregulus* of Deira; the Mercian king may have regarded him as a legitimate target. Archbishop Cuthbert, Æðelbald's own appointee to Canterbury, might have been expected to deprecate this attack on his northern counterpart; but it may be that, having lost control of the Northumbrian sees as a result of Ecgberht's elevation, he turned a blind eye to, or even approved of, the king's attack on his rival in York. At any rate, Ecgberht survived to found a famous school and library at York – and there have been archbishops at the minster ever since.

According to the entry for 743 in the *Anglo-Saxon Chronicle*, Mercian forces were also busy in the west, this time in alliance with Wessex: 'King Æðelbald and King Cuðred fought against the Welsh', is the terse entry. The enigmatic commemorative monument known as the Pillar of Eliseg[‡] records an expansion of Powysian territory during the reign of the eponymous Welsh king, which may have been the *casus belli*, or a reaction to, this Mercian/West Saxon aggression. Later Welsh tradition tells how Æðelbald also caused 'great tribulations and devastations' along the River Wye, appropriating a dozen or so churches.[7] When interests aligned, rivals might become expedient friends.

---

* The wording of the entry allows one to infer that it was 'during the reign' of King Cuðred that he fought against Mercia; not necessarily in that year.

† He, like his brother, issued coins in his own name. Sawyer 2013, 76.

‡ See above in Chapter 5, p. 152.

Æðelbald's motives in campaigning in the lands across the Humber and Severn do not seem to have included the idea of conquest or annexation. Harrying and raiding offered young Mercian warriors the chance to test and hone their martial skills and acquire booty. The testimony of the coinage at this time suggests that the European silver supply had begun to run dry and Mercian coins were increasingly being debased with copper and tin.[8] So if the king's coffers, filled over the previous two decades by a golden era of regional and international trade, were now looking a little empty, then a small war or two was the most expedient means of refilling them.

More than a hundred years later another fighting king, Ælfred, would write in his Old English translation of Boethius's *Consolation of Philosophy*:

A man cannot work on any enterprise without resources. In the case of the king, the resources with which to rule are that he have his land fully manned: he must have praying men, fighting men and working men.[9]

The praying men – and women – of Mercia and its satellites must now have been almost as numerous as the pool of fighting men – the *comites* and their *duces* – who went to war under Mercia's banner. A very rough calculation based on assessments recorded in Tribal Hidage suggests that a seventh-century king like Penda may have been able to muster an army of 3,000 men. * Those were theoretical men; in practice, the thirty *duces* who fought and died with him on the *Winwæd* in 655 may have led warbands of fewer than 100 men – perhaps even half that number. It seems unlikely, after two decades of plague in the latter half of that century, that Æðelbald's fighting strength would have numbered anything like 1,500 or 2,000 armed warriors. On the other hand, if one takes a conservative estimate of the number of

* See my calculations in *The First Kingdom*, Adams 2021a, 328–329; but see below in Chapter 10, p. 349 for a larger estimate.

active monastic and church communities across Greater Mercia in the middle of the eighth century – say, the thirty-odd identified by contemporary sources, doubled by the number of implied but unnamed foundations, it is possible to envisage a 'praying force' that matched those wielding spears in the king's name, even if the 600 or so people who comprised the religious and lay complement at Wearmouth was wholly exceptional.[10]

The number of 'working' men and women of England's midland farms – the ceorls and their households who rendered their *feorm* and were also liable for services of one sort or another – is much more difficult to estimate. Archaeologists are still a long way from identifying more than a very modest proportion of their settlements, while their cemeteries have not yet been found in anything like the numbers expected. Bede took the view that there were not nearly enough priests to fulfil their pastoral duties of preaching and teaching in every hamlet across the kingdom of Northumbria. Many people lived in inaccessible places where a bishop had never been seen, he lamented in his letter to Bishop Ecgberht in York shortly before his death in 735.[11] In the aftermath of two or more decades of plague there were fewer people and they lived further apart. Those who did survive thrived, though; and they seem to have been active in building new, more economically rewarding settlements.

In any case, as Bede so bitterly complained in the same letter, in those days it was hard to tell fighting, praying and working men apart. Cynical members of Northumbria's secular élite had, across a whole generation, insinuated themselves unworthily into the spiritual life. Such men, he wrote, had no knowledge of the monastic life but obtained lands from the king in hereditary right under the false pretext of building a religious community. Those same men lived blatantly in these so-called monasteries with their wives and children, with all the rights and privileges of territorial lordship and free of the imposition of taxes but scorning a life of spiritual virtue and service. That is to say, the closed monastic life of contemplation, scholarship, worship and pastoral care had been infiltrated, debased and soiled by the secular élite for the purpose of avoiding the payment of the *feorm*, performing public services and fighting in the king's army.[12] Bede's all-too prescient belief, that unbalancing the levers of the fighting, praying and working machinery of the kingdom would lead to disaster, was to bear fruit before the end of the century.

The 'Letter to Ecgberht', as Early Medieval scholars refer to it, is regarded as one of the most important and revealing documents of the eighth century. Its off-screen soundtrack is the pounding ring of the smith's hammer, striking the anvil of an Anglo-Saxon revolution forged by human capital from the land's wealth. Slowly but surely evidence for the existence of that much-beaten anvil is accumulating across the counties of England and nowhere more so, perhaps, than in Mercia.

Cookham* in Berkshire, the site of a Thames-side monastery founded sometime in the early eighth century, looks to the archaeologist more like a miniature version of the trading settlement at *Lundenwic*, a few days' rowing down the Thames, than it does the sanctified enceinte idealised by Bede in his *Vita* of St Cuthbert.[13] Cookham's community, presided over by an abbess, made profitable use of its location on a bend in the river close to the well-wooded Chiltern hills. Industrial workshops forged copper alloy and cast-iron items like decorative dress pins and a carpenter's axe, recovered from one of the boundary ditches. Cattle were brought to the site on the hoof and then butchered – to be salted, preserved and traded downstream, perhaps. Fragments of lava quern stones, so absolutely diagnostic of continental trading connections, have been recovered from the site and surfacing for a waterside loading area provides the tell-tale evidence for Cookham's connections along England's foremost trading artery.[14]

The Cookham estate must have been land in the direct gift of the king, for at some time during the archiepiscopacy of Cuthbert (740–57) Æðelbald donated it to Canterbury, together with a generous land grant:

---

* Celebrated as the setting for Kenneth Grahame's *The Wind in the Willows* and as the birthplace of the artist Stanley Spencer.

This monastery, namely with all the lands belonging to it, Æðelbald, the famous king of the Mercians, gave to the Church of the Saviour which is situated in Canterbury, and in order that his donation might be the more enduring, he sent a sod of the same land and all the deeds of the aforementioned monastery by the venerable man Archbishop Cuthbert, and ordered them to be laid on the altar of the Saviour for his everlasting salvation.[15]

Barking Abbey, on the Lower Thames, has produced a rich inventory of imported goods including glass and gold, while the styli found there betray the presence of a scriptorium.[16] Many of the other minsters sited on the navigable reaches of the Thames are likely to have been well-connected with *Lundenwic* and with continental ports of trade just as, at Minster-in-Thanet, abbesses were negotiating the remission of tolls on their own ship.

At Yarnton, just north-west of Oxford and no more than a mile from the Thames, a modest settlement of sunken-floored buildings was transformed in the years after about 700 into an agricultural complex with squared enclosure ditches, trackways, a substantial timber barn and – judging by patterns of close-packed post-hole structures – granaries, wells and stock-keeping areas; perhaps, even, a fowl-house.[17] The Anglo-Saxon plant and agriculture researcher Mark McKerracher interprets the structures here as elements of a grain processing and drying installation;[18] and John Blair believes that the very striking intensification of centralised farming activity at the site is a tell-tale of monastic investment – a sort of venture-capital takeover, exploiting an existing but unproductive estate close to the Thames and turning it to profit.[19]

Castor, on the north bank of the River Nene a few miles upstream from *Medeshamstede*, lies at the junction of two Roman roads: Ermine Street, which joined London to the ports of the River Humber, and an east–west road that gave access to the rich heartlands of East Anglia. The religious house founded here, on the site of a former palatial Roman villa and temple complex, by King Penda's daughter Cyneburh in the late seventh century, has produced sherds of pottery known as Ipswich Ware – diagnostic of trade in a high-value, low bulk return cargo for whatever was being exported from Castor – and of wheel-thrown continental pottery.[20]

These minsters belong to a group of sites known, because of the striking quantities of coins and other metalwork recovered by excavation and metal-detection, as 'productive' sites. If Bede was sceptical that such industry and trade was being carried out strictly in the name of the saviour, he could have pointed to a number of 'productive' sites that, while they also exhibit features of religious observance, look otherwise to be very much secular enterprises. Brandon, on the Little Ouse in Suffolk, and Flixborough, on the Trent in Lincolnshire, are textbook examples of rich Middle Anglo-Saxon sites that seem to hybridise monastic and economic investment.[21] If this secularising momentum shocked Bede, it also prefigured another period of monastic capitalism in the Cistercian movement of the twelfth century – emblematic of a Europe-wide technological and intellectual revolution that produced cam-driven fulling mills, huge minster complexes and the creative outpourings of spiritual-intellectuals such as Peter Abelard and Hildegard of Bingen. Eighth-century Mercia may have had its Abelards and Hildegards, but their written words do not survive.

The most progressive and aptly-sited Middle Anglo-Saxon monasteries might invest in the technology of watermills, of which increasing numbers have been recognised in excavations at Tamworth, at Old Windsor on the Thames and elsewhere.[22] Old Windsor excites archaeologists for other reasons, too: its Anglo-Saxon name *Windlesoran* means 'ridge with a windlass': it was known, and named, as a riverside location where boats could be drawn up either to trade or to be repaired.

Monastic entrepreneurs may also have had access to new strains of wheat and to wider expertise in livestock breeding, making use of their 'secondary products' – milk and cheese, wool and leather – as well as their meat; to the heavy plough; to grain-drying ovens and larger granaries and to systems of crop rotation and specialisation.[23] English estates were producing more food, in a wider variety, more efficiently and with a view to surplus that might be stored, traded or taxed.

Mercia had not begun the eighth century with the trading advantages of its rivals in Kent, Wessex and East Anglia; but by the middle of the century its natural resources in fertile soils and minerals like salt and iron were being exploited by promoting monastic investment in farming intensification. It traded its surplus through a variety of interconnected routes that gave its producers and traders new, profitable outlets.

In one sense at least Mercians were rediscovering key elements of the Romano-British landscape. Roman administrators, prospectors and military engineers joined *Britannia's* navigable rivers with a necklace of metalled roads criss-crossing the Midlands, linking harbours north, south, east and west, and allowing both goods and armies to flow at speed between ports of supply and markets. The roads may, by Æðelbald's day, have been in a more or less pitiable state, the centralised Roman market economy shrunk beyond all recognition; but Mercian administrators and their long-reigning eighth-century kings had begun to grasp the possibilities.

Monasteries were not the only 'productive' sites in Æðelbald's Mercia; and the heads of religious houses, genuine or otherwise, were not the only entrepreneurs in the Anglo-Saxon landscape. Away from the rivers Thames and Trent, Nene and Severn, Avon and Ouse, concentrations of eighth-century metalwork and coins have been turning up in some unexpected places. Dunstable in Bedfordshire* had been a key crossroads in the Romano-British landscape, standing at the intersection between Watling Street and the ancient (but still relevant) Icknield Way that linked East Anglia and the line of the Chiltern Hills to south-west England, crossing the Thames at the later Alfredan *burh* of Wallingford in Oxfordshire. Over the years a 'thick scattering' of eighth-century

---

* Old English *Dunstaple* – perhaps 'post-on-the-hill', as in a posting station.

silver *sceattas* has been found in the area below the Chiltern scarp,[24] prompting the idea that the crossroads was a periodic market place or fair where traders might meet – perhaps on one of the quarter days that marked the turn of the seasons.* A similar case can be made for Royston in Hertfordshire, at a crossroads where Roman Ermine Street met the Icknield Way and where, again, coins have been found in abundance – 116 of them by 2003, when the numismatist Michael Metcalf listed Royston as one of the thirty-four most 'productive' sites in Britain. Another, Bidford-on-Avon in Warwickshire (14 coins), lay within the network of Droitwich saltways. All three were places where a canny king and his reeves might see fit to extract tolls from those exchanging goods and cash.

Lincolnshire features prominently in lists of the most 'productive' sites. Lincoln itself – the Roman provincial *colonia* and diocesan seat of Lindsey's bishops – probably its kings, too – was superbly connected to the River Trent via the Roman Foss Dyke; to the Humber estuary ports via Ermine Street; and to the Wash along the River Witham. Lincoln was the terminal of the Fosse Way, through which it was connected to the Middle Anglian see at Leicester and to former Roman towns as far away as Exeter, while Ermine Street gave access to all points south. Lincolnshire, the most enthusiastically ploughed county in all of England, boasts five or more 'productive' sites[†] and a rash of finds concentrations that partly reflects present-day farming practices – which bring buried materials to the surface – and partly reveals its ancient productivity.

Archaeologist Kevin Leahy, who has made a long-term study of Lincolnshire's Middle Anglo-Saxon riches,[25] has shown how finds of imported goods, of Frisian *sceattas* and of decorative metalwork are concentrated along lines of communication that fan out from Lincoln and the Humber and Wash ports, along the Jurassic Ridge and the Lincolnshire Wolds. Garwick, now no more than a farm

* Metcalf 2003, 40; and see below in Chapter 7 for the possibility that Dunstable was the lost *Stretburg*.
† Lincoln itself plus Melton Ross, Riby, West Ravendale, Garwick and Torksey.

lying just off the main road between Holbeach and Sleaford – with its probable monastic church – has yielded more coins, particularly Frisian *sceattas*, from the first half of the eighth century than any other 'productive' site in England. It seems to have been a trading settlement channelling goods along the Roman-period canal known as the Car Dyke and the Wash-draining River Witham, connecting Lincoln with *Medeshamstede* and its many monastic satellites.[26] The Witham itself has produced tantalising evidence of Lindsey's material and artistic wealth, in a number of gilded brooches and dress pins, decorated with hybrid beast motifs.[27]

At Melton Ross, on the western edge of the Wolds where the modern main road between Scunthorpe and Grimsby runs, a particularly rich concentration of finds and evidence of an important settlement with droveways and enclosures tells, once again, a story of intensive production and the activities of traders. Despite the temptation to see all such establishments as having, inevitably, to be linked with the capitalising impetus of religious foundations, I cannot see any reason why Æðelbald's more overtly secular élite, starved for the most part of opportunities to glorify and enrich themselves in war, should not attempt to cut a piece of the mercantile pie for themselves.

Mercia was not alone in seeing an early eighth-century boom in production and trade. Northumbrian kings produced substantial coinages whose loss pattern suggests a highly productive core in Deira, between York, what later became the East Riding, and the Humber. In Kent, through which so many travellers and traders passed, economic, cultural and political links with the Frankish and Frisian kingdoms were already well-established in the previous century. East Anglia, where the extreme poverty of historical narratives and charters is an almost impenetrable barrier to serious analysis, may very well have been the wealthiest and most economically advanced kingdom in these decades. Archaeology is slowly making up some of that deficit, exposing enormously rich complexes at places like Brandon in Suffolk and at West Fen Road in Ely.[28] Mark McKerracher sees East Anglia and Essex as seventh-century hubs for agricultural innovation, with the trading settlement at *Gipeswic* providing the motive force.[29] The products of its very early and highly productive Frisian-inspired pottery industry penetrated deep into neighbouring kingdoms along the same paths that carried products to their markets.[30] A succession of well-connected and long-reigning East Anglian kings from about 663 to 749 provided the stability and royal impetus for economic and administrative success in spite of, or partly because of, the devastating effects of the plague. East Anglian trading settlements developed early and strong links with the Frisian trading sites at Domburg and Dorestad, while its kings corresponded with English missionaries like Boniface.[31] They were early adopters of gold and silver coinage. Two bishops, at North Elmham in Norfolk and at *Dommoc*,* provided administrative oversight for its churches and monasteries. Felix's *Life of St Guðlac*, and the long-reigning King Ælfwald's own correspondence with Boniface in Mainz, are indicators of a high level of literacy† in the kingdom. The striking possibility that King Æðelbald, a youthful exile from Mercia, was politically and culturally educated at the East Anglian court may go some way to explaining the sophistication of his own Mercian revolution.

Some of the seeds for that revolution were sown in the natural tensions generated by a century of monastic and diocesan foundations. Land permanently alienated from the king and held by a spiritual élite barely, if at all, separated from their secular kin, encouraged a parallel and competitive form of territorial lordship among ecclesiastics. Bede had anticipated the potential consequences; so did Mercia's kings. There is little or no sign in the charters that, before Bede's death in 735, kings had begun to reign in or withdraw some of the privileges that churches enjoyed or thought they should enjoy. Those privileges included immunity from taxation – that is to say, from the obligation to render the *feorm* or food rent to their royal or aristocratic patrons; they also

* Probably Dunwich on the Suffolk coast.
† Of the king's ecclesiastical colleagues, if not of the king himself. *St Boniface Letters* LXV.

included exemption from the public obligation to repair bridges and build and maintain fortifications, due from those who held land directly from their lords for a life interest. And they included a more obvious exemption from the obligation to serve in their lords' armies.

Bede's warning letter to Ecgberht, primarily concerned as it was with a lack of pastoral provision, signals a more widespread concern that spurious religious houses and those that were already, or were in the process of becoming, tax-free hereditary holdings, were cheating. Secular lords, thinking themselves hard done by, in turn sought to book their own lands by hereditary right: the king's *comitates* did not see why they should not enjoy a slice of the action.

In 742, two years after the elevation of Bishop Cuthbert to the archiepiscopal see at Canterbury, a synod 'assembled in the famous place that is called *Clofesho*' and presided over by King Æðelbald, for the first time ruled on matters of ecclesiastical election, on monastic independence from external interference, and on what privileges the churches of the southern kingdoms ought to enjoy:

> ... in all things the honour and authority and security of the church of Christ on this side of the River Humber be denied by no person and also all of the monasteries established within Kent should remain both free from secular services and also secure from all burdens, major or minor.[32]

As it happens, the 'famous place that is called *Clofesho*', which as far back as the days of Archbishop Theodore had been promoted as the proper venue for annual synods, has never been securely identified.* So far as church privileges go, the Council's provisions of 742 look as though its co-sponsor, King Æðelberht of East Kent, and his new 'Mercian' archbishop were seeking reassurances from their overlord in Mercia that customary rights

* See Chapters 7, p. 246, and 10, p. 339.

of independence and freedom from public exactions were to be protected; and these they duly won. It is a declaration of harmony.

However, four years later a rash of correspondence from the former missionary Boniface, now archbishop of Mainz, paints a quite different picture. An opening letter written directly to King Æðelbald and delivered by Boniface's personal messenger, Ceola, is accompanied by suitable greetings and the gift of a hawk, two falcons, two shields and two lances as tokens of affection – a form of diplomatic lubricant in return for which Boniface begs diplomatic immunity and protection for his messenger, that he might travel freely and without hindrance on his master's business.[33] Part of that business, it seems, was to gather intelligence on the state of the church in England. Given that Boniface's own royal sponsor was the Frankish ruler Carloman,* it is likely that more than mere ecclesiastical intelligence was collected on that trip.

Some months, perhaps a year later, Boniface wrote to the Mercian king again, this time armed with accusations.[34] His 'trusty messengers' had assured him that Æðelbald was very liberal in alms giving; that he was a defender of widows and of the poor; that he was active in the suppression of robbery and rapine; that his kingdom was a land of peace. But he had also heard that the king had not taken himself a wife; that he remained unmarried not through virtuous abstinence but because he indulged in the sin of adulterous fornication with holy nuns and virgins – not merely imperilling himself in the eyes of God but setting a poor example to his people. Worse, Boniface had heard that the children born of these relationships were subject to infanticide, 'crowding tombs with corpses and hell with miserable souls'.

On the matter of ecclesiastical privileges, Boniface understood that the king had revoked the privileges of many churches and deprived them of some of their property. Since the time of Pope Gregory the Great, who had dispatched St Augustine on his mission to the English in 597, the privileges of the English church were

---

* He was the eldest son of Charles Martel (ruled 718–41), Mayor of the Palace and de facto ruler of the Franks.

inviolate, says Boniface, right up until the days of King Ceolred and of King Osred of Northumbria, both of whom notoriously not only violated the nuns in their monasteries but also destroyed the monasteries themselves. Both kings met terrible ends.

Boniface's reforming zeal extended to his own provincial church. In a long, 2,000-plus-word letter from the same year, 747, to Archbishop Cuthbert in Canterbury, he told how he forbade the servants of God in his province to hunt, or to go about the woods with their dogs, or to keep hawks or falcons. He exhorted his own bishops to make the rounds of their dioceses annually; to seek out and suppress pagan rites, divination, fortune-telling, charms and 'all Gentile vileness' – including carrying arms.[35] The letter ends with this warning:

> As to the forced labour of monks upon royal buildings and other works, a thing unheard of anywhere excepting only in England, let not the priests of God keep silence or consent thereto. It is an evil unknown in times gone by.

If the intelligence gathered by Boniface's embassy in 746 was accurate, the silence of the charters on matters of church privileges before the reign of King Æðelbald must be obscuring already-simmering tensions. It may have suited nervous religious leaders in the English kingdoms that an English exile should challenge their lords so openly; it also seems to have emboldened them to exert their own pressure on the king. In September 747 a second council was held at *Clofesho*, convened by King Æðelbald with his *principes* and ealdormen and with a dozen bishops in attendance.[36] Its Acts survive in full, albeit in a very fragmentary, partially burned manuscript that barely survived the fire that engulfed the Cotton Library in 1731. Simon Keynes believes that the council was the direct response of the king and his archbishop to Boniface's stinging criticisms – a copy of which belongs in the same manuscript collection.[37] The explicitly canonical provisions recorded at *Clofesho* cover the appropriate admonitions to bishops, copied almost word for word from Boniface's letter to Cuthbert, and their duties to inspect monasteries 'if they can be called monasteries', especially those 'we know not how, possessed by secular men... by presumptuous human invention' – an echo of Bede's sentiment.[38] But the headline is a carefully worded restatement of those privileges granted to the Kentish church under King Wihtred, protecting their liberty, security and authority and exempting them from all secular services... 'except [military] expedition and building of a bridge, or castle'.[39] These critical exceptions show that while Æðelbald was prepared to take Boniface's personal admonition on the chin, his attitude to exemptions on monastic obligations towards the state had hardened since the first *Clofesho* council in 742. Æðelbald may have learned how the Frankish ruler Carloman, at the Council of Estinnes in 743, set out to use church lands to provide *precaria* – essentially the living of the land – for his warriors; it was a timely precedent.[40]

A tree-fringed mound in a Leicestershire field: all that remains of a former assembly site of the Mercian kings.

Two years after the second *Clofesho* council, in 749, Æðelbald convened a further council at a place called *Godmundesleach*. This time the site can be identified with satisfying precision. The small village of Gumley, lying on a back road between Market Harborough and Leicester, is surrounded by once-tilled arable lands now turned over to grazing for sheep and horses. A couple of hundred yards south-west of Gumley's single, house-lined street, in steeply undulating parkland, lies a natural amphitheatre containing a pond known as the 'Mot', overlooked by a prominent tree-covered mound. By general acceptance this is the site of the council of 749 and two further royal councils held in 772 and 779.[41] Its present obscurity may be misleading: it lay close to one of the sources of the River Welland, which may have formed a significant Middle Anglian boundary in the eighth century.

The *Acta* recorded at Gumley are directed to a single purpose: to free minsters and churches from all public burdens, *except* building bridges and repairing fortresses 'against the enemy', and to reassure them that small gifts – *munuscula* – through which minsters offered feasting and hospitality to the royal household, should not be demanded of them unless they were given freely, and with love.[42] Religious communities were, then, formally freed from paying the *feorm*, or food rent exacted from secular lands, from maintaining roads and from military service but they must contribute to the so-called 'common burdens'; and kings expected to be entertained lavishly once in a while. In Francia, in Mercia and in other eighth-century kingdoms, royal and ecclesiastical administrators were adapting to an increasingly competitive environment, negotiating and writing down the rules for a new social contract.

Significant trends come into clear focus from this decade of frenetic negotiations between church and state. The tensions that they aimed to resolve between public obligation, a secular élite starved of martial opportunities and protected ecclesiastical privilege had evidently been rumbling since the days of kings Ceolred and Osred in the first decade of the century. Those tensions paralleled the secularising trend towards specialisation and productivity among the monasteries best placed to profit from riverside or estuarine locations. And then, the increased royal – one might say 'state' – interest in excepting bridge-work and the repair of fortifications suggests that those 'common burdens', as they came to be called, were being deployed in governmental projects. If that is the case, who were the 'enemies' against whom fortresses were being raised and repaired; where were the bridges that needed to be built; and who, if not the dependent tenants of church lands, was providing the sweated labour to do all that work?

Archaeology, and the recognition of some very suggestive patterns in eighth-century place names, are beginning to reveal how the hand of royal control was exercised to support, certainly to control and exploit, perhaps even to initiate, Mercia's rise to economic dominance. So far just a single example of a Mercian bridge across a major river is known. North of Newark-on-Trent the tiny village of Cromwell sits in the low-lying Nottinghamshire flood plain of England's most dynamic river. Here, a canal lock bypasses a weir where the Trent narrows to something like 230 feet across: fast-flowing when rains come streaming off the Peak District hills and the uplands of Cannock Chase. Here, in the early eighth century, a remarkably sophisticated wooden bridge spanned this most daunting river.[43] Its foundations were discovered in 1882 during improvement works to the Trent Navigation. Two timber piers were encountered in the river bed and, thinking that they were Roman, surveyors measured and drew them before they were blown up with dynamite. Fortunately, sufficient of one of the timbers has survived that a tree-ring date (missing the crucial outer sap rings that would have pinned the date to a precise year) has been obtained from them, proving that they belong to the century of Æðelbald and Offa, not of Hadrian.

The bridge at Cromwell was constructed of lozenge-shaped caisson piers – perhaps seven of them, twenty-nine feet apart, spanning the whole channel and filled with rubble. A central timber baulk, which supported the long-rotted bridge superstructure above, was held in place inside each caisson by piles driven

The fast-flowing Trent drains the Mercian core lands. At Cromwell in
the eighth century a great wooden bridge spanned the river here.

through mortices into the river bed. Other timbers retrieved from
the bridge, probably from later repairs, suggest that it was still in
use in the eleventh century or later. This was a major infrastructure
project, allowing traffic moving along the Fosse Way between the
Humber, Lincoln and the Middle Anglian heartlands to cross
the Trent in all weathers. It must also have provided another
convenient toll-extraction point for the only authority who could
instigate such a project: the Mercian king. And, if the Trent could
be bridged here, in such style, how many more bridges must have
been constructed – at places like Wychnor, for example, which
overlooked a key crossing point of the Trent along Ryknield Street
close to Lichfield and to the important settlement at Catholme?[44]
Cromwell bridge was a product of refined engineering principles:
it cannot have been the first. Along the length of the Thames,
Avon, Nene, Great Ouse and Severn, where river and road traffic
met, ferries must have plied to and fro whenever the need arose;
but armies and the king's baggage train relied on speed and

security. The Watling Street line between London and Wroxeter carefully and deliberately avoids major river crossings; but the Fosse Way, its diagonal twin, crosses major rivers at Leicester, the seat of Middle Anglian bishops; at Halford on the River Stour in Warwickshire and the Somerset Avon near Bath.

At such key road-river crossings Mercian engineers – or continental engineers acquired by Mercian kings – possessed the technology to construct bridges where needed; and in reserving bridge-building as a public burden to be imposed on all estates, religious or secular, eighth-century Mercian kings were exercising royal authority to strategic purpose. Middle Anglian bishops at Leicester may have had to suffer the material burden of maintaining a dry crossing of the Soar in good working order; but the opportunity to exploit that crossing for trade and for the extraction of tolls probably made it a price worth paying.

King Æðelbald's imposition of services in kind extended to the upkeep of fortresses; but since few seventh- or eighth-century royal *vills* show any signs of being fortified by rampart, ditch, wall or palisade, archaeologists have to search more actively for clues to confirm their existence. John Blair, in his masterly study *Building Anglo-Saxon England*, reasons that the most credible candidates are the hillforts of the Iron Age, slighted and abandoned during the Roman period but now repurposed.[45] Such reoccupation has been identified through excavation at many hillforts in the North and West of Britain and they seem to have become centres of territorial lordship in their own right during the fifth century and beyond. A century after Æðelbald, King Ælfred belatedly adopted several abandoned hillforts as part of his brilliant *burh*-based defensive scheme.[46]

If archaeologists are, for the time being, unable to point directly to more than one certain Æðelbaldan *burh* or fortress\*, they can at least hold up the smoking gun left at the scene: a naming code for places that fulfilled specific functions in the Mercian administrative mind-set. It has long been known

\* See below on p. 204.

that names like Witton (*wudu-tūn*), Eton (*ea-tūn*) or Rushton (*rysc-tūn*), very rare before the first half of the eighth century, were given to places holding a critical resource or fulfilling a specific role: so-called 'functional –*tūns*'.[47] The key names, so far as Mercian administration goes,* are Burton, Charlton, Eaton, Knighton, Stratton and their variants;[†] possibly also the directional –*tūn* names Norton and Sutton, Weston and Easton.[48] Sometimes such names appear isolated, in a landscape whose rubric cannot now be decoded. But sometimes they form suggestive clusters: satellites orbiting central places. John Blair identifies one of these clusters surrounding the village of Great Glen,[‡] eight miles south-east of Leicester, where a royal charter was issued in 849.[49] Within a couple of miles are found Burton Overy, Carlton Curlieu, King's Norton, Great and Little Stretton – on the Roman road between Leicester and Godmanchester – and Newton. Great Glen itself lies five miles north of Gumley, the site of the 749 council at which Æðelbald confirmed his imposition of the 'common burdens'. These focused clusters have lain beneath the radar partly because they do not appear in the charters – at least not for another century or so; and because they leave so few tangible traces for the archaeologist. As a result, it is almost impossible to be sure whether they represent an explicit administrative policy or whether they developed organically. In any case they found their place, sooner or later, in the toolbox of royal control.

* The same code appears to be valid for both Northumbria and Wessex, too.
† Burton: the *tūn* servicing a fortress or stronghold; Charlton, the estate of a ceorl rendering services to the royal *vill*; Eaton: a *tūn* fulfilling a function at a river crossing, such as a ferry; Knighton: the *tūn* of a thegn owing a service – perhaps military protection or as a courier – to a royal *vill*; Stratton or Stretton: a *tūn* on a Roman road – perhaps a place where stabling and fodder might be found. Directional names often come in pairs and seem to have fulfilled well-understood (to their administrators) functions in the overall scheme.
‡ Old English *Glenne*.

If the *burh-tūns* are the pivotal elements in this scheme, an over-arching pattern is clear: they cluster along the key communication routes of the Fosse Way and River Trent; and they are found at locally prominent sites with strategic value for defence. Nowhere is this more apparent than in the Mercian heartlands of the Upper Trent, where Tutbury – perhaps 'look-out *burh*'* – controlled the confluence of the Trent and River Dove, while Burton-on-Trent and Stretton protected the river and the passage along Ryknield Street close to Catholme and the river crossing at Wychnor. Similar clusters are identified by John Blair further down the Trent. One is concentrated on the old Roman town of *Margidunum* – later known simply as *Burh*,[50] on the Fosse Way, whose satellites include a *brycg-ford* (East Bridgford), Toot Hill and a *burh-tūn* (Burton Joyce, on the opposite, north-west bank of the river). A second cluster lies further downstream at Littleborough, where a key Roman road linked Lincoln with the Great North Road.[51]

So far, just one certain candidate for eighth-century defensive works can be proposed: Irthlingborough, overlooking the River Nene from its north bank fifteen miles or so downstream from Northampton. Here, a timber building of about the right date reinforces the testimony of a charter confirmed here between 787 and 796. The *burh*'s relationship to an estate centre lying across the river at Higham Ferrers† implies the existence of a crossing here – perhaps a causeway or bridge spanning the broad floodplain, as many bridges do today. The physical nature of the defences implied by these *burh*- names and their satellites may, in time, come into clearer focus; but it seems that a cluster of dependent thegns bound to the king's services may have acted just as effectively as rampart or palisade, creating a protective ring around key royal centres.

In the Early Medieval mind-set, insofar as a twenty-first-century historian can evoke it, physical and imagined landscapes were inseparable worlds of seasonal cycles, hard graft, ancient memory and wonder or threat, invested with meaning and

---

* Old English *tōt* – a lookout.
† See below in Chapter 7, p. 242.

Tutbury – perhaps the 'lookout' *burh* – protected core Mercian
interests along Ryknield Street and the Upper Trent Valley.

magic. Favourable relations with God, or older gods, spirits or ancestors, mattered as much as the physical realties of a hard winter or fine harvest, fertile beasts or the wondrous movements of moon and stars. Created at the pace at which humans, horses and oxen moved, homestead and paddock, field and meadow, woodland and open wold were experienced as the machinery of more earthly constellations.

Itinerant kings saw their world as if from a higher viewpoint. They knew more lands, had met more people, enjoyed wider networks of relations and enjoyed greater agency as the physical and spiritual embodiment of their people, the *gens*. Their local world was concentric, like that of the farmer; but they also experienced landscape as a network of linear routes, intersections and progressions counted in days of travel and rest between estate centres – royal *vills* or *tūns* – where the land's surplus awaited them. Mercian kings' mental map of their kingdom might have looked something like a rough square, bounded at top and bottom by Trent and Thames, left and right by Welsh mountains and East Anglian Fens. Within that square, diagonal lines described by the Fosse Way and Watling Street connected nodes of royal power or agency: the *burhs* and minsters, patrimonial lands, trading places and customary assembly sites. River crossings and crossroads were places of risk and opportunity, of decision and doubt. Frontiers might be borders defined by treaty or zones of weaker affinity, sometimes to be defended, sometimes to be crossed with malice aforethought. Kings' fortunes, their projection of geopolitical reach and authority, pulsed like the tides. Æðelbald's interaction with these landscapes and his reinforcement of ideas about 'public' service reveal a strong sense of emerging statehood, underpinned by codified relations between territorial lords and their rights and privileges, duties and obligations.

Historians and archaeologists pondering the monumental dyke constructions attributed variously to Offa and Wat along the Mercian/Welsh border and Woden (Wansdyke) in the west country have sometimes considered Æðelbald as having conceived or had a hand in their construction. Wat's Dyke, the diminutive sibling of its grander and more famous neighbour, is now more securely dated to the ninth century and Offa's Dyke, despite doubts raised over the years, is hard to dislodge from its traditional attribution. There is no consensus over the dating of the massive earthwork complex known as Wansdyke, whose Wiltshire and Somerset components appear to reference Roman routeways including the Fosse Way south of Bath. The most impressive section, running along the downs above the Vale of Pewsey, lies close to the site of the battle fought in 715 between King Ceolred of Mercia and King Ine of Wessex. This was an ancient zone of tension and conflict. But if Æðelbald's extraction of public labour included earthwork construction, solid evidence for it is so far unforthcoming although improved dating techniques, particularly Optically Stimulated Luminescence, offer archaeologists hope for the future.

The policy of tightening royal control over the political administration of Mercia and its peripheral subject states is, nevertheless, manifest in other ways. The charter issued in 736/7 by King Æðelbald to Æðelric, his *Hwiccian comes*, for land at Wootton Wawen on which to found a monastery looks, at first sight, as though the king was founding a minster (one of several recorded in similar charters) on *Hwiccian* lands – probably former royal lands that he had appropriated. The historian Patrick Sims-Williams interprets such charters as recording rewards for loyal service: tax-free retirement gifts of land on which old comrades might become abbots of their own religious houses.[52] That may well be true in some cases:* a confirmation of the trend deplored by Bede towards a secularisation of minsters in Mercia.

But a historian with a different perspective, Steven Bassett, sees these lands as estates already belonging to secular lords like Æðelric, on which they would establish proprietorial minsters

---

* Most convincingly, perhaps, the subject of the Ismere Charter of 736, which Sims-Williams traces through to its eventual donation to the church at Worcester and which he characterises as a 'family' monastery. The grant was made with the consent of the *subregulus* and *comes* Æðelric. Sims-Williams 1990, 149.

for the sake of their souls.[53] In that view, the king's consent and blessing was required because the land in question was being, so to speak, privatised – removed from the pool of lands from which the *feorm* and military service were due. That is consistent with the extension of royal control in the reservation of the 'common burdens', and with the relegation of such men as Æðelric from the independent status enjoyed by their fathers' generation. The lack of surviving charters in the Mercian core may partly reflect the fact that kings did not need to give themselves consent to alienate royal lands there.

The witness lists of Æðelbald's surviving charters reveal another side to his grip on secular lordship. At *Clofesho* in 742, after the understandable prominence of senior ecclesiastics attesting the council's acts, comes a list of the king's senior lords: Ofa, the old *patricius* from his days in exile, who witnessed no fewer than sixteen surviving charters; Ealdwulf *dux*; then Æðelmod, enigmatically listed as the 'offspring' of the king of the Mercians; then the king's brother, Heardberht, and two other *duces*, Eadbald and Beorcol, followed by more junior *ministri*: Cyneberht, Freoðorne, Wærmund and Cuðred, and an abbot, Buna.[54] The secular signatories to the second *Clofesho* council, of 747, do not survive. At Gumley in 749 the order of preference among secular testators is: Heardberht; Eada; Cyneberht; Beorcol *patricius*; Friðuric (the king's brother-in-law); Eopa; Eadbald; Brynhelm; Mocca; Aldceorl and Alhmund.[55]

Some of these men appear only once – substitutes on a sporting team – and are never heard of again. Like shares on a stock market, orders of rank in the king's cabinet and the king's choice in who should accompany him record the ups and downs of their fortunes. Æðelbald's last known charter, from the year of his death in 757, records a grant of land at *Toccan seaga* (Tockenham) in Wiltshire in the territory of the West Saxons:

Wherefore I, Æthelbald, king, not only of the Mercians but also of the neighbouring peoples over whom divine dispensation wished me to rule without judgement on my merits, freely granting, bestow on the servant of God, Abbot Eanberht, land of ten hides, into the control of the church of Christ, for the salvation of my soul and for the atonement of my misdeeds. That land is near the wood they call Tockenham, having nearby the *tumulus* that has the name Rada *beorg*. [56]

Since Cynewulf, recently installed as king of the West Saxons, was a signatory to the charter, his subordinate relationship to Æðelbald is clear: the Mercian king had the power to gift land well beyond Mercia's traditional border on the Thames. Bishops of the *Hwicce*, of Sherborne and of Winchester were in attendance. So too were a number of familiar councillors: Heardberht, Eadbald, Eada and Æðelric.* One witness, Cerdic, bears a name that is almost certainly West Saxon – a member of Cynewulf's own *comitatus*.

The king's brother, Heardberht, witnessed four of his charters and several more under Offa. He often features alongside Beorcol, but the latter disappears as a testator after Æðelbald's death. Was Ofa, his most senior and trusted *comes*, dead by 749, or had he fallen out of favour? And, who, in this seemingly loyal band of *comitates*, was plotting to bring the king down?

The appearance in a single charter witness list of the otherwise unknown Æðelmod, 'offspring' of the king of the Mercians, raises all sorts of questions. Why not call him a son? Was he illegitimate? According to Boniface, the king did not take a wife but indulged in adulterous relations with, among others, women who had taken the veil. King Offa would recognise his son, Ecgfrið, as a frequent testator to his charters, but also as his heir, and successors were sometimes installed as their father's *subreguli* in Kent, Northumbria and Wessex. Æðelbald's reluctance to recognise an official wife and produce heirs who would be regarded as legitimate successors is inexplicable in the conventional code of Anglo-Saxon kingship. The expectation is that he would have taken a young woman of

---

* Assuming this to be the same Æðelric who was a *subregulus* of the *Hwicce*, it shows that in Wessex his stock was lower than in his home lands.

royal stock from one of the other kingdoms, and his East Anglian ally King Ælfwald would be a prime candidate for supplying a consort. Otherwise, whom did he expect to succeed him? Can one infer from Boniface's testimony that the nuns with whom he had sexual relations bore his children and that he connived in their killing as infants?

Two possibilities cannot be discounted. One is that, by agreement, a successor from another family line had been chosen. In that case, the only realistic candidates are Offa, his first cousin twice removed,[57] or the man who immediately succeeded him after his death: the otherwise unrecorded Beornred. The failure of either to appear in his charters or councils as witnesses tells against this idea. A second possibility is that he had a wife whose name was not recorded and that she died without issue or even, perhaps, during childbirth. Even in the hard-headed world of Early Medieval kingship it is not impossible that grief played a part in Æðelbald's otherwise unfathomable dynastic policy. Either way, Æðelmod's status and history are confoundingly obscure.

Mercian 'foreign' policy in the second half of Æðelbald's reign looks much more conventional. The alignment of anti-Welsh interests that had led to a joint Mercian–West Saxon campaign against the Britons in 743 did not last. In 748 King Cuðred's son Cynric rebelled against his father and was slain. Two years later one of Cuðred's ealdormen, named Æðelhun, led an unsuccessful rebellion against him. Two years after that, in 752, a West Saxon army fought Æðelbald at an unidentified place called *Beorhford**\* and 'put him to flight', according to the *Chronicle*. Beneath the headline lurks the likelihood that Mercian sponsorship had lain behind the two West Saxon rebellions. At any rate, by the time the two armies met at *Beorhford* Cuðred's rebellious ealdorman Æðelhun was fighting with him against the Mercians. The change in sides looks timely and decisive. The following year a West Saxon army was independently campaigning against the Welsh.

\* Despite the obvious attribution of this name to Burford in West Oxfordshire, the derivation is not certain.

King Cuðred's success in throwing off Mercian overlordship lasted three years, until his own death in 756. He was succeeded by a kinsman, Sigeberht; but within the year he had been deposed by a council of senior ealdormen and replaced by the Cynewulf of the Tockenham charter who, although he bore an impeccably sound West Saxon royal name, comes with no pedigree. Sigeberht was driven into exile in the Weald. Cynewulf, so insecure in the immediate aftermath of his coup that he had to witness a Mercian charter granting land in his kingdom, survived his Mercian overlord and managed to hold the throne of Wessex for nearly thirty years.

Place names fossilise Mercian royal administrators' success in binding key locations into a centralised scheme of control.

Continued political instability in Northumbria, seemingly fed by King Eadberht's preoccupation with his northern neighbours, erupted in that same year, 756, as if rebellion were an infection. He is found campaigning in August with a Pictish army, under Onuist map Vurguist,* against the Britons of Strathclyde, attacking their craggy fortress at *Alclut*: Dumbarton on the River Clyde. Factions in Northumbria took the opportunity to rebel against him.[58] Onuist was, by some assessments, Æðelbald's only serious rival for his pretensions as *Rex Britanniae*.[59] In his reign Pictish power was consolidated in the East and he was able to dominate and defeat the armies of Dál Riata and Strathclyde in the west. An enigmatic entry under the year 750 in the northern annal known as the *Historia Regum* or *Chronicle of 802* records that King Cuðred of Wessex made war against Æðelbald *and* Onuist, implying that they had formed an alliance with King Eadberht of Northumbria. The Early Medieval historian Alex Woolf has suggested that, in the aftermath of the victory at Dumbarton, a Pictish army came south into Mercian territory and was almost completely annihilated at a place called *Niwinbirig*.† I am not quite convinced; but there is no reason to doubt the alliance, albeit allusive, between the two greatest overlords of their day.

Even Æðelbald, whose extraordinarily long and successful reign is untainted by any known record of rebellion, outlasted the patience, loyalty and subservience of his *comitatus*. By 757 many of his old comrades-in-arms, like Ofa, may have been dead. The king must himself have been sixty or so years of age. He had no designated heir. The *Anglo-Saxon Chronicle* records tersely that he was murdered at *Secandune* (Seckington), near Tamworth. The *Historia Regum* offers a little more detail. He was 'treacherously killed by his bodyguard'.[60] In the aftermath, it goes

---

* Alternatively Óengus mac Fergusso.
† *Niwinbirig* would conventionally become Newborough, a name so common that candidates for the site of this disastrous Pictish defeat have been proposed on the River Tyne, in Lancashire or near Lichfield. Woolf 2005.

on, the Mercians commenced a civil war. The *Chronicle* records that the king's body was taken to Repton, some dozen miles to the north, for burial in what would become a royal Mercian mausoleum. Guðlac, his hermit patron, had been a monk there before his Fenland exile. The relief carving of a mounted warrior recovered from the excavation close to the surviving crypt may be his memorial.

The immediate beneficiary of Æðelbald's death was one Beornred, who 'ruled a short time and unhappily'.[61] An entry in the annal known as the 'Continuation of Bede' recorded that in the same year after Beornred began to reign, Offa killed him and 'strove for the kingdom with a bloody sword'.[62] Was Æðelbald's bodyguard-turned assassin the proxy for a disaffected faction at court? We cannot say. The king may now have been infirm or losing his wits: a lame duck. Was this Beornred, whose name alliteratively links him to a series of kings who would reign in the next century, a left-field opportunist or a consensus candidate? Or was the king's cousin, Offa, a chosen successor, briefly usurped in a pre-emptive strike? Seckington is an otherwise historically unremarked village tucked into the north-west corner of Warwickshire. In the Domesday Book it would be recorded as a mere *vill* with a manor of five hides. The site of the assassination makes it almost certain that *Secandune* was a royal *vill* in the eighth century, so its regicidal association may have prompted its abandonment by subsequent Mercian kings in favour of relocation to nearby Newton Regis. It is one of the choicer ironies of Mercian history that it is by such offhand means that the geography of royal estates is painfully teased out.

Æðelbald has not been remembered as a great king of the Mercians. No known church or great earthwork; no chronicle; no *Vita* survives to monumentalise his deeds. If the Repton stone was his cenotaph, it is a modest legacy. He slumbers in the shadow of his illustrious successor. A letter written around the time of his death by an unknown monk, and curated among the manuscript letters of Boniface, lists Æðelbald as one of those 'tyrants' whom he had encountered in a vision of hell, a 'seething whirlpool of fire and blackness'.[63] But if his sins marked him out for opprobrium by the more austere of his ecclesiastical contemporaries, his achievements in enforcing a pan-Mercian peace, in fending off rivals and in establishing a novel and long-lasting form of statehood built on solid economic foundations, render him worthy of some sort of redemption.

# Chapter 7

# Offa: Overlord
## 757–779

I n the heroic poetry that celebrated Anglo-Saxon traditions of migration, battle, honour and kingly might, Offa of Angel was remembered as a warrior lord who had defined and defended the frontiers of the continental Angles at the base of the Jutland peninsula: the founding father of a tribal nation. He appears in the Old English poems *Widsith** and Beowulf and, in the eighth-century Mercian genealogy preserved in the *Anglo-Saxon Chronicle*, he makes an appearance as the grandson of Woden, three generations before the dynastic progenitor Icel.[1] He was the tap root of the Mercian people's mythic past.

In *Widsith*, the 'Traveller's Song', the eponymous poet or *scop* eulogises rulers of great renown from Alexander and Caesar to Alaric the Goth and Theuderic of the Franks. These mighty warlords and their peace-weaving queens appreciated fine poetry; were generous with gifts of gold and gleaming treasure. It is regarded as one of the oldest of the Early Medieval poetic compositions to survive – told and retold many times over the generations before being written down. *Widsith's* heroic Offa, while still in his youth, 'conquered the greatest of kingdoms'; no-one was more noble in battle; and *'ane sweorde merce gemærde'* [...] 'with his lone sword he defined a frontier [...]'.[2]

The written version, as it has come down to us, bears striking Mercian overtones: the real, eighth-century King Offa was

* *Widsith*, like *The Wanderer*, the poem known as *The Fates of Men* and the so-called *Maxims*, is preserved in the manuscript collection known as the *Exeter Book*, dating perhaps to the tenth century.

remembered in later centuries for constructing the dyke that bears his name – the ultimate definition of a frontier; and the word for frontier used by the poet, *merce*, seems deliberately to allude to the tribal name. The real King Offa's sword also seems to have borne particular resonance for later generations. The 'bloody sword' with which his thirty-nine-year reign is introduced in the *Anglo-Saxon Chronicle* was still famous in the eleventh century when Aðelstan, son of King Æðelræd II, left 'the sword which King Offa owned' to his son Edmund 'Ironside'. *

The Mercian Offa was not the only royal bearer of that name: there had been an Offa among the East Saxon royal family in the first decade of that century, who abdicated and accompanied the Mercian King Cœnred on his pilgrimage to Rome in 709.[†] In praising the legendary Offa of Angeln the poets who entertained royal courts with *Widsith* and *Beowulf*, both possibly eighth-century Mercian compositions,[3] drew on popular ancestral traditions to resonate with contemporary audiences. The new Mercian king's auspicious name tapped into that mythology and reinforced its potency as part of the Anglian creation legend.[‡]

King Offa's geographical origins are obscure. One clue is offered by two grants of land in the kingdom of the *Hwicce*: at Bredon in Worcestershire, as remembered in an Offan charter of 780, and at Westbury in Gloucestershire, gifted by King Æðelbald to Offa's grandfather, Eanwulf.[4] A second clue is offered by

---

* Atherton (2005, 65) suggests the sword in question might be the Hunnish weapon gifted to Offa by Charlemagne and mentioned in a letter of 796 written by Alcuin. *Alcuin Letters* XL 40.

† See above, Chapter 4, p. 127.

‡ I have often wondered if the conspicuously 'noble' names of Anglo-Saxon kings – the dithematic Æðelræds ('noble counsel'), Ceolwulfs and Ecgfriðs – were names given at birth or conferred at an appropriate age when the youth took up arms or was inducted as heir apparent, in which case they are more likely to have been chosen to suit the individual's perceived virtues. The humbler Beonnas, Uhtreds and Eadas who attest their charters have a much more vernacular ring about them. If Offa received his name as a young adult it may have carried prophetic overtones.

one of Offa's own foundations, of a nunnery at Winchcombe in Gloucestershire, where the royal estate would constitute its own tiny shire in the tenth and eleventh centuries.[5] It seems that, like Penda, Offa's family affiliations lay in the south-west, in the heartlands of the *Hwicce*. Despite doubts over the date of his marriage and his wife's background, there is a strong possibility that Offa was an early exile at the West Saxon court and was betrothed to a royal princess there.

The brief but bloody civil war of 757, which followed Æðelbald's murder, seems almost to have shaken the natural order of things. In the same year, King Sigeberht of Wessex was deposed and in his place Cynewulf became king. In Northumbria in 758 King Oswulf, ruler for a single year after the abdication of his father, King Eadberht, was killed by his bodyguards and the throne usurped by one Æðelwold Moll, a man unconnected to the royal family. Then, in Æðelwold's ...

> ... second year [759] there was great tribulation by reason of pestilence, which continued almost two years, divers grievous sicknesses raging, but more especially the disease of dysentery... [6]

The symptoms seem incompatible with a recurrence of the plague of 664. Dysentery outbreaks were exacerbated by poor hygiene; it is possible, even likely, that this was a livestock disease spreading through the human population. Its effects on domesticated animals and people are hard to quantify but at any rate while the pestilence was still raging, in 760, there was a 'battle between the Britons and the Saxons' at Hereford in the land of the *Magonsæte*, according to the Welsh Annals. King Æðelbald had fought against the Welsh in alliance with the West Saxons in 743, although the geography of that campaign is obscure; this time, it seems, Mercia was defending a satellite territory. Æðelbald is not recorded in military action for the first seventeen years of his reign; Offa went into battle in his third year. Powysian forces may have taken advantage of both Æðelbald's recent death and the ravages of the pestilence to raid or attempt a conquest of loosely held lands at the head of the River Wye; perhaps to attack the bishop there.* Thomas Charles-Edwards, the historian of Early Medieval Wales, suggests that this is the period commemorated in the inscription on the Pillar of Eliseg, whose eponymous king 'seized the inheritance of Powys by force from the power of the English'.[7] With turmoil in Northumbria and Wessex and a new, unproven king in Mercia, Welsh kings, ambitious to test the mettle of the *Lloegr*,† seem to have chosen an opportune moment.

Historians also look to this fragment for evidence of a framework of conflict in which the origins of Offa's great earthwork project might be sought. In truth, the complexities of Anglo-Welsh relations in this period are hidden from view. No English source recorded Mercia's conflict with Powys or the other Welsh kingdoms and the best proof of that long-running antagonism comes from the Dyke itself, an overtly military enterprise designed to dominate, intimidate and threaten.‡

Then, after plague and sword: ice. In the winter of 763–4...

Deep snow, hardened into ice, unlike anything that had ever been known to all previous ages, covered the earth from the beginning of winter till nearly the middle of spring, by the severity of which the trees and shrubs for the most part perished, and many marine animals were found dead.[8]

The Chronicler called it 'the hard winter'. The compiler of the *Annals of Ulster* recorded...

* In 760 Welsh ecclesiastical practice was still divergent from Roman orthodoxy on matters of the date of Easter and the tonsure; there is no evidence that the Welsh recognised the authority of any Anglo-Saxon bishop. Alignment came eight years later, according to the Welsh Annals, perhaps under Mercian pressure.
† Old Welsh *Lloegr* meant literally, perhaps, 'borderer', but was applied as a generic to English people.
‡ See below in Chapter 8, p. 260.

A great snowfall which lasted almost three months; a great scarcity and famine; an abnormally great [summer] drought.[9]

In an age when winter survival depended on storage of a surplus from the previous harvest and on slaughtering animals not required for future breeding, long, hard winters might have devastating effects – psychological as well as physical. The Old English word *wintercearig* means 'winter-anxious' and in those dark days it might seem like spring would never come. Snowstorms 'fettered the earth' in the poetry of exile and loneliness, while in Bede's metaphor of the mead hall, winter and darkness symbolised the infinite unknown before birth and after death in a heathen world, in opposition to the warmth of the hearth, the light of the candle and a life lived in the knowledge of salvation in the Christian world.[10] Winter and adversity were synonyms.

And after the ice, with the onset of a widespread drought, came fire. The *Historia Regum* recorded the catastrophe that followed…

Many cities, monasteries and vills in various parts and, moreover, kingdoms were suddenly laid waste by fire; for example, *Stretburg*, *Venta civitas* (Winchester), *Hamwic*, *Londonia civitas*, *Eboraca civitas* (York), *Donacester* and many other places suffered under that calamity…[11]

It seems as if the drought created conditions, like those of 2022, in which timber and thatch buildings, desiccated by heat, were as kindling to wildfires caused by stray sparks or wilful arson. Three of the named settlements were major trading posts, whose close-set buildings and concentrations of people and goods rendered them particularly vulnerable to the ravages of fire. *Gipeswic*, modern Ipswich, seems to have escaped their fate. London, York and *Hamwic*, reaching their greatest extent around 750,[12] may already have been suffering from the generally prevalent economic decline indicated for the second half of the eighth century; but perhaps the fires of 764 were partly responsible.

The enigmatic name *Stretburg* is a reminder that even settlements that must have been prominent in their day can defy modern identification. The compilation of the *Historia Regum* belongs to the beginning of the ninth century, so this entry preserves a rare early record of a *–burh* name. The *Stret-* prefix is an indicator that it lay on a major Roman road, and the complete name is a unique and direct Anglo-Saxon equivalent to the city of Strasbourg on the Franco-German border – formerly the Roman *Argentoratum*, which lay at an important crossroads. The Frankish historian, Gregory of Tours, writing at the end of the sixth century, knew it by both names.[13]

Where, among the Anglo-Saxon kingdoms, might one find a suitably important central settlement on a crossroads or major route? Of those settlements whose identities are not already secure, the most likely candidates are former Roman towns still functioning in the eighth century. While London, York, Winchester and other settlements razed by the fires of 764 recovered and were rebuilt, the loss of the name *Stretburg* suggests that it did not recover – at least not in time to retain its early name. A few candidates readily suggest themselves: St Albans in Hertfordshire and Dunstable in Bedfordshire both lie on Watling Street. St Albans, site of the shrine of a great Romano-British martyr visited by St Germanus in 429, sits on a hillside above the Roman town of *Verulamium*, which survived the end of the empire in some form. Later medieval legends attribute the revival of the shrine to King Offa. But a charter dating to the 990s refers to the settlement there as *Wætlingaceaster*.[14] Dunstable – Roman *Durocobrivis* – has a better claim, perhaps: it was sited at the point where Watling Street crossed the Icknield Way and, as one of the most prominently productive of the 'productive' sites, it has mid-eighth-century form as a market. If Dunstable is a good candidate, so is Royston, sited at the crossroads of the Icknield Way and Ermine Street, some thirty miles to the north-east – and for the same reasons. Towcester, fortified in 917 by Edward the Elder, had been Roman *Lactodorum*, sited on Watling Street where it joined the road heading south towards

Dorchester-on-Thames. A last speculative possibility is the site of *Venonis*, the small Roman fort situated where Watling Street crosses the Fosse Way between Hinckley and Lutterworth. Now only a hotel and two farms indicate its former importance as a critical junction in Early Medieval England. No settlement of the period has yet been discovered here but there are tell-tale signs of a former role in the Mercian network of *burhs*. Four miles to the west of *Venonis* lies Burton Hastings and there are *Stret-tūn* names on two axes of the crossroads here.* Archaeology may yet recover the evidence for an eighth-century fire in such a plausible location. If *Stretburg* was, in fact, a Mercian *burh*, the jigsaw of strategic Mercian royal settlements would acquire a satisfyingly useful piece.

Early Medieval kings were itinerant by habit and need. Although there are signs that Offa's court spent significant amounts of time in London and Tamworth, it is very likely they spent much of the year on the move. His 'winter quarters', occupied when travel was difficult and provisions to cover the months between harvests would be dwindling, are unknown – Tamworth is a strong candidate in the Mercian heartlands. Using just the main arteries of Watling Street and the Fosse Way the court must have covered considerable mileage – the most demanding schedule of any Anglo-Saxon kingdom. The need to plan accommodation and provisions across such an extensive kingdom may be a more significant driver of Mercian kings' progressive, not to say revolutionary, establishment of *burhs* and *burh-tūns* than the need to defend against enemies otherwise unidentifiable. The historian of Anglo-Saxon towns Jeremy Haslam has pointed out that many of the Mercian *burhs* seem to have been located at river crossings, where the same common burdens that provided labour for their defences might also be called on to construct new bridges or causeways – at Oxford, for example – or refurbish old Roman structures.[15]

* Stretton-under-Fosse to the south on the Fosse Way and Stretton Baskerville to the west along Watling Street.

He offers tentative evidence for Mercian bridges at Hereford, Northampton, Bedford and Tamworth.

*

Notwithstanding the weakness of stumbling dynasties in Northumbria and Wessex, King Offa's immediate political initiatives were directed at Kent. Overlordship of that ancient kingdom delivered valuable prizes: in the fruits of continental diplomacy, in cross-Channel trade and in the power to appoint or influence Canterbury's archbishops. Offa's bloodless coup in Kent can be traced to the years following the death of King Æðelberht II in 762, after a reign of more than three decades. In 764, the year of the great winter, one Sigered is recorded granting land as *Rex Cantiae*, with the consent of Æðelberht's son and co-ruler King Eadberht II.[16] The same year sees the only other mention of Eadberht, confirming remission of tolls on two ships at Sarre and Fordwich, granted previously by kings Æðelbald and Offa to the community of nuns at Minster-in-Thanet.[17]

East Kent (roughly the diocese of Canterbury) and West Kent (the diocese of Rochester), divided by the River Medway, had always been separable entities, periodically united under a single ruler.[18] Æðelberht's death seems to have precipitated its fracturing once more and Offa took immediate advantage. A third Kentish charter of 764, issued at Canterbury, grants twenty *sulungs** of land at Islingham to Bishop Earwulf of Rochester.[19] The donor is King Offa; the two principal witnesses are Archbishop Bregowine and the otherwise completely obscure King Heahberht, who must, one supposes, have asserted himself as the ruler of West Kent.

A year later Kent's complex politics had delivered another ruler, Ecgberht, apparently from a previously unrecorded dynasty. He granted land at Rochester to its bishop in a charter also

* A unit of land assessment unique to Kent; probably a rough equivalent of Bede's *familiae* or the hide.

confirmed by King Heahberht;[20] but a confirmation was added
by King Offa, indicating that his authority was required for the
grant to be validated. Around that time Sigered issued his last
known charter, now more modestly styling himself *'rex dimidae
partis provinciae Cantuariorum'* – king of half of Kent.[21]

Mercian control over Kent was no more than indirect in these
years. In 764 Archbishop Bregowine died and Mercian interests
dictated that a candidate be elevated from the ranks of Mercian
bishops, as his predecessors Nothelm and Tatwine had been. But
King Ecgberht retained sufficient independence to promote the
interests of Canterbury's own candidate, Jænberht. He was a true
Man of Kent* – a monk, later abbot, of St Augustine's monastery
in Canterbury and a kinsman of one of King Ecgberht's reeves.[22]
Relations were, nevertheless, sufficiently cordial that King Offa is
recorded as making direct grants of modest estates to Jænberht
in the next decade, exercising the soft power of patronage. The
charter whose witness list survives to attest one of these grants,
made in 774, has King Offa attended by the recipient, Archbishop
Jænberht, by his Mercian bishops, and by his *principes* and *duces*:
Brorda, Berhtwald, Esne, Eadbald – perhaps formerly a minister
under Æðelbald; otherwise, a fresh generation of Mercian
nobility with their own particular interests and loyalties to the
king.[23] The occasion seems more secular than ecclesiastical, the
modest grant more likely the lubricant in some political deal than
the principal subject of the assembly.

This seems to have been a high point in co-operative relations
between the two kingdoms. *Imperium*-wielding overlords and
their tributary kings responded to political undercurrents,
pressures and opportunities, each testing the other's mettle on
the field of political combat while exploring and exploiting novel
tactics and, occasionally, trying out new weapons. A charter of
799 issued by Offa's successor, King Coenwulf, at Tamworth,
records how substantial lands in Kent, which King Ecgberht

---

* That is, from East Kent; those hailing from West Kent are traditionally
called Kentishmen.

had conferred on Christ Church, Canterbury,* in return for the 'recompense of great riches', had been taken off that church by King Offa and then distributed by him to his *comites*. The king's reasoning was that King Ecgberht (Offa's 'minister' in the charter text) had given these lands away without his lord's (Offa's) permission. King Cœnwulf was now returning these lands to the church, in return for a payment of 100 mancuses, or 3,000 silver pence.[24]

King Offa's 'rights' as overlord of Kent were evidently circumscribed by the limits of his political and economic influence there. King Ecgberht does not appear in the lists of witnesses to Offa's Mercian charters, as the *subreguli* of the *Hwicce* had been expected to do. He was able to issue a rather elegantly designed silver coinage in his own name without any apparent Mercian interference, although only in very small numbers.[25] He also seems to have emerged the victor after a battle fought between Mercians and Kentishmen at Otford in West Kent in 776.[26] It is a classic site for a set-piece encounter, where the valley of the River Darent, running north towards the Thames at Dartford, crosses the line of the ancient Pilgrims' Way – a natural east–west route whose origins lie deep in prehistory, running along the north edge of the Vale of Holmesdale between the North Downs and the Wealden ridge.

Battle was, ultimately, the decisive means of resolving tensions between subject kings and their putative overlords even if, in victory, the liberation might be short-lived. In the event, for nearly eight years afterwards Offa seems to have been unable to impose his overlordship on King Ecgberht or his successors. Kent's independence as a kingdom had always relied more on its trading wealth, its possession of the pope's vicar among the Anglo-Saxons and on its relative geographical isolation from the 'great' powers of Northumbria, Mercia and Wessex. During the second decade of Offa's reign that independence was reasserted, with some success.

* That is, Archbishop Jænberht's cathedral.

The Mercian king had softer targets to aim at. According to the *Historia Regum* he invaded the kingdom of the South Saxons in 771 and 'subdued the *Hastingas* by force of arms'.[27] On the face of it this seems an unlikely campaign. The *Hastingas*, as the name suggests, were a people who inhabited the narrow coastal strip of East Sussex between the Pevensey Levels and the marshlands around Rye in Kent. A Roman road had once reached this far off the beaten track; but its state in the 770s is unknown. Even today Hastings is hardly the most accessible of coastal towns. An invading army must pass through the formidable hilly forests of the Weald – and return – and it is far from clear what riches or glory might be won on such a campaign. How the raid caught the attention of the northern annalist compiling the *Historia Regum* is also a mystery. The result of the raid is clear: Offa was able to establish Mercian overlordship over all of Sussex.

The names of no South Saxon kings are known after this period and, in 772, the Mercian king was able to grant land at *Bixlea* (Bexhill) in Sussex, to the Bishop of Chichester, for the founding of a minster church.[28] Among the witnesses were Kings Ecgberht of Kent and Cynewulf of Wessex – testators as much to Offa's *imperium* as to the ceremony of the turf and book. Alongside familiar names from Offa's inner circle are four *duces* of the South Saxons: Oswald, Osmund, Oslac and Ælfwald. Of these, Osmund had enjoyed the title *rex* in a charter of 770, witnessed by Offa and his senior councillors, while Oslac was ruling jointly with the otherwise obscure King Ealdwulf in about 765, according to a South Saxon charter of that date supposedly also confirmed by Offa.* These, then, seem to have been brothers and sometime co-rulers of the South Saxons: now diminished, like their counterparts in *Hwicce*, to the rank of *dux* after the imposition of Mercian *imperium* over their lands.

---

* Electronic Sawyer S49; S50. The attestation of Offa's consort, Queen Cyneðryð, and their son Ecgfrið, are almost certainly later interpolations, since Ecgfrið is unlikely to have been born at that date.

The mechanisms by which eighth-century overlordship was asserted or accepted have to be reconstructed: the rule book was not written down but was part of the unstated political culture of tribal kingdoms. Subordinate kings collected tribute in the form of cattle and treasure and rendered it to their new lord; they were obliged to fight in his wars, attend his councils and offer up marriageable daughters for his male heirs. Overlords must reinforce their authority through gifts and patronage, by confirming charters and, sometimes, through force of arms or the threat of punitive violence.

On the accession of a king through inheritance, coup, murder or battle, his predecessor's *imperium* no longer held – it had to be re-established. The senior nobility of a kingdom would expect to attend their new lord, if they were not already part of his *comitatus*; submit to his authority; swear an oath of loyalty and pledge military support. Those more peripheral lords, like the former kings of *Hwicce*, who believed it to be in their best interests to seek the new king's protection and patronage, would also submit, probably voluntarily. Hostages and gifts would be exchanged; marriageable offspring betrothed to cement the alliance. Even those lords fallen from their own thrones and relegated to the role of *dux* or *princeps* were highly sensitive to rank, proud of their ancestry and possessed of a keen sense of their place in the hierarchy. In an Irish milieu such relations were regarded as either 'honourable' or 'dishonourable'; sometimes the distinction must have been very fine.

Alternatively, tributary kings might shake off their former allegiance and attempt to establish or re-establish their independence, as Kentish kings were able to do in the late 770s; as Offa himself was able to do in the face of West Saxon overlordship in the early years of his reign. Sometimes, battle must decide; but warfare was a high-risk venture: the loser humiliated, perhaps wounded or killed, the flower of his people's warrior élite cut down with him or left vulnerable to internal plotting. The rule of many kingdoms, it is evident, was shared by more than one king, either from historical tradition or through division among siblings. Blood feud, diplomatic misunderstanding, bitter personal hatred and the traitorous intent of consorts and hostages might tip the delicate balance of *imperium* at any time; famine, ice, plague, drought and the other capricious fates, against which no man or woman was immune, might alter the trajectory of history.

The Welsh Annals record that in 778 a Mercian army 'devastated the South Britons' – that is, the Welsh: a more rational target, seemingly, than the *Hastingas* of East Sussex. Frictions between Welsh kingdoms and their Mercian and West Saxon antagonists are a constant backdrop to the eighth century. King Cynewulf of Wessex (reigned 757–86) 'frequently fought great battles against the Welsh'[29] and the sparse record of Mercian military intervention there is surely an accident of historical survival rather than a reflection of undisturbed entente.

Inter-tribal tensions were also played out along the line where territorial control was disputed between Mercia and Wessex: the Thames corridor. King Æðelbald had numbered the new King Cynewulf among witnesses to one of his last charters in 757. After Æðelbald's murder the balance of *imperium* was tilted the other way: charters issued by King Cynewulf in the 760s bear King Offa's name as a witness, indicating that in his early years the young Mercian king lay under tribute to Wessex.* At some time, probably in the 760s, Cynewulf was able to assert West Saxon control over Berkshire, which had periodically lain under Mercian rule.[30] But by 772 Cynewulf was present at the Mercian court as a charter witness,† a stark reversal of fortunes.

A year after King Offa's attack on Wales in 778, Mercian and West Saxon armies met at Benson, on the Thames near an important river crossing at Wallingford in Oxfordshire. The bland account in the *Chronicle*, which records that King Offa 'took the village', underplays its significance: for with control of such a strategic site,

---

* Electronic Sawyer S265, for example, a gift of land to St Peter's Minster at Bath.
† See above: Electronic Sawyer S108.

just downstream of the contested episcopal see of Dorchester-on-Thames, came territorial rights over a swathe of land along the middle Thames Valley, replete with lucrative productive sites and riverside minsters strung out along the river towards London.

An ongoing legal conflict over possessions on the Thames brings these tensions into sharp focus. The records of a synod held at the unidentified *Clofesho* in 798 follow the resolution of a long-running dispute over the Thames-side monastery and trading site at *Cocham* – Cookham, in Berkshire.[31]* Its eighth-century fortunes tell a remarkable story of theft, of royal chicanery and dynastic politics; and they introduce the figure of Offa's queen Cyneðryð. According to the *Clofesho* account, King Æðelbald had given Cookham to Christ Church, Canterbury. But, after Archbishop Cuthbert's death in 760, two of his pupils, named Dægheah and Osbert, stole the deeds and improbably delivered them to King Cynewulf in Wessex. The king from then on treated the monastery as his own. Cookham, we know from recent excavations, was part monastery, part trading and production settlement – a glittering and profitable jewel on the precious silver necklace of the Thames and a highly attractive royal or archiepiscopal asset. This blatant act of theft had been the subject of complaints by subsequent archbishops; but their pleas for restitution had gone unanswered. Then, King Offa...

> ... seized from King Cynewulf the often-mentioned monastery, *Cocham*, and many other towns [*alias urbes quamplurimas*] and brought them under Mercian rule.[32]

The likeliest date for the original theft is in the early 760s, when Cynewulf was able to assert West Saxon power over a weak King Offa along the Thames corridor. Sometime later in that decade, Offa gained the upper hand and was able to seize Cookham for himself. King Cynewulf was subsequently persuaded, through

* See above in Chapter Six, 6, p. 165 and below, Chapter 10, p. 343.

'tardy penitence' and the threat of 'anathema', to return the deeds to Canterbury, along with a great sum of money in compensation for the original theft. Nevertheless, Offa retained *de facto* possession of Cookham and its revenues and subsequently bequeathed it to Queen Cyneðryð who, at the time of the synod in 798, was still in possession as its abbess. The archbishop now agreed formally to assign her the deeds of the monastery in return for 110 hides of land in Kent, 60 hides at Fleet in Hampshire, 30 more hides at *Teneham* (an unknown location) and 20 hides at the source of the River Cray in West Kent – a high price that reflects Cookham's economic and symbolic importance to the Mercian royal family. But the deal also highlights the staggering landed wealth that Offa's queen was able to trade in this mutually advantageous end to a long-running dispute. The then archbishop, Æðelheard, also granted Cyneðryð hereditary possession of a monastery at *Pectanege*, not certainly identified but perhaps Patney on the Wiltshire Avon.

Cyneðryð's career, and her historical visibility, mark a step change in Mercian dynastic strategy – a break from three successive kings' reluctance to marry, provide heirs or deploy royal kin as diplomatic assets. Offa's policies were much more orthodox in these respects, if radical in others. Following contemporary thinking on the continent, he regarded dynastic legitimacy as an imperative; and that meant ecclesiastical legitimacy. If he took lovers, especially from among those noblewomen who had taken the veil, he was discreet about it.

Queen Cyneðryð was the first Anglo-Saxon woman to mint her own coinage, her stylish silver pennies graced with a portrait that looks remarkably like that on the coins of the contemporary, if distant, Empress Irene of Byzantium.[33] Both were inspired, it seems, by Roman coins bearing the image of Faustina, wife of Antoninus Pius, or those bearing the portrait of Helena, mother of the Emperor Constantine.[34] Cyneðryð also appears as a senior witness to Offa's charters, apparently fully integrated into the machinery of Mercian royal government.[35] In this she is virtually unique among Anglo-Saxon consorts before the tenth century. The first charter in which she appears as Cyneðryð *regina*

*Merciorum* in 770 records a gift of land by Uhtred, *subregulus* of the *Hwicce*, to one of his ministers, confirmed by her husband.* She must, obviously, have married Offa by then; their son, Ecgfrið, was installed as his father's successor by 787 and this is unlikely to have happened before his fifteenth birthday at the earliest.[36] The couple also raised four daughters, three of whom married, or were at least betrothed to, scions of other Anglo-Saxon royal houses – with varying fortunes.

Comparisons inevitably arise between the very real Cyneðryð and the powerful royal consorts in the *Beowulf* poem, which may have enjoyed a period of popularity in the royal Mercian courts of the eighth century; may, indeed, have been written down and read at Offa's court.† Like *Widsith*, *Beowulf* contains references to the ancestral hero of Angeln bearing the name Offa – King Offa's own putative ancestor. The *Beowulf* Offa[37] was 'the best of all mankind', a brave spear-wielder and wise governor whose fame crossed the seas:

> *Forðam Offa wæs geofum ond guðum, garcene man,*
> *wide geweorðof, wisdome heold eðel sinne.*[38]

Ringing tales of the greatness of the heroes of old entertained and flattered the Mercian court of Offa's day, for whose *scops* praise of one's patrons was a stock-in-trade. But the *Beowulf* poet offered much more equivocal praise of the legendary Offa's queen. His consort Ðryða had borne a reputation for cruelty and murder before her marriage to the strong and virtuous Offa, after which she became a good and loyal wife. The more celebrated consorts

---

* Electronic Sawyer S59. The addition, as subscribers, of their children Ecgfrið and Ælfflæd, who can only have been infants, is either a diplomatic courtesy or a later interpolation, unless Offa's marriage can be back-dated to the very beginning of, or even before, the start of his reign in 757. The latter seems very unlikely, given that Ælfflæd was married to King Æðelred of Northumbria as late as 792.

† For the argument that the epic was first written down, in more or less the form that we have it, at the court of King Offa, see Whitelock 1958.

in *Beowulf* are by turns peace-weavers, gift givers, bearers of the mead cup and inciters of war: Freware, Wealðeow and Hygd.[39] The real Cyneðryð enjoyed a long and influential career as a royal consort. The poetic imputation against her namesake may have been the poet's scurrilous fancy, since Ðryða's story seems to belong to a stock of conventional queenly tropes that circulated in European mythology across the centuries. Cyneðryð's husband gave her power and wealth and she, in turn, provided both a political counterweight in the royal council and the dynastic assets that should have ensured generations of Offings at the centre of Mercian power.

Queen Cyneðryð's background is nowhere made explicit but her name has a West Saxon flavour, like those of her seventh-century predecessors as Mercian consorts and princesses: Cynewise, Cyneburh and Cyneswið.* If King Cynewulf was her father, or brother, it would be consistent with a marriage date around or after 769, by which time Offa had shaken off West Saxon overlordship and asserted his own dominant position among the southern kingdoms – for by the ineluctable rules of Anglo-Saxon diplomacy the gift of a daughter was a token of submission.† The *Beowulf* poet's jibe at Ðryða's expense may have been aimed at her West Saxon background.

Offa's military and political dominance over the South Saxons may have been the lever by which he was able to assert authority over Cynewulf in the late 760s; but one suspects some other, invisible force at work. The fortunes of the community and production site at Cookham, so pointedly lying on the traditional Thames frontier, acquire a heightened significance in this context, for Cynewulf had stolen King Æðelbald's gift to Canterbury;

---

* Pauline Stafford, who has made a detailed study of Cyneðryð, believes they were both members of the Mercian nobility. Stafford 2001, 36; I have argued above in Chapter 2 that Cynewise, Penda's queen, was of West Saxon origin.
† Overlords also married their daughters to subject kings as a means of cementing alliances. Diplomatic niceties must have determined the terms and protocols; and failure to observe them might end in trouble, as Offa would discover himself when dealing with Charlemagne. See Chapter 9.

Offa had taken it from him by force and, it seems, given it to his queen – likely Cynewulf's sister or daughter – while the deeds still lay in the possession of the West Saxon king. Did Offa seize it after his victory at Benson in 779, as a great prize, or was Cookham part of a marriage deal – the dowry – in 769–70, with Cynewulf keeping the deeds as surety? The latter has a satisfying political logic to it, even if Canterbury's lawyers would later paint it as naked theft.

Offa's political elevation of his queen to such an exalted and visible status is a mark of his now-immense political capital and the radical nature of his thinking about royal authority. The administration of his possessions, trading settlements, diplomatic relations and finances required a sophisticated apparatus of both literate clerics and capable ministers and in these he involved his whole family: his queen, sister, daughters and son.[40] Various of them are signatories to his charters, especially later in his reign. One of these, dated to 787,* confirms privileges to the minster at Chertsey on the Thames, subscribed thus:

> *Cynedritha regina mea et pro Egfrido filio meo et filiabus meis Ethelburge* [Æðelburh] *Abbatisse et Æthelflede* [Ælfflæd] *et Edburge* [Eadburh] *et Æðelswithe.*

Cookham was by no means the only minster in which Offa and the women of his household enjoyed interests. A privilege of Pope Hadrian I (772–95) records the possession of many monasteries by Offa and Cyneðryð, all in the name of St Peter; but the text of that document does not list them.[†41]

Offa's charters open a fascinating window onto the land's resources and the ways in which it was exploited. One of those witnessed by Queen Cyneðryð, gifting land to a minister,

---

* Electronic Sawyer S127. Some commentators, while accepting the basic authenticity of the charter, judge that it has been copied and interfered with.
† But see below in Chapter 9, p. 310.

Ealdbeorht,* and his sister, Abbess Seleðryð, describes land at Palmstead in Kent and its appurtenances,† which included meadows, rights to woodland in *Andredesweald* for pasturing swine, 100 loads of wood in Buckholt and cartloads of wood for fuel, to 'cook' salt or as building materials.[42]

Rights to the fruits of woods and wood pastures held by or donated to sometimes distant communities were highly valued assets – for fuel, charcoal and construction timber, but also for fodder and for pasturing pigs in autumn to fatten them for slaughter – the practice called pannage. King Offa's loyal and hard-working ministers expected such rewards; and historians use the evidence of their holdings to construct complex relationships between widely dispersed land holdings that provided the full suite of resources from which to manage extensive estates.[43] Possession of settlements in favourable trading locations on navigable rivers or coasts; rights to the remission of tolls on ships; privileges to extract salt, iron, lead, fuel and, perhaps above all, access to summer pasturing for livestock, made up the portfolios from which they might become wealthy.

In previous centuries, lords had to live off the estates – the contiguous *vills* and shires – of their ancestors. What one might call the squirearchy of the seventh century, in which a very large number of minor lords possessed lands that their fathers had enjoyed for a life interest, morphed in the eighth century into an acquisitive élite of greater lords holding extensive portfolios and enjoying a spectrum of natural and economic resources – much more like the picture painted at the time of the Domesday Survey of the late eleventh century. In turn, the patronage they received from the king allowed him to extend Mercian royal influence and power, tentacle-like, from his own core territories into all those lands over which he was able to exert *imperium*.

Judging by his appearances in so many charters as a senior

---

* Witness to eight charters between about 786 and 805.
† In the context of bookland, 'appurtenances' means dependent settlements – that is, the component *vills* of an estate.

witness, King Offa's most powerful minister – his *patricius* – was a man called Brorda, whose career can be traced fitfully between 764 and the notice of his death in the *Historia Regum* thirty-five years later. Ironically, he nowhere appears in an Offan charter as the recipient of a grant: his holdings must have lain in the old core lands of Mercia, for which charter records are non-existent. Like the king himself, Brorda's career appears only in silhouette: he witnessed forty-six charters, either with the king as the principal sponsor or on behalf of Uhtred, *subregulus* of the *Hwicce*.

At the Synod of Gumley in 749 King Æðelbald had irrevocably asserted the state's right to impose the burden of public works on land holdings. Military service, the repair of bridges and the construction and maintenance of fortifications – those enigmatic *burhs* – were imposed universally, in theory at least. King Offa, while parcelling out estates to his favoured *comitates*, grasped the potential of the common burdens as a powerful tool of royal power. Building on his predecessor's establishment of burghal clusters at strategic sites, he expanded and reinforced it, creating a pan-Mercian network from which power could be exercised at a regional level. Somewhere, in this web of control and patronage, is to be found the geography of men like Brorda.

Half-way between Lincoln and Leicester, both seats of Mercian bishops, the Fosse Way runs close to the south-east river terrace of the Trent Valley for some dozen miles in south-east Nottinghamshire. Two Roman towns straddled the Fosse Way here and both were sites of important river crossings by ferry, causeway or bridge from the Roman period onwards. Newark,* which may have been among the possessions granted to *Medeshamstede* at its seventh-century founding, was the site of a small town known in the Roman period as *Ad Pontem*: 'At the Bridge'.[44] Ten miles to the south-west along the Fosse Way stood the Roman settlement of *Margidunum*, close to the medieval Trent crossing at East Bridgford. During the second

---

* Literally 'New building' in Old English: *nūwe (ge) weorc*.

century it had been provided by Roman civic engineers with earthwork ramparts; then, later, with a wall.[45] A tell-tale cluster of surrounding place names identified by John Blair marks *Margidunum* as the site of a Mercian *burh*.[46] *Margidunum's* early twentieth-century excavator, Felix Oswald, suggested that the town was an important site for trans-shipping lead out of the Peak District in the lands of the *Pecsætan* – the Peak dwellers. Across the river to the north lies Burton Joyce, where a polygonal earthwork enclosure at Lodge Farm betrays the sort of structure provided for by the provisions of the 'common burdens'. There is also a Carlton close by, and two places called Toot Hill are clues to look-out points on the approaches to *Burh* – the name given to *Margidunum* in later centuries.* A hill named Spellow, which means 'Speech mound', lying two miles or so south-west of *Margidunum*, indicates an assembly site.[47] So *Margidunum* bears striking similarities to the cluster of names and sites around Great Glen in Leicestershire that include the site of the important council held at Gumley in 749.

Merchants rowing or punting their barges down the broad plain of the swift-flowing Trent, past another crossing near Newark and then beneath the timber bridge at Cromwell, would come in time to another Roman town, named *Segelocum*, where the Roman road from Doncaster crossed the river heading for the great *colonia* at Lincoln. Littleborough, as it later became, may have been known in the seventh century as *Tiowulfingacaester*, named by Bede as the site of a church where Paulinus, having first converted *Blæcca*, the reeve of Lincoln, baptised the people of Lindsey.[48] Nearby are Gate Burton,† West Burton and two Sturtons (*Stret-tūns*). Here, as at *Margidunum*, the imperative of military strategy in securing major routeways and crossroads was combined with control of trade, trans-shipment and

---

* It also bore the name *Eald (ge) weorc* – 'Old construction', according to Blair.
† Also known, in the medieval period, as *Theothinga*, from *Þēod-Þing* – an assembly place. Blair 2018, 206.

toll-extraction points, to assert and exploit royal authority. These places are likely also to have played a key role in providing accommodation and a change of horses for the king's messengers. Such *burhs* must, one imagines, have been maintained under the control, perhaps the very lucrative control, of senior *ministri* in the Mercian royal household. It would be highly gratifying to be able to associate such places with individuals like Brorda, Berhtwald, Esne, Eadbald and Ceolmund: the king's loyal men, trusted with responsibility and rewarded, like satraps, with the fruits of prestige and revenue.

As it happens, the first surviving charter to hint at the process of establishing *burhs* was confirmed at Gumley in 779, the year in which King Offa fought King Cynewulf of Wessex at Benson.[49] Here, in the company of five of his bishops and his senior councillors – including Brorda, Berhtwald, Esne and Eanberht, the Hwiccian *subregulus* – the king granted his minster, Dudda, four hides of land on either side of the River Windrush at *Sulmonnes burg*: Salmonsbury in Gloucestershire. Salmonsbury, described as an *urbs* in the charter, is the site of a substantial, oddly rectangular late Iron Age hillfort on the edge of the village of Bourton-on-the-Water in the heart of the Cotswolds. From here, Ryknield Street led north to the Mercian heartland around Lichfield and Tamworth, intersecting with the Fosse Way and close to the line of three salt roads that led from Droitwich to the Thames and across to the Chilterns.[50] To the south-west on the Fosse Way lay Spelstow – another assembly site – and John Blair points to a number of associated enclosures and meeting places in the area that became prominent in later centuries.[51] The Bourton or *Burh-tūn* next to which Salmonsbury lies speaks for itself.

Dudda, almost the first case of a secular lord being given bookland* independent of any religious house, witnessed two other Offan charters, both in the lands of the *Hwicce*. In one of these, dated to the same year as the Salmonsbury grant,

---

* Effectively freehold, the provision of perpetual ownership formerly confined to monastic foundations.

he is described as *pincerna* – the king's butler – an important functionary of the royal household.[52] Holding the fort, so to speak, at the *burh,* Dudda benefitted from the resources of the river, no doubt extracted tolls on the king's behalf along the salt roads and at the crossroads of the Fosse Way and Ryknield Street and ensured that the regional authority of the king's writ ran in Mercia's south-west province. The grant of four hides was hereditary and John Blair suggests that it captures the precise moment in the eighth century when a *burh* and its accompanying *burh-tūn* were created, both as functional sites and as name types.[53] But it is unlikely to have been the first, if the establishment of the earliest *burhs* can be aligned with the establishment of the common burdens in the 740s.

Salmonsbury was held by a relatively junior minister like Dudda. *Margidunum* and Littleborough are good candidates for *burhs* that might have been held by senior king's men like Brorda along the strategically vital River Trent. But the Mercian system as conceived by its eighth-century kings required that trusted councillors held critical locations right across Greater Mercia. Where to look for the others? One might first address the thought that the five or six Mercian bishops who sat in council with the king – at Hereford for the *Magonsæte*, Worcester for the *Hwicce*, Lincoln for Lindsey, Leicester for the Middle Angles, London for the East Saxons and Lichfield for Mercia proper – held episcopal sees that in many respects resemble ready-made *burhs*. Leicester, Lincoln and Worcester had all been Roman towns with defences, visible and probably still functional in the eighth century, while Lichfield and Hereford seem to have assumed some of the functions of nearby Roman power centres. Bishops were substantial landowners with a great many resources at their disposal. They belonged to their regional élites and it seems unlikely that, on secular matters considered by the king's councillors, they offered no opinions; had no interests. We know that some of them enjoyed rights to remission from tolls and that they played an active part in commerce, perhaps issuing coinage. They were territorial lords as much as they were spiritual lords;

farmers and traders as well as fishers of men. But they owed their wealth and status to the king's patronage.

King Æðelbald had specifically, even painfully and at considerable political risk, refused to exempt church lands from the public burdens of bridge and defence work. At Hereford, excavations on narrow sections of the medieval defences have shown two phases of pre-conquest ramparts, the first of which, consisting of a gravel bank, dates to the late eighth century.[54] Hereford's immediate hinterland shows none of the characteristic *burh-tūn* or *Stret-tūn* names that would argue for a burghal cluster here but, given that Hereford was close to the site of a battle in 760 and, given the Offan attribution of the great dyke constructed some twenty miles to the west, the idea that bishops of Hereford were more than merely God's spiritual vicars among the Mercians and their satellites has much to commend it. A coin of Offa, and excavated evidence of grain drying and iron working, reinforces Hereford's credentials as a production centre. Although the cathedral lies at the centre of the city's defensive walls, the original minster of St Guðlac was located in the south-east corner of Hereford's defensive circuit. John Blair suggests that Offa (or his bishop) employed the site as a provisioning centre for the construction of his great dyke before relocating the church to the site of the later cathedral.[55] Hereford commanded a crossing of an important river, the Wye, which may have been navigable to this point in the Early Medieval period[56] and it lay close to a major Roman road connecting Gloucester with the heartlands of southern and central Wales. Not so far away from Hereford, at Marden, a rare Middle Saxon watermill has been excavated.[57]

Worcester is known to have been fortified by Ealdorman Æðelred and his consort Æðelflæd, King Ælfred's daughter, at the end of the ninth century,* and excavations have located the relevant period of its defences, which were revetted in stone and seem to have been surmounted by a timber rampart.[58] No earlier

* See Chapter 11.

defensive line has yet been identified beneath these defences, but there is a strong case for suggesting that the *Hwiccian* church and episcopal seat were constructed within existing Roman earthwork defences, a 'D'-shaped enclosure on the east bank of the River Severn, some fifteen acres in extent. The tenth-century Æðelredan-period defences expanded this enclosed area to the north. Across the river to the west lay a cluster of *wīcs* (in this case productive farms rather than trading emporia) that belonged to the minster, with the lucrative and strategically important Droitwich brine pits lying to the north-east on the Salwarpe, a tributary of the River Severn.[59] Worcester's control of a major Severn crossing and its siting on a western branch of Ryknield Street gave it an importance equivalent to Lincoln, Hereford, Leicester, London and Lichfield – key centres of Mercian ecclesiastical administration. If Worcester was not a *burh*, it was very like one.

Lincoln, too, has its *wīc*, the trading settlement of Wigford, across the River Witham and overlooked by the Roman *colonia* and minster on the hill to the north. There is also a tell-tale *burh-tūn* immediately to the north-west: Burton-by-Lincoln. Leicester, despite its having been more extensively excavated than any other Roman town in Britain, has yet to produce evidence for a refurbishment of its defences before the Norman Conquest; perhaps they remained in serviceable order into the eighth century. But with the cluster of Strettons and Burtons around Great Glen, and the assembly site at Gumley to the south-east, Leicester has its burghal hinterland and, as a result of its strategic location on the River Soar straddling the Fosse Way, Middle Anglia's bishops were likely key players in the Offan network.

London, it almost goes without saying, was a crucial Mercian asset. The *wīc* brought in revenue and controlled access from the Upper to the Lower Thames and vice-versa. It linked Mercia directly with the continent and provided a control centre for Mercian interests in Kent – not just secular control but also through powerful and wealthy religious houses like Minster-in-Thanet. London's bishops, their seat lying imperiously within the old Roman walls, enjoyed remission from tolls on their ships, reflecting both a privileged ecclesiastical status and, perhaps, a role alongside the *wīcgerefa* – the king's own functionary at *Lundenwic* – in the lordship of the former Roman capital. The old city, having lain more or less empty in the immediate centuries after Roman withdrawal, would become *Lundenburh* under King Ælfred. Royal *vills* located upstream at Chelsea and Brentford were frequently visited by the royal household, judging by the number of charters promulgated at those sites in the second half of Offa's reign.

It is increasingly clear that Mercian royal authority was focused on key regional cores and that these were both heartlands of former satellites – *Hwicce*, *Magonsæte*, Lindsey, Middle Anglia and the lands of the Middle and East Saxons – and the locations of Mercia's bishops. King Offa had personal affiliations in *Hwicce* but increasingly, also, in the old Mercian territories around Lichfield, Tamworth and at Repton, with its royal mausoleum where his immediate predecessor was buried. Mercian assets were strung out along the Roman road network, especially on Watling Street, Ryknield Street and the Fosse Way and along arterial rivers: the Trent, Thames and Severn. But even with the insights provided by recognition of the eighth-century Mercian *burhs* and their complex geography, locating the regional power of Offa's governing class has so far proved beyond reach.

I find my attention frequently drawn to another cluster of Mercian interests that does not seem to fit with the most obvious royal or episcopal interests of Offa's day. From the royal monastic foundation at Weedon, at the head of the River Nene where it crosses Watling Street, downstream as far as *Medeshamstede* (Peterborough) and on into the Fens around Crowland, runs a string of high-status sites. Four miles downstream from Weedon, Kislingbury is a *burh* site of unknown foundation date sited on the gravels of the river valley – perhaps at an important crossing point and possibly a successor to the Iron Age hillfort just downstream at Hunsbury. A Roman town located at what is now the western Northampton suburb of Duston lies along a Watling Street spur from *Bannaventa* (Whilton Lodge). That road probably crossed the Nene just upstream of a significant tributary opposite Hunsbury and continued east-south-east towards the River Great Ouse at Bedford. Less than a mile south of Duston a single Anglo-Saxon sunken-featured building, excavated in the 1960s in advance of road construction, hints at a much more substantial, high-status settlement in the area. A weaving room complete with box-seat benches and rows of loom weights, its only know counterpart is an almost identical, if smaller, structure at the royal Northumbrian township of Yeavering.[60] In Northampton itself, immediately adjacent to the site of St Peter's church, stood one of the largest timber building complexes of the period. And Northampton lay in the ancient hundred of Spelhoe – a 'speech mound' name.[*] The extraordinary basilican church at Brixworth – often touted as the site of the lost *Clofesho* – stood a few miles further north.

Downstream from Northampton – merely *Hamtun* in the earliest records – lay Wollaston, site of the 'Pioneer'[†] warrior burial, close to the Roman town of Irchester. On the opposite, north bank stands the town of Wellingborough, another (undated) *burh* site and, evocatively, two or three miles to the north-east, Finedon: Ðing-denu – 'Assembly-valley'.[61] Finedon lies close to the *burh* of Irthlingborough, a site known to have been visited by King Offa since he confirmed a charter here between 787 and 796.[62] The *burh* here lies conveniently close to Rockingham Forest, very likely a royal hunting ground even before the Norman Conquest. The *tūn* belonging to this *burh* can be identified with Burton Latimer, three miles to the north-west in the valley of the River Ise: the valley to which the *denu* in Finedon refers. Irthlingborough overlooks both the Nene to the immediate south-east and, on the opposite bank downstream from Irchester, a unique Middle Saxon settlement at Higham Ferrers. A little further downstream again lay the Middle to Late Saxon settlement of Raunds; then the Wilfridan riverside monastery at Oundle in the district of

---

[*] The speech mound in question, a hundredal and probably tribal assembly site, can be identified on the ground with Spellow Close in Weston Favell, now a suburb of Northampton.
[†] See above, Chapter 1, p. 25.

*Inundalum*. The remarkably dense cluster of monastic houses in the immediate hinterland of *Medeshamstede*, including Guðlac's hermitage at Crowland, ends the sequence.

The River Nene is Britain's tenth-longest river, at 105 miles. The upper reaches of this Wash-draining valley seem to have supported more than their fair share of élite Middle Saxon sites. Falling just 200 feet between Northampton and its outflow into the Wash, the Nene is navigable today as far upstream as Northampton* and it seems likely, although incapable of proof, that it was also navigable from the Roman period. The Nene Valley was a centre of the Romano-British pottery industry and the easily drained gravels and light soils of its flood plain were equally attractive for settlement in the centuries following the end of the Roman Empire.

A minster seems to have been founded at Northampton by the beginning of the eighth century, perhaps one of the many religious houses established by Wilfrid but unidentified by his hagiographer. Just as likely, perhaps, it was one of those royal foundations referred to in the privilege of Pope Hadrian I (772–95).† Northampton may have been only the latest incarnation of a trio of successive administrative centres, beginning in the Iron Age with Hunsbury hillfort, then the Roman town at Duston, before relocating to the Anglo-Saxon and medieval site.[63]

Two early churches, St Peter's and St Gregory's, stood on a bluff overlooking the Nene close to its confluence with the tributary known as the Brampton Nene, which flows in from the north.[64] Between the two, at about the turn of the eighth century, perhaps a little before, a very substantial timber hall was built – of double-square rectangular form with squarish annexes at either end and opposed central entrances. Its foundations lie within a few feet of the excavated remains of a prehistoric ring-ditch enclosure, suggesting that the site may have possessed ceremonial significance across the millennia. In scale and design the Northampton hall almost matches

---

* With the provision of locks to keep levels sufficiently high; but flash weirs are known from the pre-Conquest period.

† Referred to above on p. 233.

the great seventh-century structures at Yeavering and Sprouston in Northumberland, at Atcham near Wroxeter in Shropshire and one or two others – at the peak of Middle Anglo-Saxon architectural scale and pretension.[65] The social context would seem to be royal – indeed, the principal royal *vill* of a king – and yet, sandwiched between two early churches and devoid of any direct association with a known Mercian king, there is something to be said for the thought that Northampton's great hall belonged to a religious house.

The closest architectural parallel from the eighth century comes from a site at Brandon on the Little Ouse in the Breckland of the Suffolk–Norfolk border. Here a thriving riverside monastic and trading site, laid out on a carefully gridded axial plan, blurs the boundaries between enclosed religious community and secular production – a small encapsulation of the spectrum of eighth-century communities. Excavations at the settlement, its two cemeteries and church – a structure closely similar to the 'hall' at Northampton – yielded evidence for textile production, a *scriptorium* and the importation of exotic goods from the continent, including drinking glasses. From here, some 24,000 sherds of Ipswich Ware were recovered.[66] Had the missing 'hall' at the monastic-productive site of Flixborough on the Lincolnshire Trent been recovered to give structural context to the extraordinarily rich finds recovered from excavations there, it would have reinforced archaeologists' sense of an emerging coherence and consistency in such settlement architecture.[67]

Questions about the status of Northampton's grandiose building become even more intriguing in light of the fact that it was replaced directly, around the end of the eighth century, by an even larger, more substantial stone-built hall, unique in Mercia or any other Anglo-Saxon kingdom in the period. Neither is mentioned in any contemporary source and neither survived into the medieval period. The remains of several mortar mixers were found in the immediate vicinity. These are circular pits in which slaked lime and sand were mixed by means of a paddle driven around an upright pivot, a *gin*, by means of human or animal traction. They are a marker of very high-status royal and ecclesiastical sites

The south wall of the tower at All Saints', Brixworth. Mercian royal investment in great churches monumentalised its self-conscious grandiosity.

like Jarrow in Northumberland; a hint at ecclesiastical patronage. Unfortunately the material retrieved from excavations here is not illuminating: none of the *styli* or sculpture that would indicate a monastic function; none of the trading tell-tales – Ipswich Ware pottery or *sceattas* – redolent of a *sele* or trading hall. *

Archaeologists have, understandably, looked some six miles to the north to the parish church of All Saints', Brixworth – the finest standing Middle Anglo-Saxon church in England – and seen the two buildings as twin expressions of Mercian power and wealth. Brixworth is mysterious, unique: an enigma. All Saints' church stands on a spur of sandstone at a little under 400 feet above sea level. A small stream splits the village below in half, and the prominence of this cleft in the hill, together with the magnificence of the church, have led to a cautious identification of Brixworth with *Clofesho*,† although why this site should have been chosen as a suitable location for church councils by Archbishop Theodore, in the late seventh century, has not been satisfactorily answered.‡

The church here was constructed, like Northampton's stone hall, at the end of the eighth century: a great, solid basilica with arcaded aisles whose rounded arches were formed using recycled slim terracotta Roman bricks laid radially on edge – later filled in and now visible in the fabric of the exterior walls.[68] Above them, smaller, round-arched clerestory windows exaggerate the height of the nave to impressive effect. At the east end, a very rare ring crypt was built and at the west end stood five compartments, later modified to form the base of a tower with an external turret staircase. Looking down into the nave from the triptych of round-arched windows that pierce the tower is to sense another time and place – there is something evocatively Carolingian about the

---

* The ceramic specialist Paul Blinkhorn informs me that a few sherds of Ipswich Ware have now been recovered from sites in Northampton, increasing the likelihood of direct trading links downstream with the Wash ports.

† Old English *clof, clofa*, 'a fold, a cleft [...] used of a valley' and *hoh*, 'a heel or spur of land'. Keynes 1993,14; Parsons 1977.

‡ See below, Chapter 10, p. 339.

Brixworth's round-arched arcades, fabricated from re-used Roman tiles, are deliberate nods to an imperial past, the materials scavenged from local towns or villas.

whole conception, with the altar nestling in the apse beyond a triptych of round-headed arches, again decorated with terracotta Roman tiles set on edge. The plastered and whitewashed walls bring out the rich texture of the arch decoration.

For clues to the milieu in which Brixworth was conceived historians have speculated on the nature of the relics for which the crypt was probably built, even though little remains of its original form after a major reconstruction in the nineteenth century. David Parsons, noting that there was once a three-day festival held here to mark the feast day of St Boniface, has suggested that some of his relics may have been brought here. Boniface was murdered in Frisia in 754 and his only known connection to the Mercian royal house was his unflinching moral assault on King Æðelbald. It seems unlikely that the church was erected to venerate him, unless in expiation of that king's sins.

Questions have also been asked of the recycled Roman material in the walls and arches – where did it come from? Seventh-century church builders often used *spolia*, stone and carved mouldings from Roman sites – crypts at Hexham and Ripon, both built by Wilfrid, are good examples. The nearest Roman town from which such materials might be scavenged for Brixworth is Duston, just outside Northampton; but a villa stood about half a mile to the north of the village.[69] The building stone for the bulk of the church fabric has been shown to have originated in quarries near Leicester, some 25 miles to the north.[70] Either way, the scale and evident expense of the church here strongly implies royal, possibly episcopal patronage: it is public architecture of the highest order.

Simon Keynes has emphasised the importance of the Middle Anglian bishops of Leicester as charter witnesses and attendees at church councils. Brixworth may have been a bishop's showpiece, perhaps even the core of an episcopal palace complex. Excavations to the west of the church have revealed the presence of an impressive ditch, possibly part of a large enclosure in which the church sat. John Blair wonders if the church was the set-piece centre of an imperial-style minster estate, inspired by the burgeoning connections between Offa's court and that of the Frankish King Charlemagne at Aachen.[71] It is equally striking that a natural line of communication joining Northampton and the Middle Anglian diocesan centre at Leicester runs through Brixworth, the assembly site at Gumley and the burghal centre at Great Glen: links in a chain of royal and ecclesiastical power. There is a suggestion that Brixworth belongs to a group of eighth-century minsters sitting inside defensible enclosures – John Blair has identified examples at Bampton in Oxfordshire and at Stamford; Bath may have been another. The sharp academic distinction between minsters and secular *burhs* is, as Blair argues, overstated.[72]

Archaeology has delivered more tractable results further down the Nene Valley. Excavations between 1994 and 2001 at Higham Ferrers, a small town sitting on a limestone ridge overlooking the river terrace from the south, revealed a substantial eighth-century settlement: a large, keyhole-shaped enclosure redolent of the great corral at Yeavering. Inside, an organised cluster of large timber buildings provide evidence for grain processing and livestock management and indicate productive relations beyond the region.[73] Cattle, sheep or goat, pig and chicken bones were found here in abundance, and the ages at which they were slaughtered indicate that while some were reared expressly for their meat, others were kept for their 'secondary products': for milk, butter and cheese and eggs, for their wool and as traction animals. The range of buildings shows a capacity to store large amounts of the estate's surplus – the sorts of products listed in the surviving description of an Offan food render. A malting oven, essential in the preparation of ale, was found almost intact, complete with stone-lined pit and flue and charred barley grains.[74] Not just barley, but wheat, oats and rye grains were also retrieved from ditch and pit fills on the site, alongside the sort of vegetable staples recognisable to any modern gardener: peas and beans.

Leafing through the pages of a necessarily technical excavation report, one's eye is caught by Paul Blinkhorn's pottery study, in which he describes the assemblage of Ipswich Ware pottery from the site as 'one of the largest known from the inland areas of the South-east Midlands'.[75] Ipswich Ware, coil-built but finished on a turntable, was the first post-Roman industrially made pottery from Anglo-Saxon England and a tell-tale of regional trade – the footprint, so to speak, of its now-missing contents, carried by merchants along Eastern England's routeways. Its presence here in such large quantities raises the likelihood that Higham Ferrers was both producing and importing goods along the river during the mid- to late eighth century. By conspicuous contrast Northampton, at the same period and a mere dozen or so miles upstream, was not seeing Ipswich Ware in quantity, although that picture may yet change. The other key artefact assemblage that raises eyebrows here is of Niedermendig lava fragments, a highly characteristic continental import used in making high-quality quern stones for grinding corn.[76] Higham Ferrers was well-connected.

That Higham Ferrers was a farming estate established to support the *burh* across the river at Irthlingborough is an obvious conclusion to draw. What is much less clear is the economic, social and political milieu that governed relations along the length of the Nene from Northampton down to *Medeshamstede* and beyond. Oundle had its Wilfridan house. Northampton, Brixworth and Weedon seem to have enjoyed both monastic and royal associations and there is a plausible connection with the Middle Anglian episcopal seat at Leicester.

The seventh-century monastic foundation at *Medeshamstede* and its satellites at Castor, Peakirk and Crowland was wealthy and powerful and had enjoyed royal patronage since its inception in the days of Peada and Oswiu. *Medeshamstede* retained more than merely local significance during Offa's reign. Two abbots, Botwine (764–89) and Beonna (789–805/7) were frequent charter witnesses in the second half of the eighth century. In a charter dating between 786 and 796 Abbot Beonna is the donor, granting a lease of ten hides of land to a Lindsey *princeps,* Cuthbert, in return for 100 shillings and annual hospitality.[77] These were men who could, and did, sit as senior members of the king's 'cabinet' – literate bureaucrats and independent territorial lords as well as spiritual and political advisers. That *Medeshamstede's* influence might extend upstream as far as Higham Ferrers, with

Queen Cyneðryð: the first Englishwoman whose face appears on royal coinage. She played a key role in forging Mercian royal dynastic power.

its river-borne Ipswich Ware, is conceivable. Northampton may have lain beyond its political reach, while a possible demarcation point between the lower and upper Nene might have existed at Thrapston, half-way between Oundle and Higham Ferrers. I am told that downstream of this point the river name is pronounced 'Neen'; upstream it is 'Nenn'.

With the Thames, the Severn-and-Avon and the Trent, the Nene Valley seems to form a core regional powerhouse of patronage, investment, wealth and trade. Is this the heart of Middle Anglia; and, if so, was it ruled directly under a Mercian royal prerogative; was it, like the lands of *Hwicce*, the great lordship of bishops and former petty kings; or a fief of the abbots of *Medeshamstede;* or of someone of the status and stamp of Brorda, the *patricius*? One might even speculate whether King Offa, like the Mercian kings of old, had set his son Ecgfrið up as the king-in-waiting in the south-east Midlands. If nothing else, Ecgfrið's all-too-brief career as a historical footnote to his father's *imperium* would explain why his name has not been stamped indelibly on this Outer Mercian territory.

The *burhs* and their *tūns*, the pendulum swing of periodic *imperium* over the southern kingdoms, the wealth and pretension of public works and great religious houses speak of an active, progressive royal administration – of a capable, imaginative and literate élite exploring and then developing new ideas of statehood and cultural confidence. If it has taken historians, historical geographers and archaeologists until now to fully appreciate the scale and ambition of the Offan project it is, paradoxically, partly because of its more obvious expressions: an outstanding tradition of literacy and sculpture; an overtly imperial style of coinage and, not least, the massive earthwork that bears his name. These achievements were built on solid statehood and the stability of eighth-century Mercian kingship and they allowed Offa to project an unprecedented image of Anglo-Saxon royal power at its height.

# Chapter 8

# Offa: The face of power
# 780–787

❖

*Offa's portrait coins • The Dyke • Landscapes of
the Marches • The art of imperium •
Lichfield and Canterbury*

In about 780, perhaps in the aftermath of Mercia's successful campaign against the West Saxons at Benson, moneyers began to strike a series of broad, thin, silver pennies portraying busts of *OFFA REX* in profile. In some portraits the king wears a diadem over close-cropped hair, like the mounted warrior on the Repton stone; in others his hair is overtly, almost ludicrously curly. The niceties of such nods to imperial Roman precedent – the diadem associated with victory and regality; the curled hair with supernatural or divine status[1] – may have been lost on the Mercian populace, but the royal authority that underpinned the economics and iconography of his coinage was not. And when the king was shown with his eyes looking upwards as to a vision, with a symbol that might be a sword or cross, or both, floating before him, the more educated of his subjects – the clergy, his ealdormen and their consorts, merchants and rival kings – could be in no doubt of the king's motivations. Offa wished himself to be compared to illustrious forbears: to Emperor Constantine, who had seen a vision of the cross before his great victory at the Milvian Bridge in 313; to King David of the Old Testament, a potent model of righteous Christian kingship; perhaps also to a more recent royal exemplar, King Oswald, whose own vision before a victorious battle in 634 had evinced divine favour.

The Offa portrait coins were issued, appropriately, from a mint in London, capital of the Roman province of *Britannia*. Offa's innovation of the same period, granting Queen Cyneðryð the right to mint coins in her name, may also have a special significance. The numismatist Gareth Williams suggests that the

The face of power: King Offa's outstanding coinage projected his image as overlord in the imperial tradition of Roman emperors.

small issues bearing her portrait, minted by a single Canterbury moneyer named Eoba, may be the surviving material expression of the gift of a 'licence to mint', intended to provide her with an independent income, perhaps alongside her potentially lucrative lordship of Cookham.[2] King Offa's projection of power thus extended to the dynastic status of his close family, co-opted partners in Mercian royal authority.

For most Mercians and for inhabitants of its subject kingdoms, royal power was remote and theoretical, manifested immaterially through anecdote and family tradition. The liquid magic of pure silver, elementally transformed from the living ore and stamped with the emphatic and authoritative die of the moneyer, rendered the fleeting existence of human life – even that of a mortal king – eternal. Like stone buildings, sculpture, fine metalwork and the written 'booking' of charters backed by a sod of very real earth, Offa's silver coins, with their Latin inscriptions and portraits, materialised and 'petrified' that which had formerly been fluid: the negotiable value of bullion and scrap metals; oral history and custom; organic media like wood and textiles; transient, personified kingship.[3] The king had by now ruled over a whole generation of Mercians, as his long-lived predecessor Æðelbald had. Reviving the iconography of imperial Rome had been a tool of Anglo-Saxon kings since the days of King Edwin in seventh-century Northumbria – Edwin, who had progressed through his lands preceded by an officer bearing a *tufa* or imperial standard; whose *mund* or royal authority had stretched from sea to sea. The authority of King Offa's bishops and the immutability of his land gifts also came, ultimately, from Rome.

Power projected through coinage carrying the king's name was loaded with more than just symbolism. As every economist knows, the value of a currency is based on trust – and for silver coinage that means consistency in weight and precious metal content. Coinage undermined by debasement – a form of inflation – or by a reduction in weight, compromised the confidence with which merchants might trade regionally and between kingdoms; and that also compromised the ultimate authority of the king. Kings and moneyers might profit from 'recoining' – drawing in as many coins as they could through tolls and taxes, then reminting them either in smaller numbers (raising their value) or, more often, with less silver: devaluing them, until the balance of trust was lost and the coinage became effectively worthless. For numismatists, weight and silver content are key markers of a king's authority and of economic health.

The eighth-century Mercian economy was by no means market-driven: the concentration of coins in the hinterlands of the trading settlements, along major river routes and at 'productive' sites shows that most people's economic lives were still circumscribed by the taxes in kind rendered from their fields and pastures, rather than the amount of cash in their purses. Coinage operated in the borderland between customary gift exchange and hard-nosed trade. Those gifts and that trade were overwhelmingly dominated by an élite whose collective power mediated that of the king.

In King Æðelbald's day Mercian coins had carried neither the king's name nor a portrait (there are no ÆÐELBALD REX coins), an indication that, if he took an interest in coinage, that interest was pecuniary rather than as a projection of royal authority. The economic boom of the early decades of the eighth century had been followed by shortages in the supply of silver bullion – a continent-wide problem. Almost all silver was recycled from previous coin issues and, as economic confidence declined and trade shrank, so coins with higher silver content tended to be hoarded, ceasing to be actively available as currency; in turn, new coin issues were progressively debased until a point in the 740s where *sceattas* were minted substantially with non-precious metals.[4] King Eadberht of Northumbria (reigned 737–58) seems to have initiated attempts to stop the rot in that decade, enforcing a controlled weight standard, excluding the use of all foreign coins and, literally, stamping them with his authority as king, *EOTBERTUS*, with a stylised and fantastical beast on the reverse. Where Eadberht led, East Anglian and Kentish kings followed in the same decade.

On the continent in the 750s Pippin, increasingly expanding his territorial and political authority as king of the Franks, issued a *novus denarius* at a fixed weight of 1.30g. These thin but broad pennies with a high silver content offered more scope for the art of the die-cutter to experiment and express himself, with an abbreviated royal title on the obverse and carrying the name of the mint on the reverse. They maintained a consistent purity and seem, by the early 760s, to have been the inspiration behind King Offa's first issues bearing the abbreviation *OFRM* (*Offa Rex Merciorum*) with the name of his moneyer, Mang, on the reverse.

Moneyers based in *Gipeswic*, London and Canterbury were emphatically not royal functionaries in the late eighth-century Anglo-Saxon kingdoms. Their names – Lul in East Anglia, Ibba in London, Eoba and Mang in Canterbury among others, tantalisingly expand historians' knowledge of individual careers outwith the church or the immediate households of kings. Most likely belonging to an élite merchant class able to pull the levers of both regional and international trade, they enjoyed a high status and political independence; were able to strike coins for more than one king and in a range of designs – often novel and exuberantly artistic – militating against the idea that they lay under strict royal authority. Nevertheless, Mercian kings sought increasingly to control the flow and quality of the currency and to influence its designs as expressions of that control. London was Offa's town: its trading profits and privileges were substantially his, to fund pan-Mercian projects or to give as reward or incentive for services rendered. He may have begun to see its moneyers as *his* moneyers.

When King Offa's coinage was first studied in detail by numismatic scholars in the 1960s, a mere sixty-odd of his coins were known. By the early 2000s, excavation and the finds reported to the Portable Antiquities Scheme had vastly increased the known coinage to more than 600 specimens, with perhaps a dozen new coins emerging every year. Each new coin, with its subtle variations in wear, its tell-tale die marks, alternative designs and provenance, increases the potential for the outstanding work of numismatic scholars to interrogate Offan coinage, revealing its economic and social reach; its expression of his agency. Numismatists can now say with confidence that these 600-odd coins must represent a cumulative output, between about 760 and the end of Offa's reign, in the many millions of silver pennies, produced by about thirty moneyers at three southern mints.[5] Just two surviving gold coins issued by Offa, probably worth the equivalent of thirty silver pennies each and referred to in a charter of 799 as mancuses, may represent a much larger number of gold coins intended to serve within a rarefied, élite economy of high ecclesiastics and members of his inner circle – perhaps also as diplomatic gifts or land bonds.[6]

*

In taking a wide view of the Offan project and the trajectory of the defining events of his reign from such few fragments as survive, it is tempting to point to the assertion of Mercian *imperium* over Wessex after 779 and, perhaps, to the lacuna in his relations with the Kentish kings from 776, and see a shift of interest towards his ambitions in the west. A hundred years and more after the event, King Ælfred's Welsh biographer, Bishop Asser, recorded that:

> There was in Mercia in fairly recent times a certain vigorous king called Offa, who terrified all the neighbouring kings and provinces around him, and who had a great dyke built between Wales and Mercia from sea to sea.[7]

In 780, after twenty-three years on the Mercian throne, King Offa cannot really be said to have terrified all his neighbours. He had gained, then lost, superiority over the kings of Kent; had won a significant strategic victory over the West Saxons south of the Thames and was able to exact tribute from the former kings of the South Saxons. He had continued his predecessor's policy of relegating the former kings of the *Hwicce* and *Magonsæte* to mere ealdormen. Middle Anglia had been entirely absorbed as part of the greater Mercian realm and East Saxon kings accepted Mercian overlordship. But the kingdoms of East Anglia and Northumbria retained their independence and Wessex remained a periodic threat to the security of Mercia's interests along and south of the Thames Valley.

Mercian relations with the kings of Powys and of the southern Welsh were periodically hostile, as the thin record of occasional raids and battles implies, and Mercian armies were sufficiently powerful to launch destructive campaigns either pre-emptively or in retaliation for Welsh incursions; but Offa had not subdued them; had not subjected their kings to the humiliation of trailing around in his itinerant household, attending his councils and witnessing his charters. There was, so far as one can tell, no patronage network of Mercian interests much beyond the River Wye. And later kings of Powys, erecting the Pillar of Eliseg, looked back on Offa's age as a period when they had reasserted their identity and resisted Mercian expansion. Even so, historians and archaeologists have, by and large, accepted Asser's testimony that King Offa caused a great monument to be built and that, in conception at least, it does span the more than 150 miles between the south side of the Dee estuary and the mouth of the River Wye near Chepstow.

When exactly the Dyke was built and why; how it was designed, surveyed and constructed; how it functioned; what it meant for those living in its shadow and how it fits into the narrative of King Offa's reign are much darker corners than the brightly illuminated politics of his coinage. As an outstanding engineering and military achievement, the Dyke is unique and, if Offa did not cause it to be made, there is no other candidate who might have bent such enormous resources to what may in the end have been something of a vanity project. Few doubt that it speaks, and loudly, of power and the desire for monumental legacy. As a symbolic if not physical barrier between peoples and cultures, it endures more than a thousand years later. But who can say whether, on his death bed, Offa might have reflected as Shakespeare's Earl of Warwick reflected...

> Why, what is pomp, rule, reign, but earth and dust?
> And, live we how we can, yet die we must.[8]

Earthy cenotaph or not, Offa's Dyke – *Clawydd Offa* in Welsh – is the most substantial monument to survive in Britain's landscape from the centuries before the draining of the Fens and the age of canals and railways. Like the celebrated silver pennies that bear portraits of King Offa or Queen Cyneðryð and the exuberant and prolific artistic manifestations of the age, the Dyke is a brightly fluttering banner celebrating the culture of royal might, framing the silhouette of its enigmatic sponsor.

The pioneering surveyor of the Dyke, Sir Cyril Fox (1882–1967), who studied, excavated and mapped its ditches and banks over a twenty-year period, called it a 'travelling earthwork'.[9] The modern tourist, even more so the dedicated walker, can still experience something of its breathtaking ambition, its sensitivity to the complex intricacies of folded hill, winding valley and broad plateau; its indefatigable pursuit of purpose. Above all, to trace the Dyke mile after mile and day after day through those borderlands is to follow in the footsteps of an enterprising, intelligent and sophisticated mind: visionary and ambitious, determined but adaptable.

The Dyke has kept its relevance to people either side of it – and on it – across the intervening centuries: as a frontier separating England and Wales; as a physical barrier to movement and as a symbol of identity: Cymry on one side, Angles on the other. In 1928 Fox, then deep into his great survey of the Dyke, asked a Montgomeryshire informant what the Dyke meant to the people of his community. The man replied:

> 'You put your head inside the back door of Bob Jones's cottage there; tell him he was born the wrong side of Offa's Dyke, and see what happens.'[10]

The Dyke, as it survives in the twenty-first century, is by no means a continuous, single, heterogeneous earthwork; it probably never was. Sometimes its line is no more than implied; sometimes it is substituted by a river – the Severn or the Wye – and it may never have physically reached the estuary of the Dee; there is no general agreement. If it defines territorial boundaries between, say, Powys and the *Wreocansæte* or Glywysing and the *Hwicce*, it does so imperfectly. Welsh-named settlements lie to the east of it, English-named settlements to the west. It was not 'fortified', in the sense that Hadrian's Wall was, with a system of forts, turrets and milecastles. Nowhere has the Dyke been shown to have been revetted with a wall or topped with a palisade – although one or the other may yet be discovered. There are no obvious monumental 'entrances' that would control military passage or act as toll points although they, too, may have been built and used, even if they cannot now be identified.

However, there is a slowly emerging consensus that the Dyke did not stand in isolation. Although there is no bespoke road to supply garrisons directly, as there is in the Stanegate for Hadrian's Wall, evidence for an accompanying infrastructure – a hinterland of *weards*, *burhs* and 'functional' *tūns* – is beginning to come into hard-won focus. No serious scholar now doubts the overarching hand of design, of the 'idea' of the Dyke, even if its purposes or functions were much more nuanced than simply that of a defensive frontier against a hostile enemy: the protective barrier of the Mercian state against its British antagonists.

The southernmost terminal of the Dyke is to be found at Sedbury Cliffs on the tidal reach of the Severn. Here it cuts off a promontory, Beachley Point, where the Severn meets the outflow of the River Wye at Chepstow in the Early Medieval kingdom of Gwent. Chepstow is *cēap stōw* – 'market place'[11] – in Old English. Frank Noble, who studied the Dyke for many years and pioneered the Offa's Dyke long-distance path, believed that here the purpose of the Dyke may have been to restrict the passage of traffic and to force trade through the settlement.[12] The nearby place name Tutshill* also suggests a defensive 'watchtower' element at this

To walk the length of Offa's Dyke is to appreciate both its strategic logic and its surveyors' mastery of topography.

* Old English *tōt* – a lookout.

strategically important river confluence, where the remains of a Roman bridge were recently revealed by an unusually low tide.

The Wye flows from the north along a steep-sided, twisting gorge and for the most part the Dyke follows it from a height, hugging the contours on its east bank as far as Redbrook, three miles downstream from Monmouth where the Wye is joined by the Monnow. From here, for many miles north as far as Hereford, it is the twisting, dynamic course of the Wye rather than any earthwork that forms the 'frontier' and, even allowing for possible partial demolition by construction and farming, there evidently never was a major earthwork here. That does not mean that no provision was made to control the Wye valley: hillforts perch above the river, matched by later Norman castles. Control of these lands came from dependent lords acting as wardens of the march.

King Offa's interest in the *burh* site at Hereford was reinforced by its status as an episcopal see. As the name suggests,* it was an important crossing point of the Wye upstream of its confluence with the River Lugg and a strategic lynchpin of Mercia's westernmost margins. Hereford lay on the Roman road – the *here-pæth* – between Gloucester and Wroxeter, a branch of which penetrated deep into the Welsh mountains past the Roman town of *Magnis*. This was the territory of the *Magonsæte*. Its seventh-century rulers had included the Merewalh whose daughters became celebrated abbesses of Much Wenlock and Minster-in-Thanet. The broad Wye valley is fertile here, a land of cider orchards, of fine, well-drained pasture and arable fields. King Offa is believed to have had a royal *vill* at Sutton,† on the east bank of the River Lugg and two watermills were excavated close by at Marden in 2006.[13]

Some six miles to the west, and upstream of Hereford, the Roman road runs through *Magnis* and on towards Hay-on-Wye. Here, on the ridge of a prominent local landmark called Garnons Hill directly overlooking the Wye, the Dyke resumes its northerly

* Old English, probably *Here-pæth ford*, 'Army road river crossing'.
† See below, Chapter 9, p. 302.

course – perhaps, as the archaeologist Gordon Noble suggested, marking the limits of *Magonsætan* territory.[14] But it survives only in short sections as far as Kington, the small town that controls east–west access along the valley of the River Arrow. Despite suspicions that intermediate sections have been lost, it seems that the Dyke was discontinuous here, each short section serving a specific local purpose in cutting across, and blocking access through, a succession of narrow valleys. In principle there is no reason why other forms of demarcation might not also have served here, including cairns of stones or strings of ancient burial mounds; but this may have been an area of generally looser control; of more nuanced affinities. Kington – *Cyne-tūn* – was a royal *vill* and its location, a mile or so upstream and west of the line of the Dyke across the Arrow Valley (i.e. on the 'Welsh' side), is a clear demonstration of how the Dyke marked a zone of control as much as it formed a physical boundary or barrier.

From Rushock Hill north of Kington, to Knighton on the River Teme, the Dyke is more or less continuous, following the complex, sometimes steep and twisting topography of deeply incised east–west-running valleys: the Hindwell Brook and the Lugg. This landscape posed serious challenges for surveyors as well as engineers and for those whose obligations to their lords included the sweated labour of ditch-digging.*

North and west of Knighton the Dyke runs between the summits of hills rising to more than 1,300 feet above sea level; and in places in these uplands it is still positively monumental, with a broad bank that must have lost some of its height to erosion in the intervening years, and a silted-up and now grassy west-facing ditch. Sometimes, from a distance, the line of the Dyke is etched against a hillside by a line of trees along the bank, or the shadow of a fringe of gorse. In the low light of winter, under snow, its serpentine line is unmistakeable. Down into and across the valley

---

* A term, *well-geworc*, noted in ninth-century Worcester charters by Cyril Fox, may be an explicit reference to the duty to work on earthen fortifications. Fox 1955, 283 n 1.

of the River Clun below the village of Newcastle, ditch and bank run their course; then back up over higher country – the hills and folds of Clun Forest in south-west Shropshire – before descending into the Vale of Montgomery; but always maintaining a northerly direction. From here the Dyke runs dead straight, north-west across the gently undulating plain for mile after mile with military precision.

Just north of Montgomery the Rivers Camlad and Severn and the Afon Rhiw meet in a broad floodplain. The Dyke blocked or controlled access along the Camlad Valley, which leads directly to the later Mercian *burh* of Shrewsbury, heartland of the *Wreocansæte* and, somewhere nearby perhaps, unlocated Pengwern, site of the legendary Powysian seat of Cynddylan. From this meeting point of strategically important natural routeways, the Dyke rises again on its northerly line and runs above the east bank of the Severn flowing north–north-east to Buttington and Welshpool. For five miles, as far as Rhos, the Severn takes the place of the Dyke; then it strikes north once again across the Oswestry uplands, riding ridge and vale, past the massive thirteenth-century Chirk Castle as far as the River Dee and beyond through Flintshire towards the coast.

The Dyke was generally constructed as a ditch with an upcast bank on its east side. Cyril Fox's generous illustrations of profiles and excavated sections show how it varied in height and breadth to suit topography and visibility. In the Shropshire uplands near Careg-y-Bîg the earthwork, as it stood in his day, measured seventeen feet from the top of the bank to the base of the ditch, and this is typical of what Keith Ray and Ian Bapty, authors of the most recent authoritative study of the Dyke, call its 'massive' form, where segments run for hundreds of metres across what must have been open country.[15] When constructed, the bank must have been higher and the ditch deeper, with sharper profiles than the gentle 'U' seen now. In other areas, especially in the southern counties where the line is more sinuous, what they call a 'weaker' form prevails – more a demarcation than a physical barrier.

There are other reassuring certainties. The Dyke 'faces' west: it can almost always be seen, more or less prominently, from the Welsh side, either cresting a ridge or running along the contour of a west-facing slope, chiselled against a hillside. On occasion it cuts directly across one of the many river valleys that flow east out of the mountains, blocking or controlling passage downstream. At the largest scale it very roughly parallels the course of a Roman road joining Gloucester in the south with Chester in the north – sometimes closely, sometimes at a distance of more than fifteen miles.

Most strikingly, perhaps, the Roman road that runs almost due north from Wroxeter (*Viroconium Cornoviorum*) to Whitchurch (*Mediolanum*) is absolutely paralleled by a twelve-mile north–south section of the Dyke. This was constructed on a gigantic scale, designed to link the most southerly point of the course of the River Dee near Pont Cysyllte with the most northerly point of the courses of the Afon Vyrnwy, at Llanymynech, and River Severn at Derwas, a few miles west of their confluence. Here Welsh mountain meets Shropshire plain and here the great hillfort of Breiddin looms from over a thousand feet to the south-east.

It is conceivable that this section provided the blueprint for the whole conception: a bulwark of Mercian power in the face of Powysian resurgence in the old marches defining the lands of the *Wreocansæte*. At the centre of the Dee–Vyrnwy gap, less than three miles east of the Dyke, sits Oswestry: site of King Penda's bloody triumph over the Northumbrian King Oswald in 642, and a huge Iron Age hillfort, Old Oswestry. The Dyke's surveyors made clever use of broad north–south ridges and west-facing slopes with commanding views to negotiate a series of deeply incised river valleys. If it was not intended as an impenetrable barrier here, it functioned nevertheless as a line of control enabling the physical blocking of valley bottoms against cattle-raiding and large-scale attack.

Keith Ray and Ian Bapty emphasise both the broad imperative of the Dyke's north–south line and its adaptation to the immediate exigencies of local topography.[16] The crossing of the Afon Vyrnwy at Llanymynech, just south-east of the small town of Pant, is dominated by a fort on Llanymynech Hill, and the Dyke follows the contours below the hill's crest. The fort, directly overlooking the confluence of the Afon Vyrnwy and Afon Tanat, has not been excavated, but its apparently deliberate inclusion in the Dyke scheme gives it an eighth-century relevance and perhaps marks a 'start' point from which the Dyke could be surveyed to the north. While the overall north–south line is remarkably consistent here, the use of local terrain is always clever, taking advantage of steep slope and broad plateau, stream and pass.

A very similar strategy was adopted further south, near Presteigne in Radnorshire. Here, the River Lugg is joined by its southern tributary, the Hindwell Brook. Five miles upstream from the confluence a narrow gorge squeezes the river between two hills. Herrock Hill, on the south side, carries the Dyke along its western flank and then down a steep slope, directly at right angles across the valley. On the north side the gorge is overlooked by a long, steep-sided hill from which the ramparts of Burfa hillfort command access from east to west. The Dyke follows a contour below the ridge on which the fort stands, references another fort, Pen Offa, to the north-west and then resumes its more northerly course down into the valley of the Lugg. Looking west from Burfa Hill up the valley, as it broadens out onto productive farmland, one's scanning eye takes in the site of a Roman fort, a motte and bailey castle of the Norman period, many prehistoric *tumuli* and standing stones. This is a landscape of productive wealth and symbolic richness, steeped in memory and bristling with the tensions of territorial rivalry. The imperative of the Dyke here is not to possess those lands, but to demarcate a corridor, a zone of control both east and west in which Mercian power was expressed; on which it was imposed.

Eighth-century surveyors – and no-one can say if they were a professional cadre recruited for the purpose or whether such knowledge was, like farming, part of the common stock of cultural wisdom – knew how to plot a line of direction and join the dots. The Pole star may primarily be useful to sailors but all travellers and surveyors knew its value; the full moon hangs due south at midnight and, what is more, illuminates the land. Beacons lit on intervisible hilltops facilitate surveying over long distances and lines of sight can be accurately marked with posts, flags and plumb lines. In this way the Dyke was engineered in segments, sometimes joined up to form a continuous bank with west-facing ditch, sometimes left unconnected. Mostly the segments were straight, with angles created where they met.[17]

In a world lacking compasses, maps or detailed topographic gazetteers, any engineer or surveyor entering this country must have relied heavily on local knowledge, as the Domesday Book inquisitors would and as Ordnance Surveyors did more than a millennium later. Lords knew their lands: what they owned and where the limits of their lordship and powers of taxation and legal enforcement lay. Men like Dudda who was granted an estate near Salmonsbury, themselves subject to obligations devolved from *subreguli* and *patricians* like Brorda, must submit to, and enforce, the burdens of public works: bridge repairs, fortifications and military service. They knew the imperatives of lordship in the Marches: watchfulness; the transhumant logistics of pastoralists, routes taken by raiders, traders and armies; choke points, escape routes and key strategic lines of control; hidden valleys, clefts and folds where stolen cattle might be concealed.

How far individual segments of the Dyke were devolved to such marcher lords and the extent to which they advised and negotiated its route and construction is unclear. They can hardly have been unfamiliar with resolving local disputes regarding straying cattle, abduction, boundary lines and outright theft.* The landscapes through which the Dyke ran were not blank natural canvases: this country is littered with prehistoric monuments, ancient references to pastoral rights and the burials of ancestors – proprietorial sensitivities that the Dyke's surveyors may have felt able to ignore, or were obliged to placate with compensation. Long-distance seasonal pasturing and access to woodland, fertile

---

* And see below in this chapter, regarding the *'Dunsæte* Ordinance'.

arable land, hunting grounds and minerals must have complicated negotiations – if negotiations there were. Perhaps, at the local level, the Dyke reflected or redefined existing boundaries. The line of control may have been porous in parts, impassable in others.

The ways in which archaeologists interpret the function, longevity and political and military context for the Dyke depend on how, or whether, it was patrolled, manned and supplied; and that depends very much on modern insights into the strategic conception – the imagination – that lay behind it. Offa's Dyke: white elephant; vanity project; offensive military platform or visionary national frontier? Cyril Fox believed that the Dyke was a negotiated compromise – a triumph of Anglo-Welsh détente – but the consensus has turned against this idea.[18] Mercian interests such as strategic river crossings, fertile farmland and settlements, are found lying to the west. A lack of evidence for structural remodelling to bank or ditch and, for want of a better word, 'furniture' (one thinks of dozens of phases of adaptations and alterations to Hadrian's Wall) seems to militate against the idea that the Dyke went through a maturation period or evolved in response to military attack, change in strategy or political fortune. But a lack of obvious garrison structures on the Dyke itself is countered by the as yet tentative case that can be made for a number of pre-existing enclosures and Iron Age hillforts along its line to have been reoccupied.[19]

There is nothing in King Offa's long reign that marks him as a dilatory man of passing vanities. A hundred years after his death Bishop Asser related how hard-fought had been King Ælfred's battle to persuade *his* thegns to co-operate in the grand scheme of *burh*-creation and public defence that finally won victory against the Great Heathen Army of the Vikings. King Offa's Dyke project was carried through with determination, as well as style. It has been said that the Dyke was 'performative' – as potent in the planning and execution as it was in any resulting political or military advantage.[20] If the average day labourer can shift about a tonne and a half of earth – archaeologists have some experience in this – and if an 'army' of several thousand dependent farmers took part – their numbers swelled, perhaps by Welsh prisoners of war – then this was an event approaching a 'national' endeavour – a feat of common purpose carried through with the huge political and material capital on which King Offa could call.

Even so, in probing the mind-set of Mercia's royal administrators, including those formerly independent rulers of the subject territories that lay on Mercia's western flank, archaeologists and geographers now seek confirmation along the Dyke zone of the sort of thinking that lay behind the *burh-* and *burh-tūn* network through which Mercian kings imposed their will on the core lands of the Trent, Thames and Nene Valleys. Long ago, the place-name scholar Margaret Gelling drew attention to both the occurrence of many 'English' settlement names to the west of the Dyke, and to the distribution of *burh-tūn*-type names with seeming relevance to it. Gelling was, indeed, the scholar who first drew attention in the late 1980s to the *burh-tūn* names as possible indicators of a Mercian 'system' of defence pre-dating the Viking invasions in the ninth century.[21] She pointed to distinctive concentrations, not just of *burh-tūn* names but also of the *burh-weard* form in which Old English *weard* is a suffix indicating the office of a warden – a military official charged with defensive duties. Gelling's educated hunch is now accepted as an inspired insight. John Blair argues that the Offan infrastructural scheme encompassing the lands through which the Dyke passed was implemented with canny skill and detailed local knowledge.

Climbing north-west out of Knighton from the valley of the Teme onto the broad ridge whose western edge the Dyke marks in massive form, the walker sees the landscape to the west open out spectacularly: the river valley immediately below; steep and narrowly incised hills with wooded valleys beyond; isolated farms and occasional hamlets blending into the far distance – pasture lands, for the most part. Nearer to hand, just across the valley, is Knucklas, a prominent spur of high ground on which sits a small earthwork hillfort, Cnwclas Castle, at a height of about 900 feet above sea level. Looking back at the Dyke from here

it is possible to see in a direct line an earthwork called Trebert\*
and, beyond the Dyke, Treverward: the *tūn* of the warden or
guardian of the *burh*.† In this scheme, settlements in the charge
of a Mercian functionary watched over the critical valley bottom
royal *vill* at Knighton. Blair believes that Knucklas was the *burh*
in question.[22]

If this pattern was repeated elsewhere along its line, then the
Dyke was provided with *burhs* and wardens – men who held their
land in return for the service of *burhweard* – an integrated system
of surveillance. Gelling identified no fewer than seven *burhweard*
names across the marcher lands of Cheshire, Shropshire and
Herefordshire, the hinterland of the zone of control between
the Roman road and the Dyke corridor.[23] Ray and Bapty have
identified more elements in this apparent scheme: *burhweard*-
type names at Burwardsley and Brewer's Hall in Cheshire, a
Burwardstone on the Cheshire–Flintshire border and Broseley
and Burwarton in Shropshire; a pair of *worthing*-type names –
Worthenbury and Worthen – that may indicate the presence of
earthwork enclosures close to the Dyke; several Burtons along
the east side of the Dyke line and a number of *-wardine* suffix
names in Shropshire and Herefordshire including Lugwardine,
just east of Hereford.[24] The occurrence of some of these names
to the west of the Dyke reinforces the sense that the Dyke was a
zonal frontier, rather than a 'stop-line'.

More so than Knighton, Hereford, where Roman roads met
and then deviated from the line of the Dyke, bore a key significance
in the scheme – perhaps as an administrative and logistics hub for
the whole project. Other key river crossings may have fulfilled
similar focal purposes on a more modest scale: Presteigne, where
the Lugg meets the Hindwell Brook, with a 'Warden' name on
the hill behind it, a Norton controlling the north side of the
Lugg Valley and the great ramparts of Burfa camp guarding its

---

\* *Tref-burh* – a direct Anglo-Welsh hybrid equivalent of *burh-tūn*.
† *Tref-burhweard* Blair 2018, 207–8. A Shropshire equivalent is
Burwarton.

south-western approaches along the beck; while Kington has its Bredward upstream along the Arrow Valley. Oswestry, with its great hillfort, perhaps a fortress of King Penda in the seventh century, sits behind the Dyke at the dead centre of the critical stretch between the Dee and Severn.

The Dyke, then, seems to have been conceived, and functioned, as a consistent component of the eighth-century Mercian policy of extensive and intensive royal control. It stamped a Mercian presence along a line of significant military relevance throughout the reigns of all Early Medieval Mercian and Welsh rulers. Eighth-century conflicts – raids, campaigns or battles – were recorded in 743, 760 (at Hereford), 776, 778 and 784, the last two of which were recorded in the Welsh Annals as 'devastations' by King Offa.

The extent to which Mercian administrators actively interfered in the regional politics of the border is suggested by a raft of names that tantalisingly evoke the smaller territorial entities of Tribal Hidage. The *Rhiwsæte* are identified with a group living in the vicinity of Wroxeter. The *Meresæte* – 'boundary-dwellers' – may have been their neighbours to the north on the far side of the Severn in the area south of Oswestry centred on Maesbury – a hundredal court at the time of the Domesday Book. Equally shadowy groups whose names survive only in late documents include the *Temesæte*, in the lands along the River Teme that flows through Knighton; and the *Steplesaete* in Herefordshire north of the River Wye.[25] Such names have a flavour of antiquity about them: peoples named from natural topographic core lands. Early names they may have been, but their use as administrative epithets, if it cannot be tied down to the eighth century, is at least congruent with the idea of Mercian administrative interest in knowing who lived where, under whose lordship; and how those dwellers might be harnessed to the greater purposes of the Mercian state.

A West Saxon document, probably dating to the tenth century and known as the *Ordinance concerning the Dunsæte*, addresses matters of mutual interest to dwellers on either side of a river to be identified with the River Wye between Hereford and Monmouth.

This area was once called Archenfield, the Anglicised name of a small kingdom, *Ercyng* in Welsh; the dwellers in question are assumed, or imagined, to be Welsh on one side and English on the other. The *Ordinance* addresses real and potential disputes that might arise between any neighbours:

> Ðæt is gif man trode bedrifð forstolenes yrfes
> of stæde on oðer.

That is: if anyone follow the track of stolen cattle from one river bank to the other, then he must hand over the tracking to the men of that land, or show by some mark that the track is rightfully pursued.[26]

Much the same provisions are known from the more recent Anglo-Scottish 'Debatable Lands' of the sixteenth century. The composition of the *Dunsæte* 'court' is intriguing: six men from each side were to sit in judgement; and English and Welsh livestock were valued at an equal price.* If this late Anglo-Saxon document reflects the practices of earlier administrators who embraced the value of the aphorism that good fences make good neighbours, then such provisions may date as far back as the Dyke's construction, or beyond.

It is all the more infuriating, in this context, that despite many excavations of sections across the Dyke and the efforts of generations of landscape archaeologists, it defies attempts to date it. It is not Roman. It was there when Bishop Asser wrote his *Life of King Ælfred*; therefore, its construction lies floating uncomfortably in a half-millennium window; and the lack of any sort of precise dating leaves open the question whether the Dyke was the product of a single phase of construction or a cumulative process of development, perhaps beginning in the reign of Æðelbald and continuing after Offa's death in 796. A number of other linear earthworks that might belong to or supplement

---

* For more detail on the *Ordinance*, and for its contemporary setting, see *Ælfred's Britain*: Adams 2017, 324–27.

the Offan 'system' occur in Herefordshire – the Rowe ditch, for example – and on the largest scale there are substantial earthworks that might demarcate Mercia's northern and southern frontiers: the Wansdyke in Wiltshire and Somerset and the Roman Ridge dykes of Yorkshire.[27]

More than eighty physical investigations of sections of the Dyke have been carried out over the years[28] and none has produced the sort of 'slam-dunk' evidence that archaeologists dream of: a coin sealed beneath the base of the rampart; layers of organic silt in ditches containing charcoal fragments that might be roughly dated by carbon-14 assay; a buried and miraculously preserved baulk of timber from a palisade that would offer the ultimate proof: an unequivocal dendrochronology date for the felling of a tree used to build it. There is hope, however. A technique for dating buried soil horizons is proving increasingly successful in establishing earthwork construction dates with a high degree of accuracy. Optically Stimulated Luminescence (OSL) relies, like carbon-14 dating, on the steady decay of radioactive isotopes but in this case measuring minute light emissions from a number of radioactive elements. The many supposedly Early Medieval earthworks that pose such chronological problems for archaeologists are now within range of a concerted programme of dating: the massive Wansdyke in Somerset and Wiltshire; the Cambridge dykes and elsewhere. Not least, archaeologists are concerned to establish a chronological relationship between Offa's Dyke and its 'junior' sibling, Wat's Dyke.

The forty-mile-long linear earthwork known as Wat's Dyke runs, or ran, on a parallel course to the east of Offa's Dyke, between the Afon Vyrnwy near Maesbrook and, beyond the Dyke's northern extent, as far as the Dee Estuary at Basingwerk. At a relatively consistent distance of about two to three miles east of Offa's Dyke, Wat's Dyke occupies low ground, often linking key fortified sites such as Basingwerk, Old Oswestry hillfort and Bryn Alyn north of Wrexham – in long, smoothly engineered, uniform lengths with a west-facing, V-shaped ditch. Like its more famous counterpart, it scrupulously connects the southernmost bend in the River Dee

Wirksworth parish church, Derbyshire. From King Offa onwards,
Mercian power was heavily invested in its queens and their children,
monumentalised in its churches.

to the northernmost point on the Severn. It passes very close to
a Mercian royal estate centre at Maesbury, south of Oswestry,
and a Welsh royal residence at Bagillt. Often, Wat's Dyke is less
substantial than its more famous neighbour to the west, or has
been swallowed by farmland or urban development. Where
Offa's Dyke cuts east–west flowing rivers at right angles, Wat's
Dyke joins lengths of north–south flowing rivers and streams.
It belongs, very evidently, in a settled and farmed lowland
landscape.[29] Occasionally, one might find oneself travelling
along a small country road in those parts and experiencing a
disconcerting, if gentle, hump as one passes over its line.

275

There has always been a temptation to think of Wat's Dyke as a Powys-facing prototype for its larger, more substantial and overtly offensive neighbour – a natural political and military prototype that saw its magnificent fruition under Offa. But Wat's Dyke has now yielded an apparently reliable date from an excavation at Gobowen, a couple of miles north of Oswestry. Here, a long section of the earthwork was excavated in 2006 and, from a series of samples, OSL processing has dated the foundation of the bank to the early ninth century.[30] If Offa's Dyke is truly Offa's grand design, then Wat's Dyke belonged to one of his successors. It seems counter-intuitive, unless one sees Wat's Dyke as a second line of defence against Powysian aggression. If that is the case, then the very fragmented record of Anglo-Welsh conflict that survives in the Welsh Annals may need to be substantially reassessed, the kings of Powys credited with much more effective military potency.

In the next few years, Offa's Dyke may offer up its own secrets. For now, I can only tentatively suggest a plausible window during which Offa may have bent his energies and political capital to his western frontier: after his victory at Benson in 779 and before he was able, once more, to reassert his *imperium* over Kent's independent-minded kings and its truculent Archbishop Jænberht, after about 785. Looked at another way, one might ask if Offa's preoccupation with the massive frontier project distracted him from his ambitions in Kent. Those few charters that record the Mercian court's location when a grant was made show that the king was regularly in Worcester, in Gloucestershire and at Tamworth between 778 and 781.

From this time Tamworth emerges as the principal Mercian royal centre.[31] It lay close to the episcopal seat at Lichfield and, beyond, the hunting grounds of Cannock Chase. To the north lay the *burhs* and Burton of the Upper Trent Valley and the mausoleum at Repton. Watling Street and Ryknield Street gave speedy access for the king and his household to travel to the west, south and east. If Hereford was the principal militarised *burh* supporting the frontier in the south, then Chester, Whitchurch and the royal township at Atcham may have served similar functions on Mercia's Powysian frontier.

Atcham is known only from a striking aerial photograph, taken in the very dry year of 1975, of a cropmark close to the old Roman town of Wroxeter and sitting close to the junction of the rivers Tern and Severn. Like Yeavering and Northampton, it shows evidence of two large axially aligned timber halls with annexes. Other similar complexes identified at Itchington and Hatton Rock in Warwickshire and at Sutton Courtenay, near Abingdon on what was once the Oxfordshire/Berkshire border,[32] were architecturally imposing symbols of Mercian royal clout.

Reconstructions of such halls – for example, a later Viking period hall at Fyrkat in Jutland – are impressive enough; but their symbolic, economic and political impact would have been much greater at those times when the royal wagon trains arrived to spend a few days or weeks at these sites. Then, like the tented cities of music festivals, they would bristle with human energy, with social, material and political transaction, concentrating the aura of royal magnificence in their latter-day Heorots. And if such sites, apparently undefended by earthworks or palisades, look vulnerable, it is to be remembered that the jewels of the Mercian state were defended in depth. Atcham has its Burtons and Charltons, Nortons and Suttons. Not for nothing would King Ælfred's indomitable daughter Æðelflæd choose a site close by, in a tight bend of the River Severn, as the site of a *burh*, Shrewsbury, in the early tenth century.

If Mercia's, and King Offa's, power in southern Britain was expressed in his political *imperium* and military successes, monumentalised in his coinage and in the great frontier dyke, it was also expressed more subtly in a distinctly Mercian cultural milieu. So little survives of Mercian manuscripts, metalwork and textiles that it requires a feat of the imagination to fill the halls with banners, embroidery and glinting shields; the empty churches with bling, illustrated gospels and gaudily painted sculptural reliefs; to dress bare skeletons in tunics finished with patterned, tablet-woven cuffs and hems, adorned with brooches and pins. But what little

A triptych of holy figures, symbol of Breedon's wealth and status
as a favoured Mercian monastery.

remains in scattered collections, libraries and museums, in market
squares and churches, shows that Mercian artists and craftspeople
were eclectic consumers of Pictish, Irish, Frankish, Germanic,
Byzantine and Lombard art, of their designs and decorative motifs
– often carrying overtly classical allusions. From this pot pourri
they created a vibrant, confident Mercian visual culture.

The churches at Brixworth, Wing in Buckinghamshire and Repton
stand alongside Offa's Dyke and the *burh* network as monuments
to Mercian lordship over the landscapes of the Midlands; and
excavations at Cirencester have shown that a great basilican
church stood there from the beginning of the ninth century.[33] But
the glory of contemporary arts is to be found on a more modest
scale. Eighth-century stone sculpture has been preserved more
by accident than design, to be studied and admired in friezes,
crosses, slabs and assorted panels found scattered across Mercia.
In their post-Reformation plainness – what Richard Bailey called
'tasteful reticence'[34] – lacking original garish painted decoration,
the vine scroll motifs, delicately draped figures and fabulous beasts
only draw the eye when they are well-lit; more so when they
are decoded. Outstanding collections can be found at the early
Mercian foundation of Breedon-on-the-Hill, Leicestershire, with its

friezes depicting exotic beasts such as griffons, lions, centaurs and peacocks; at Wirksworth in Derbyshire; in the candy-twist columns of the mausoleum/crypt at Repton, and in Peterborough Cathedral – Bede's *Medeshamstede*. The dedicated student of Mercian sculpture must seek out many others, often recovered from Victorian church 'improvements' and cemented unsympathetically into walls.[35]

At St Mary's parish church in the Derbyshire town of Wirksworth the effect is almost avant-garde: a dizzying tessellated jumble of figurative, decorative and zoomorphic motifs. The thoroughly vernacular lead miner figure known as 'T'owd Man', with his bucket and axe-hammer, is a reminder that art celebrates the artisan as much as the patron. A coped grave slab frieze portraying the four evangelists and the Ascension, embracing the themes of judgement, glory, sacrifice and humility,[36] is so packed with figures and symbols as to evoke a great biblical crowd scene.

Set into the walls of Wirksworth's church are dozens of friezes and figurative designs now divorced from their original context. The friezes portray plant scrolls and geometric interlace, badly weathered but recognisably well-executed. Figurative relief carvings, with echoes of the formal round-arch framing

Wirksworth parish church, Derbyshire. A coped slab from a sarcophagus showing events from the life of Christ – and showcasing Mercian sculptors' skills.

of contemporary church architecture, manuscripts and classical allusions, portray the Virgin Mary, apostles and an Archangel Gabriel reminiscent of a fragment recovered from Lichfield Cathedral.* That there was a school of Mercian sculpture – literate, sophisticated and sensitive to neighbouring traditions in Northumbria and elsewhere – is certain.

It goes almost without saying that such a well-established tradition leant on a wide range of resources: not just quarries for suitable stone but transportation, trained craftsmen and the provision of well-tooled workshops. It embraced and fine-tuned a repertoire of theological and liturgical symbolism that resonated with universal concerns: cycles of life and death, devotion and sacrifice, divine intervention – even if its message was for the most part restricted to a social élite, since the democratising parish church movement had not yet been conceived. How much has been lost remains unknown.

Among the probable gems produced in late eighth-century Mercia, the small[†] whale's bone, brass-bound, house-shaped box known as the Gandersheim Casket is outstanding. Its panels are minutely carved with animal and plant scroll interlace designs in square panels. An inscription on the base, known to be relatively modern but possibly a copy of an original, records that it was made by Æda as a chrismale – for containing a flask of holy oil.[37] Its closest artistic parallels are to be found on the badly weathered cover or 'roof' of the Hedda Stone sarcophagus in Peterborough Cathedral and it may well have been crafted there.[38] The casket's swirling interlace designs are not merely the product of skilled craftsmanship; they encode 'a complex cosmological iconography, gridded and grounded in sacred numerology'.[39] The casket, which is paralleled in Northumbria by the Franks Casket and by a number of wooden reliquaries, and on a larger scale

---

* See below.

† About five inches long and the same in height, and less than three inches deep. Wilson 1984, 65. The casket once belonged to Gandersheim Abbey, but is now in the Braunschweig collection in Lower Saxony.

The Gandersheim casket – a rare survival of eighth-century
Mercian craftsmanship.

by house shrines and sarcophagi like the Hedda Stone, is both
a portable jewel and a miniature monument in its own right, a
brilliant celebration of Anglo-Saxon art and mythology that taps
into worlds almost beyond imagination.

The sumptuously illustrated Barberini Gospels, now in the
Vatican, seem likely also to have been a product of the creativity and
intellectual endeavour nurtured at *Medeshamstede*.[40] Wonderfully
coloured with a wide palette of pigments, including a rare and
highly prized dog whelk-derived purple, they evoke perfectly the
blending of continental inspiration with a repertoire of Insular Celtic
scrollwork and Anglo-Saxon zoomorphic symbolism. The better-
known Lichfield Gospel book, dating perhaps to the middle or last
decades of the eighth century, was a product of the same confident,
eclectic and exuberant pan-Mercian artistic milieu.* With affinities

* The Hereford Gospels may also date to the last quarter of the eighth
century.

to both the Lindisfarne Gospels and the Book of Kells, it contains a dazzling carpet page, featuring delicate interlace and fantastical beasts. Portraits of St Mark and St Luke are graphically bold and handled with great skill.

Alongside a competent and well-managed *scriptorium*, Lichfield must also have been able to draw on all the physical resources required for such an elaborate and prestigious work. The Lichfield folios are large, each pair fabricated from the skin of a single calf with very few blemishes such as worm holes – the tanners had ample, good-quality material to choose from.[41] Pamela James, who has studied the manuscript, even suggests that cattle belonging to Lichfield and reared specially for the production of vellum may have carried a distinct branding mark, visible in illustrations of calves in the gospel book.[42] She believes that the limited palette of colours used in the Lichfield Gospels was sourced locally, in contrast to some of the more exotic imported colours, like Afghan lapis lazuli, employed in Northumbrian and Irish manuscripts and in the Barberini Gospels.

The Lichfield Gospels' history also offers a tantalising glimpse into Anglo-Welsh tensions and the motivations of cross-border raiders. Pamela James argues that the gospels* were created at Lichfield as a companion to the shrine of St Chad. According to Bede, the venerated holy man's original house-shrine was made of wood, provided with a hole through which pilgrims might reach in to touch the bones of the founding father of the Mercian church.[43] In 2003, excavations in the nave of Lichfield Cathedral yielded large fragments of a sculpted limestone panel, carved in relief with the figure of an angel – probably the Archangel Gabriel – still crisply sharp, having been preserved from weathering. Stylistically this may date to the end of the eighth century and by analogy it probably formed part of a shrine chest.[44] If this was a replacement for the original wooden shrine of St Chad, it may have been commissioned on the centenary of his death – that is to say, in about 772:

* Now containing just two of the four gospels but perhaps originally one of two volumes.

Portrait of St Luke from the Lichfield Gospels: the great glory of
Mercian manuscript production.

a plausible date for the gospel book.[45] Because of its long, untouched earthly interment, the sculpture still bears traces of the paints used to decorate it: red, yellow, white, black and probably gold, according to the historian of Anglo-Saxon art, Leslie Webster.[46]

Sometime in the ninth century the Lichfield Gospel book was to be found in the Welsh monastery of St Teilo (perhaps Llandeilo Fawr). We know this because at the top of page 141 a marginal entry has been added to the script, in Old Welsh, recording the fact that one Gelhi had bought the book *'pro illo equm optimum'* – for the price of a good horse[47] – and then donated it to St Teilo's church for the sake of his soul. It would have to have been an extraordinary horse indeed for the cathedral at Lichfield to contemplate trading such a prize possession. More likely, the gospel book had been stolen from the shrine at Lichfield during one of the cross-border raids mounted by the Welsh in the first decades of the ninth century and recorded in the *Brut y Tywysogion*:[48] a great trophy, sold cheap. At any rate, by the tenth century the book was back at Lichfield – ransomed, perhaps, like that other great literary treasure, the *Codex Aureus*.[49] Both Gospel books and sculpted stone shrines may have played a part not just in celebrating the relics of St Chad and the founding of the Mercian church, but also in a grander ecclesiastical occasion. For the 780s would see the culmination of another of Offa's political ambitions: the elevation of the see of Lichfield to an archdiocese.

\*

In 784 King Offa's forces 'devastated' Wales, according to the Welsh Annals – whether pre-emptively or as reprisal is unclear. A year later, the king was at his royal *vill* in *Celchyð* – Chelsea, on the Thames west of *Lundenwic* – from where he issued a charter granting a substantial estate in Kent to his minister Ealdberht and Ealdberht's sister Seleðryð.\* That Offa was able to grant

---

\* Electronic Sawyer S123. The first of two grants made to Ealdberht and his sister Seleðryð. See Chapter 7; and below, Chapter 9.

land here in his own name without recourse to King Ecgberht suggests that the latter had by now died or been deposed. A King Ealhmund is known from a brief reference in the *Anglo-Saxon Chronicle* under the year 784. From then until Offa's death in 796 the names of no Kentish kings are recorded. At least one Kentish noble, eligible for the kingship, was forced by Offa into the priesthood and thence into exile on the continent.[50]

Offa may now have begun issuing coins from Canterbury moneyers. Two more grants of land in Kent in the following few years cement the idea that Mercian overlordship over the kingdom was now complete. But the Mercian king's relations with Archbishop Jænberht in Canterbury were increasingly fraught. A year after Offa's first independent grant of land in Kent in 764, an occasion attended by both Mercian and Kentish ealdormen and by Heahberht, the then Kentish king, the archbishop had been anointed at the Mercian court.[51] He attended royal councils in 780 and 781 during the years after King Ecgberht's reassertion of Kentish independence.* But Canterbury's autonomy was itself now under threat; Jænberht had lost his royal friend and patron and, late in 786, King Cynewulf of Wessex, the last potential break on Mercian power in the south, died after a bitter family feud.† It was in this new, opportune political climate that Offa confiscated from Canterbury estates that had been granted to it by King Ecgberht and his reeve – on the grounds that they had been granted without his (King Offa's) leave.[52]

---

* At the council held at Brentford in 781 Jænberht had witnessed the resolution of a dispute between King Offa and Bishop Haðored of Worcester in which the bishop had been forced to 'restore' to the king substantial estates, which Worcester believed to have been granted to the church by King Æðelbald (Electronic Sawyer S1257; English Historical Documents 77). Such demonstrations of royal authority can have left Jænberht in no doubt of Offa's political clout.

† The *Anglo-Saxon Chronicle's* very long and detailed account of this feud, contained in the annal for 755 (correctly 757), is unique, and remarkable.

Into these political currents, in the months before King Cynewulf's death, sailed a legation sent by Pope Hadrian – the first to visit these shores since the days of Archbishop Theodore a hundred years before. Like Boniface's embassy in the reign of King Æðelbald, the papal mission of late autumn 786 was as much motivated by Frankish curiosity about the state of the English political scene as by purely theological or canonical concerns for the Anglo-Saxon church: the legates were accompanied by a Frankish abbot and courtier, Wigbod, acting as an 'assistant' to the legates.[53]

Since 768 Francia had been ruled by Charlemagne, son of Pippin. In nearly twenty years he had conquered Lombard Italy and Brittany, forced the submission of Aquitaine and Gascony, and suppressed rebellion in Saxony. He knew much of English affairs, partly because he was prepared to offer the hospitality of his court to Kentish and West Saxon exiles and partly because he had recruited to his household the Northumbrian scholar Alcuin.* Alcuin's correspondence provides a key source for the last years of King Offa's reign and he was, as it happens, a member of the papal legation.[54]

The report on the legation's mission has substantially survived.[55] It relates how, on arrival, the legates,† carrying various letters and admonitions from the pope, were met at Canterbury by Archbishop Jænberht. From there they travelled to the court of King Offa, who received them with 'immense joy and honour'. The letters were read at the Mercian court and Offa then convened a council with King Cynewulf of Wessex (presumably, also, his bishops) to address the 'vices' that Pope Hadrian had identified. Pope Hadrian later recalled, in a rather pointed letter to King Ceolwulf in 798, that King

---

* c.735–804. Alcuin was a student of Archbishop Ecgberht in a celebrated school at York. He met Charlemagne (for the second time) and impressed him, in 781, during a visit to Rome whose purpose was to confirm York's archiepiscopal status. He was invited to join the palace school in Aachen and in 782 became master of the school.

† Bishop George of Ostia and Bishop Theophylact of Todi.

Offa had at this time promised to send to the Roman church a gift of 365 mancuses every year[56] – perhaps part of his special gold coin issues of which just two examples survive. At this opportune moment it seems likely that Offa believed himself to be purchasing both papal support for the installation of an archbishop in Lichfield and a status befitting the overlord of a great Christian kingdom.

One of the legates now went into Wales,* while the other went to the Northumbrian court of King Ælfwald. A series of decrees arising from these councils was drawn up and witnessed by the kings, archbishops and bishops. The matter of the decrees is primarily liturgical and canonical, concerned with dispute resolution, marriage, the consecration of bishops and archbishops, and the anointing and legitimacy of kings. The latter were of more than passing interest to King Offa, because neither his father nor grandfather had been kings. He had already taken steps to assert the rights of his dynasty by the overt acknowledgement of his queen, Cyneðryð, and formal recognition of his son, Ecgfrið – in overt contrast to King Æðelbald's notorious licentiousness and failure to marry or produce an heir.

The decrees of 786 also contain fascinating asides: condemnation of the use of ox-horn as a material for making chalices and patens; bare legs not to be shown when celebrating mass;[57] just and equal weights and measures to be used in assessing tithes;[58] pagan mores in tattooing, dress and feasting, including the eating of horse-flesh, to be repudiated.[59] Significant legal and moral imperatives were also addressed: powerful lords were not to favour the rich, hold contempt for the poor or take bribes.[60] The capitularies, read out to the assembled council in both Latin and in the vernacular 'so that all could understand', were attested by the archbishop and all the southern bishops; by King Offa, his *duces* Brorda and Brihtwold and his *comes* Eadbald. In surviving across the

---

* Whose 'British' bishops had finally accepted Roman orthodoxy regarding the calculation of Easter in 768.

centuries, like coinage and sculpture they monumentalise a set of relations between state, church and people: a social contract for eighth-century England.[61]

Whether by long-nurtured plan or because of this fortuitous opportunity, King Offa now sought from the distant pope the formal anointing of his son as king, knowing that Charlemagne's sons had been anointed in person by the pope and wishing to emulate his Frankish rival.[62] Perhaps anticipating that Archbishop Jænberht would refuse, or extract too heavy a price for his complicity, Offa may have mooted during his councils with Hadrian's legates (and with his generous papal gift in mind) the idea that Mercia ought to be provided with its own archbishop. Canterbury's primacy was a historical artefact; ecclesiastical power ought to reflect political power; ought to reside in the Mercian royal heartlands. Mercian bishops should belong to the jurisdiction of a Mercian metropolitan. Offa was at the peak of his political powers; who could gainsay him?

The *Anglo-Saxon Chronicle* entry for 787 records that:

> ... there was a contentious [*geflifullic:* 'stormy'] synod at Chelsea and archbishop Jænberht gave up a part of his jurisdiction, and Hygeberht was appointed [archbishop of Lichfield] by King Offa. And Ecgfrið was consecrated king.[63]

The jurisdictions in question are not defined in surviving contemporary documents but the twelfth-century historian William of Malmesbury, who sometimes had access to sources no longer in existence, provided a list of the bishops who now became subject to Lichfield: Worcester, Leicester, Lindsey, Hereford, *Dommoc* and Elmham (both in East Anglia). Canterbury was left with authority over Rochester, Sherborne, Winchester, Selsey and (perhaps surprisingly) London.[64] The pope duly sent Hygeberht a pallium, the band of woollen cloth that symbolised the office of a metropolitan,[65] and Lichfield's new archbishop duly anointed Ecgfrið as King Offa's legitimate heir. The emasculation of Kent's kings and its archbishop was, seemingly, complete. King Offa, like a puppet master, could tug at military, administrative and ecclesiastical strings, from his western frontier to the shores of the English Channel, and all the puppets would dance to his tune. He commanded the stage: Mercia had become his own monument.

# Chapter 9

# Offa: War by other means
# 788–796

✦

*The great game • Viking raids
• Brides and grooms • Exiles • A trade war
• Home farms • Succession*

Between the Legatine Mission of 786 and King Offa's death a decade later, the pace of history seems suddenly to accelerate into a fast lane of civil war and murder; diplomatic crisis and an international trade war; shocking pirate raids; marriages and exiles; gossip and dastardly plot. The pendulum swing of political fortune is given new impetus. It is as if the wick on a guttering lamp has been turned up: ghostly personalities come alive; silhouette and profile now have features; history is tangibly within reach.

The resumption of a bloody and protracted civil war in Northumbria in 788 was a reminder to the Mercian court – enjoying unprecedented stability from the reigns of two very long-lived kings – of the ultimate frailties of royal power. The unchanging rules by which Early Medieval kings gained, maintained and bequeathed their thrones were brutally simple and uncompromising. Those eligible to rule a people by virtue of their lineage faced competition from collateral members of increasingly fragmented dynasties steeped in warrior-myth and in legends of their glorious forbears – eager to gain influence, wealth and power by tilting for the highest prize. A successful bid for the kingdom must be backed by a powerful fighting force, the *comitatus*, physically able to impose itself on rivals. Aspiring kings must be 'accepted' or 'elected' by a core group of élite secular lords and legitimised by the sympathetic endorsement of their spiritual counterparts: bishops, abbots and saintly hermits like Guðlac. Those ecclesiastics debated among themselves the nature of good (and bad) kingship and, in turn, tried to educate and influence their secular patrons, occasionally necessitating that they dine with the devil, supping with a long spoon.[1]

A prince in exile fought for his hosts and in return expected their military support for his bid to seize the reins of power when the chance arose. If successful, he must reward supporters with land, treasure and regional power, or else he might find that their loyalty was expedient. He must find a wife among the daughters of allies or subordinate kings and produce healthy, worthy heirs. In time he must judiciously arrange for the marriage of his own offspring and balance the delicate levers of generosity, punishment and the wisdom of advisers: navigating treacherous waters to end his days in battle or in bed, but in any case with a son to succeed.

Mercian kings had proven themselves able players in the grand board game of political strategy: creative; innovative if not revolutionary in exploiting new tools of governance and influence; able to take as it were a bird's-eye view and at the same time address the minute implementation of their policies. They monetised land and trade, delivered infrastructure, infiltrated patronage networks and judiciously reserved their executive arm to deliver decisive victory on the battlefield. They, like more recent tyrants, understood the performative value of monumentalising themselves in grand buildings, mighty earthworks, shiny coins and epic poetry. They knew, also, the value of staying alive long enough to win at the long game. King Offa saw eight Northumbrian kings come and go during his reign.

Northumbria's Idings, the Bernician warlord dynasty who had established rule over all the English-speaking lands north of the Humber in the late sixth century, were still able to compete for the kingdom at the end of the eighth; but Northumbrian power had fragmented, split into factions whose geographies are obscured by a lack of detailed narrative sources. After Eadberht (737–58) a succession of short-lived kings ruled the North, none of them lasting a decade. Eadberht's grandson, King Ælfwald (769–88) was assassinated by an ealdorman, Sicga. His successor, Osred II, was forcibly tonsured and expelled a year later[2] – elbowed out, it seems, by a returning exile, Æðelred.* The fact that the Northumbrian scholar Alcuin, embedded at the court of Charlemagne since 782, arrived back at York in the year of Æðelred's return suggests that the restored king's exile may have been spent at the Frankish court under Charlemagne's protection; and that he persuaded Alcuin to join him. Alcuin was still in Northumbria in 792 when Osred returned from his own exile on the Isle of Man with his *comitatus* and, with the support of an 'Iding' faction, attempted a reverse coup. Æðelred was forewarned of the attempt. He had Osred captured and killed at an unidentified place called *Aynburg*.[3]

Alcuin was keenly aware, as Bede had been, of the need for strong, legitimate kings. Bede's idealised *reges* fought just wars to defeat paganism and bring salvation to the *gens* – their people. Alcuin's patron, Charlemagne, came from the same mould: his defeats of pagans in *Germania* and Muslim Arabs in the south aligned him squarely with the Old Testament's King Saul. Christian rulers supported and protected the church, benefiting from the counsel and blessings of its senior clergy. They ruled justly and spread the wisdom of the Old Testament. Their dooms were written down by literate clergy in indelible ink on vellum for the ages to consult. Alcuin was deeply interested in Northumbrian affairs, acquainted with many powerful Anglo-Saxon interests through his participation in the Legatine Mission of 786 and eager, on behalf of his Frankish master, to promote dynastic stability. It may have been he who helped arrange King Æðelred's marriage in 792 to the Mercian princess Ælfflæd, sealing an advantageous – if subordinate – alliance with Mercia's overlord.

Alcuin had previously written to the Mercian heir-apparent, Prince Ecgfrið, urging him to be worthy of his illustrious parents and invoking the Old Testament's authority in his sermonising. Surviving letters also reveal a warm friendship between Alcuin

---

* Æðelred had been installed on the Northumbrian throne, seemingly as a child, between 774 and 779, before being expelled in favour of Ælfwald.

## The Mercian–Welsh border

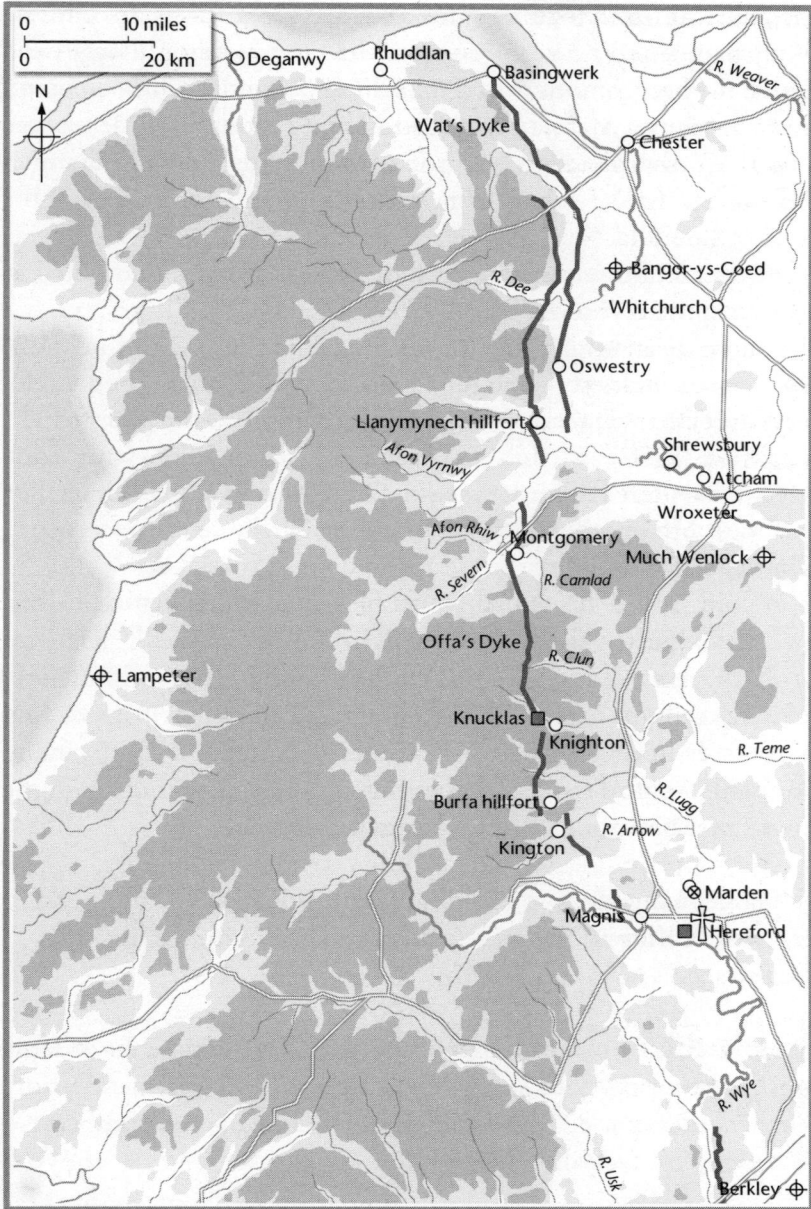

and Queen Cyneðryð. He promoted relationships between Mercia's royal women and Charlemagne's consort Liutgard through the exchange of gifts and messages. It seems entirely likely that such a powerful consort as Cyneðryð should have been active in the promotion of her children's interests. The *Beowulf* poet, perhaps a Mercian contemporary,* portrayed royal women as active political agents and he may have had Cyneðryð in mind. Alcuin referred to her as '*dispensatrix domus*' – controller of the royal household.[4]

Married daughters acted as diplomats and perhaps, to an extent, spies at foreign courts. They led privileged lives, too, but the currency in which they sometimes paid the price of privilege was local animosity – one might call it xenophobia. They might, like Æðelred's Queen Osðryð, find themselves targets for the assassin's knife or the poisoner's draught. King Beorhtric's Mercian Queen Eadburh was said, in later tradition, to have fled the West Saxon court after her father Offa's death and sought Charlemagne's protection.[5]

Alcuin found his Northumbrian homeland much disturbed and the king's attitude 'not as I hoped or wished'.[6] He was probably referring to the ignominious murder of King Ælfwald's sons, taken by force from the archbishop's church at York in 791. An insider at Charlemagne's Aachen court and an eye-witness to the immense powers accumulated by Frankish kings through the might of their armies, Alcuin lamented that...

Scarcely anyone is found now of the old stock of kings, and I weep to say it; the more obscure their origin, the less their courage.[7]

He was very much in sympathy with the sentiments of Bede who, in his often-cited letter of 734 to Bishop Ecgberht – Alcuin's teacher and himself the brother of an Iding king, Eadberht – deplored the fact that Northumbria's warrior élite were wont

---

* See Chapter 7, p 231.

to book themselves land under the false pretext of monastic foundation, weakening the military resources of their overly-compliant kings...

> Such places [...] are [...] useless to God and man because they neither serve God by following a regular life nor provide soldiers and helpers for the secular powers who might defend our people from the barbarians.[8]

Bede's words were prophetic. In 793, a year after Æðelred's marriage to Ælfflæd, and with the murder of royal princes fresh in the memory, fiery flashes in the sky and a great famine seemed to portend imminent disaster. Then, like 'stinging hornets', pagans came out of the North in their swift ships and fell on the holy island of Lindisfarne, laying it waste, plundering its riches, killing or enslaving its brethren.[9] The following year Bede's own monastery at Jarrow, on the south bank of the River Tyne, was attacked along with *Portus Ecgfridi*, the nearby royal harbour and trading settlement. The monastery on Rathlin Island off the coast of Antrim was attacked in 795 and Iona, the holiest of all the Insular monasteries, suffered periodic deadly raids from 802 onwards. No Irish, Pictish or Anglo-Saxon king was able to stop these raids or protect their coastal and estuarine monasteries from plunder.* The Viking Age had begun.

Alcuin, by now back at the Frankish court, received the news with horror and foreboding. Writing to Bishop Higbald of Lindisfarne in a tone of commiseration and admonition and, perhaps, with Bede's words ringing in his ears, he asked...

> What assurances can the church of Britain have, if St Cuthbert and so great a company of saints do not defend their own? Is this the beginning of greater suffering, or the outcome of the

---

* The first recorded raid on the English mainland was noted in the *Anglo-Saxon Chronicle* entry for 789, on the coast of Wessex, probably at Portland. The story of the Viking Age is taken up in *Ælfred's Britain*.

sins of those who live there? It has not happened by chance, but is the sign of some great guilt.[10]

For the time being Mercia was insulated from these raids. King Offa and Queen Cyneðryð, diplomatically triumphant in securing a Mercian archdiocese at Lichfield and consecrating their son Ecgfrið as anointed successor, seem to have been preoccupied with securing their own dynastic legacy. The Northumbrian marriage of their daughter Ælfflæd to King Æðelred followed that of another daughter, Eadburh, to King Beorhtric of Wessex in 789. A third daughter, Æðelburh, was dedicated as a bride of Christ, becoming abbess of the royal minster of Fladbury in the *Hwiccian* heartlands and another correspondent of Alcuin. *

It was entirely natural that the most powerful king in Britain should also wish to enter into a dynastic alliance with his illustrious neighbour across the Channel. The Mercian and Frankish courts had been in diplomatic contact since at least the time of the Legatine Mission of 786, from which Charlemagne gained useful knowledge of the state of the English kingdoms. He knew that Offa was a great overlord; that Mercia was pre-eminent among the English kingdoms. He also knew of Francia's ancient relations with Kent, as its sometime overlord. He knew of the primacy of Canterbury's archbishops. In 796, at the height of a diplomatic crisis, Alcuin was to refer to an 'ancient pact' between Offa and Charlemagne.[11] Such treaties might be reinforced by the same sort of political generosity that worked to engage regional power. A charter issued at Tamworth, increasingly important as a Mercian royal centre, made provision for the exemption of certain tolls on land in London to be granted to the church of St Denis in Paris – an ancient royal foundation and pilgrimage site[†] – at the same

* She is identified as an abbess and daughter of the king in a charter, S127, of 787.
† The charter as it survives (Electronic Sawyer S133) is a forgery; but probably based on a genuine original.

time confirming previous grants to the church for land along the South Coast.

The cosy idea that Carolingian influence – in politics, the arts and economics – flowed one way towards the English kingdoms has very recently been shattered by the discovery of a unique coin dating to 793 and bearing the name of Charlemagne's third 'dear and very lovely wife the queen', Fastrada, who died the following year. She is the only queen so named on a Carolingian coin and the numismatist Simon Coupland confirms that the issue must have been inspired by the Mercian exemplar of Queen Cyneðryð, dating to the 780s, rather than the other way around.[12]

It seems to have been Charlemagne who took the first cross-Channel initiative on the subject of marriage, in or before 790, employing as his envoy Gervold, abbot of St Wandrille on the River Seine and collector of the king's tolls and taxes at various ports, including *Quentovic*:

> [Gervold] many times discharged diplomatic missions to [...] King Offa [...]; finally one on account of the daughter of the aforesaid king, who was sought in marriage by the younger Charles [Charlemagne's son by his second wife, Hildegard]; but [...] Offa would not agree to this unless Bertha, daughter of King Charles the Great, should be given in marriage to his son...[13]

We cannot be sure which of Offa's four daughters was the subject of so great an honour, since the marriage never happened; probably it was Æðelswiðe, mentioned in the charter of 787.* For Charlemagne the initiative was an orthodox diplomatic move: Charles the younger was unlikely to succeed his father;† a formal Mercian alliance would cement ties of friendship and mutual

---

* See above, p. 233.
† Charles later became Duke of Maine, while Francia was ruled by Louis the Pious, a younger son.

recognition without costing more than a minor dynastic asset. On the Mercian side, Offa and Cyneðryð would acknowledge the superiority of the Frankish king: gift of a daughter to a junior Frankish royal was an acceptance of inferior status by Offa that, nevertheless, would draw Mercia more intimately into the club of élite European kingdoms.

But a bigger prize might be within reach for Offa and Cyneðryð. A marriage between Prince Ecgfrið, the Mercian heir-apparent, and Princess Bertha would have brought significant diplomatic advantages, reflecting a much more equal relationship between the two kings. King Offa's suspicions about Charlemagne's relations with other Anglo-Saxon kingdoms may also have motivated his desire to bring a significant Carolingian hostage into his court.[14] Significantly, perhaps, Bertha had been the name of the Merovingian princess given to King Æðelberht of Kent in the days of Augustine's mission to convert the Kentish people. Jo Story, who has made a study of the relations between Carolingian kings and their Anglo-Saxon counterparts, offers the insight that Offa may have intended Ecgfrið to rule as his *subregulus* in Kent while Offa still lived. A Frankish queen named Bertha would sooth Kentish sensibilities.[15]

In the end, neither marriage took place. It may have been Archbishop Jænberht's opposition to Ecgfrið, as much as a perceived diplomatic insult on the part of the Frankish king, that scuppered the deal and engendered a period of frosty, not to say openly hostile, relations between the two kings – even if Einhard, Charlemagne's biographer, believed that the Frankish king simply wanted his daughters to stay at home.[16]

More complex political undercurrents can be detected. Sometime between 784 and 791, plausibly during the diplomatic comings and goings over Ecgfrið's inauguration and the promotion of Lichfield to an archdiocese, King Offa's *missi* – envoys – arrived at the court of Charlemagne, warning him of wicked and false rumours concerning, and later conveyed to, Pope Hadrian in Rome. The pope himself later told Charlemagne that...

... some enemies of himself and yourself had brought to our apostolic notice that the same King Offa was suggesting to you [Charlemagne] that, with his advice and encouragement, you should (which God forbid) remove us from the holy seat and [...] place another upon it from among your own people...[17]

Charlemagne had taken Offa's warning seriously and dispatched both the Mercian *missi* and his own ambassadors to Rome to counter the rumours and assure Pope Hadrian that these 'most wicked, hostile' rumours were 'quite certainly false'.[18] The pope assured the Frankish king that he found the 'filthy assertions' and 'unheard-of deceit' *incredibilis*.[19] Nicholas Brooks[20] and Jo Story suggest that the conspiracy was fashioned to discredit King Offa at the papal court and that its instigators may have been Archbishop Jænberht and an anti-Mercian faction in Kent.

The distinguished Early Medievalist Jinty Nelson, who has studied Frankish history over many decades, sets this 'dastardly plot' in a context of periodic, very real plots against Rome's popes among rival factions in Ravenna and Rome throughout the eighth century. Accepting the possibility that both Offa and Charlemagne were acting in good faith in bringing such rumours to Hadrian's attention in order to assure him of their honest and supportive intentions, she suggests that they may also have taken advantage of such expedient intelligence to exert their own pressure on him. Modern diplomats might be surprised to find that their stock-in-trade of plot, counter-plot and media spin were also standard, if under-recorded, features of eighth-century European diplomacy. Royal courts were quite as manipulative as their successors in state governance.

As it happened, Archbishop Jænberht died in 792 and Offa was able, for the first time in nearly thirty years, to impose his own candidate on Canterbury: Æðelheard, former abbot of the Lindsey monastery at Louth[21] and thereafter bishop of Winchester. Archbishop Hygeberht of Lichfield presided over his inauguration – a strategic *coup de main*, so far as King Offa was concerned. From that point the new archbishop of Canterbury is to be found attesting the Mercian king's charters *after* the signature of his Lichfield colleague. In a rank-sensitive world, such niceties were emphatic statements of precedence.

While the cross-Channel diplomatic stand-off continued, diplomatic logic may have underlain the proposal that Offa's and Cyneðryð's as-it-were spare daughter marry King Æðelberht of East Anglia in 794.[22] Almost nothing is known of East Anglian history in this period, except that King Offa's ability to have coins in his name minted by East Anglian moneyers – probably at *Gipeswic* – is a strong indicator of at least theoretical Mercian overlordship; and Æðelberht's minting of coins in *his* own name is an equally eloquent indication of his periodic assertion of independence. A dynastic marriage would seal Æðelberht's submission to Mercian *imperium*. But something went very badly wrong. A blandly chilling entry in the *Anglo-Saxon Chronicle* for 792 (correctly 794) records that King Offa ordered King Æðelberht's head to be struck off. Brief as it is, the *Chronicle's* account does not hint at assassination by a Mercian agent infiltrating the East Anglian court; rather, it suggests that Æðelberht was in attendance at the Mercian court; that he was imprisoned and then executed while under his lord's protection. This sort of extreme executive action sometimes occurred when kings were found to have repudiated their wives. But East Anglia had also, in the days of St Guðlac, offered sanctuary to Mercian exiles. Had Offa's daughter reported some disturbing intelligence of a plot against her father? An allusion to this shocking event and to Offa's motivations may be contained in a letter later written by Alcuin to a former minister at Offa's court whom, he wrote, knew 'well how much blood the father shed to secure the kingdom for his son'.[23] Later tradition, and a cult that developed in medieval Hereford, had it that Æðelberht was murdered at the royal *vill* at Sutton Walls, site of an Iron Age hillfort overlooking the River Lugg.[24]

Royal exiles posed political challenges for all parties. They had to make themselves useful to their surrogate lord, fighting for his interests with their *comitatus* while keeping an eye on developing opportunities back home. The death of a rival offered opportunities; demanded swift action and the support of their surrogate lords. On the flip side, those lords who sheltered exiles at their court might face diplomatic consequences, incurring the displeasure of an ally. One Mercian exile caused particular problems for Charlemagne. Sometime after the installation of the Mercian Archbishop Æðelheard at Canterbury, Charlemagne wrote to him requesting his intervention with King Offa on the matter of a Mercian warrior named Hringstan, exiled at the Frankish court with his followers. There is no indication of Hringstan's background, or if he posed a genuine threat to the Mercian régime; but Jo Story suggests that he may have been a man of Lindsey, Æðelheard's own homeland.[25] One might even speculate that he had first tried his luck in East Anglia, with the disastrous consequence that Offa took executive revenge on its king.

The archbishop was to understand that Charlemagne's motives were entirely benign – affording Hringstan hospitality only in the hope of brokering a reconciliation; but the awkward fact was that Hringstan had died. Charlemagne now sent the remnants of his *comitatus* – 'these miserable exiles' – to the archbishop in the hope that he might broker their peaceful return to their own country. If not, then they should be sent back to the Frankish court – better to remain exiled than die in their own land.[26]

Charlemagne's disingenuous shedding of these unfortunates did not prevent him from accommodating other Anglo-Saxon exiles. None was more threatening to the existing order than Ecgberht, a West Saxon prince who had been driven out of his homeland and into a thirteen-year Frankish exile by kings Beorhtric and Offa, after King Cynewulf's death in 786.[27] That he was also said to be the son of Eahlmund, briefly king of Kent in about 784, is an indicator of both his potency as a future threat and a dynastic conundrum for historians. Later West Saxon genealogists gave Ecgberht an elaborate pedigree that included such illustrious forbears as King Ceawlin, the mythological progenitor Cerdic, and Woden.[28] That is hard to square with a father who was king of Kent, unless Ealhmund himself had a West Saxon mother. Ecgberht's fate matters because he would return to seize the throne of Wessex after Beorhtric's death in 802; would reign for thirty-seven years; and would decisively conquer Mercia. He was also the grandfather of King Ælfred.

A third exile of interest to Mercia was a cleric of dubious credentials named Odberht in Frankish sources and Eadberht *Præn* 'the Priest', in his native Kent.[29] The circumstances of his arrival in Francia, and his reasons for requesting Frankish protection from supposed Mercian enemies, are obscure; but he constituted a thorny issue for Aachen's diplomats. A letter from Charlemagne to Offa assures the Mercian king that this man had been sent on to Rome to resolve his issues with the Mercian king.[30] One may read from this[31] that Eadberht, a pretender to the Kentish throne, had been forcibly tonsured by Offa and expelled from Kent; and that Charlemagne did not wish to warm his hands on such a diplomatic hot coal. Let the pope intervene if he wished.

The quarrel over royal marriages that so exercised Alcuin when he was at York rumbled on. In a letter of about 790 to the abbot of Clonmacnoise in Ireland, Alcuin promised his friend some oil that, he lamented, was now almost unobtainable in Britain because traders on both sides of the Channel had been embargoed – the first time that such a trade war appears in the Anglo-Saxon sources.[32] The value of that choked-off trade is increasingly clear from the range and quantity of coins now known from the period and from the nature of coinage reforms implemented under direct royal control. It is conspicuous that the most important of these was initiated by Offa in about 792, in the midst of the trade embargo that had seemingly been imposed unilaterally by Charlemagne on cross-Channel traffic.

Offa's so-called 'Heavy' pennies, a consistent 1.45g in weight,[*]

---

[*] Up from 1.30g under the previous reforms initiated by Pippin in Francia and subsequently adopted by Offa. Sawyer 2013, 77.

were produced at *Gipeswic*, London and Canterbury, and have been found across most of England outside Northumbria. A conspicuous lack of contemporary Frankish coins in England might indicate that Offa was able to effectively prohibit the circulation of Charlemagne's coins from his ports of entry – perhaps recycling them through his own moneyers. *

The new Mercian standard was widely adopted across England. The style was consistent, with a common obverse design inscribed OFFA in a central panel, with REX below and, above, a monogram uncial M for Mercia, with the moneyer's name on the reverse. Within a year or so Charlemagne responded with his own coin reforms, raising the Frankish penny to an even heavier 1.7g.[33] This was the prosecution of war by other means.

Charlemagne's order that no ship from Britain was to land on his shores so concerned Abbot Gervold of St Wandrille, the king's chief toll-collector on the Channel ports, that he was said to have tried to persuade his master not to implement it.[34] Nevertheless the embargo was at least partially enforced until the middle of the decade. But needs must. The idea that Britain's coasts could be made impermeable to those wishing to cross it for the sake of pilgrimage, refuge or commerce seems to persist despite all evidence to the contrary. Charlemagne rebuked the Mercian king with an accusation that some enterprising English traders had disguised themselves as pilgrims in order to evade tolls; and it seems that a number of them had been arrested.[35] From Northumbria, Alcuin wrote to a correspondent in Francia in confident expectation, to request that he be sent five pounds of silver for bartering or selling, some three-ply garments of goat-hair and wool for the boys under his care, linen for himself, black and red goat-hair hoods (if his correspondent could find any), plenty of paints of fine sulphur and dyes for colouring.[36] Years later, after the embargo was lifted, he showed his own generosity, sending his friend Eanbald, by then

---

* As Sawyer notes, some numismatists believe that cross-Channel trade was so small in volume that few Carolingian coins would find their way here in any case.

archbishop of York, one hundred pounds of tin to clad the roof of the archbishop's belfry at York.[37] The tin had very likely been imported into Francia from Cornwall. The result must have been splendid. Ultimately, economic pragmatism combined with skilful diplomacy would prevail.

Notwithstanding Viking raids* and the diplomatic stand-off with Charlemagne, the Mercian court maintained its domestic policies of judiciously calculated grants, the suppression of native dynasties and strategic support of the church. Offa had already reinforced his patronage network in Kent after reimposing Mercian *imperium* there in about 785. In a charter of 786 the king granted to his minister Ealdberht and to Ealdberht's sister, Abbess Seleðryð, a generous estate of fifteen *sulungs* in Kent, the minutiae of which are revealing of the sorts of portfolio that élite lords might assemble, in plundering the wealth of a subordinate people. It came with...

> ... the swine-pastures in the weald of the Limen people [Lyminge district] and the weald of the *burh* people [Canterbury] and in Buckholt and in Blean. And in Hardres 100 loads of wood and the entry of two wagons in winter and summer, and in Buckholt, timber [for] building and fuel-collection; an urban estate at Curringtun in Canterbury, said to be located to the north of the market-place, with everything belonging to it. And fishing for one man in Pusting weir, and his salt-works above it, and wood-collecting in Blean for this [i.e. fuel for the salt-boiling].[38]

---

* Notice of a raid on Francia comes from 799, against the island monastery of Noirmoutier off the Aquitaine coast, not far south of modern Nantes, mentioned in a letter by Alcuin in which he describes 'pagan ships'. He might be referring to Moors from the south (Letter 65, to Arno, bishop of Salzburg. Allott 1974, 79) but in any case it seems highly probable that vulnerable monastic targets along the North Sea and Atlantic coasts of Continental Europe were subject to the same predations as the British and Irish kingdoms. The *Royal Frankish Annals* describe the English Channel in 800 as being 'infested' with pirates.

A less informative charter of 788 granted to Offa's minister Osberht, and to Osberht's wife and children, a single *sulung* of land at Duningcland in the Kentish district of Eastry.[39] This plot had, perhaps, been seized as punishment for some act of disloyalty on the part of its minor lord; or he had died intestate. It is compelling evidence of how deeply penetrative Mercian *imperium* in Kent could reach.

In 792, in the aftermath of Archbishop Jænberht's death, King Offa responded to a request from his new archbishop, Æðelheard, granting by charter at *Clofesho* confirmation of the privileges previously given to the churches of Kent by King Æðelbald and by their own King Wihtred...

> I, Offa, king of the Mercians, release these monasteries from all the burdens of royal taxes or secular service [...] either in the king's refreshment and alimony, or in the works for the royal *villae*, or in [...] feasts; nor in hounds, nor hawks, nor horses [...]; nor in the pastures of the plains, nor in the woods, shall they be in any way liable...[40]

The sting was in the tail. Mercian kings, since 747, had reserved from their ecclesiastical grants the three so-called common burdens or *trimoda necessitas*. No doubt with very recent raids on Lindisfarne and Jarrow still fresh in the memory, King Offa now imposed on all the churches of Kent the same reservations: for the construction of bridges and fortifications and expeditions against the heathen at sea. If this looks a little like the sort of friendly visitation of a racketeer, it is consistent with the Mercian policy of embedding religious institutions in ideas of a common national purpose; of public good; and, not least, keeping a tight rein on spurious monastic proliferation. Offa was in a position to enforce such demands.* Eighty years later a West Saxon

---

* A second charter from late in King Offa's reign also explicitly reserves these burdens in a grant to the *Hwiccian* Ealdorman Æðelmund (S139 of 793–6).

counterpart, King Ælfred, barely surviving the overrunning of his kingdom by those same heathen pirates, found it very difficult in practice to extract such burdens from his countrymen.

How were the monks, priests and lay tenants of the monasteries supposed to fulfil these obligations? Saints Cuthbert and Wilfrid, both born of noble Anglian stock, had served in the armies of their kings before taking holy orders. A former king of East Anglia, Sigeberht, had been dragged from monastic retirement in the 640s, sent into battle brandishing only a staff – and duly martyred. But eighth-century bishops, abbots and their dependants were hardly likely to follow suit. Who, then, supplied the labour for bridge-work, the repair of defences and active military service? One clue is to be found in a much later document, the Domesday Survey of 1086. By the middle of the eleventh century every five hides of land was required – at least in theory – to provide a thegn to fight in the *fyrð*, the regional or national muster. In his masterly analysis of the Domesday Book the Victorian scholar of legal history, Frederick Maitland, saw that...

> ... many of the prelates [in Domesday] have thegns, and for the creation of thegnlands by the churches it would not be easy to find any explanation save that which we have found in the territorialisation of military service. [...] The thegns of the churches [...] have been endowed [...] in order that they may do the military service due from the ecclesiastical lands.[41]

Monastic proprietors had by then adopted a practice, already known in Francia, of sub-letting parts of their estates to men of weapon-bearing rank who would render the minsters' military obligations on their behalf. The origins of this practice would appear to lie in eighth-century Mercia. As for the heavy labour required for ditch-digging and bridge construction, it addresses a broader question about the labour that made minsters so productive; that provided them with substantial enclosures and the buildings within. When they were endowed with lands by generous patrons, those semi-free or unfree tenants who belonged to the land – its *rustici* and *villani* – became dependants who must serve their ecclesiastical lords and render them the fruits of their toil just as they had their secular predecessors. How reliant Anglo-Saxon lordly farmers were on populations of slaves – þeowas – cannot be known. *

King Offa was at pains not to alienate key members of his ecclesiastical élite closer to home. In 781 he had come to an arrangement with Bishop Heaðored of Worcester for the restoration of a monastery at Bath to the royal portfolio – it having been granted to his grandfather, Eanwulf, by King Æðelbald.[42] In return the king conceded to the bishop a number of other contested estates and their food rents for three years: 'that is, six entertainments'. In 794, probably at the same *Clofesho* council at which he confirmed Kentish church privileges, he restored to Bishop Heaðored a five-hide estate at Aus (Cliff in Gloucestershire) that had been illegally seized by his *comes* Bynna; the bishop had been able to produce a charter of King Æðelbald to prove his title to the land.[43]

Offa and Cyneðryð were also active patrons of a number of minster foundations on their lands. Pope Hadrian, perhaps acknowledging Mercia's value as a friendly state after the exposure of the (real or otherwise) plot against him, had confirmed their possession of a number of churches dedicated to St Peter. But where were they? The Annals of Winchcombe Abbey claim Offa as the founder of a convent there in 787. Winchcombe was also a royal *vill*, later acquiring the status of a shire whose boundaries might reflect its original extent as an estate.[44] A second foundation may have been sited in what is now Hertfordshire. A late Anglo-Saxon

---

* As the reference in a charter of King Æðelbald shows (see above, p. 168) slaves were embedded within territorial lordship rights. They might be victims of raids or war or punishment for crime; or by birth. In later Anglo-Saxon period references to their manumission are common – Bishop Wilfrid had freed more than 200 on his Sussex estates in the seventh century. The Old English word for a slave was *þeow*; the perjorative *wealh* was also sometimes used. Pelteret 1995.

charter of 1007 records a grant of King Æðelred II to St Albans Abbey that enclosed the shrine of the Romano-British martyr of *Verulamium*. Part of the charter reads...

> Offa, in former times king of the Mercians, held a portion of these lands by royal right [*regali iure possedit*] and he granted it to the aforesaid monastery for the love of such a great martyr.[45]

The St Albans historian Matthew Paris (c.1200–59), writing *Lives of the Two Offas* in the belief that the Mercian king had been the abbey's founder, was able to draw on the abbey's own traditions as his source, even if a lacuna in the abbey's fortunes during the Viking period gives modern historians cause for scepticism. Paris claimed that Offa had discovered the British martyr's bones and built a shrine to hold his relics; but this is suspiciously conventional and not to be taken at face value. Bede knew of a Roman church there, still visited by pilgrims in his day. King Offa may have founded a minster to replace that church and house the shrine.[46] Even if Offa was not the *de facto* founder of the abbey, he seems at least to have been a generous patron and, as so often with the king's policies towards the church, he did so with a broader strategy in mind. St Albans commanded the approach to and from London into central Mercia along its most important highway, Watling Street. Paris believed that Offa had built a *burh*, Kingsbury, a short distance to the west of the church. Today's Kingsbury area of the town* lies a little way north-west of the abbey, overlooking the Roman town of *Verulamium* across the River Ver; but Rosalind Niblett, who has studied the St Albans of the Offan period, believes the *burh* lay above the Roman forum across the river: the centre of a royal *vill*.[47] If she is right, it is an important piece in the Offan burghal jigsaw. In time, more of these supposed royal minsters and *burhs* may yet be identified.

* Shown on the Ordnance Survey Second edition.

The spectrum of charters issued under Æðelbald and Offa over a period of four-score years reveals a cumulative process of investment in productive landscapes. Bishops and abbots assembled portfolios of resources, sometimes scattered across wide areas but often focused on prime riverside sites or on specific resources like brine springs or the rich sheep pastures of the Cotswolds. Remission of royal tolls and the booking of land in perpetuity allowed the more entrepreneurial ecclesiastics to invest their accumulated wealth not just in fine trappings for their churches but also in the land itself. Unlike peripatetic kings, who had to move between their *vills* and estates consuming produce on-the-hoof, ecclesiastical communities were inherently static: abbots, abbesses, bishops and their communities needed to be fed year-round at the centres of their estates. The telltale imposition of gridded axial layouts and buildings capable of storing large quantities of surplus, together with increasing evidence of riverine trade, a partially monetised economy and the importation of foreign goods, shows itself in long-lived, stable settlements with ample material culture yielded by excavation. Farming settlements display increasing signs of specialisation and interdependency. English kingdoms won a reputation abroad for fine and abundant textile goods and for their scholarship – which itself was a high-value export. The rivers Nene, Trent, Severn and Thames were highways for the transportation of those goods; but also for information and ideas.

The archaeologist Duncan Wright has proposed the emergence during the eighth century of the 'home farm': a dependent agricultural settlement acquired by a minster, which functioned to support the community's less 'productive' members – its clerics, monks and the household of the abbot or abbess. He identified two sites in East Anglia and another in Gloucestershire where the archaeological evidence seems to support the idea of active relations between minsters and such farms. The Cowage Farm complex, a dependency of Malmesbury Abbey, significantly lies upstream of it on the Somerset Avon; that at Fordham in Cambridgeshire lies upstream of the minster at Soham to which it seems to have been attached – so that produce literally flowed to the place where it would be consumed or traded onwards.[48]

In Mercia the cluster of economically active sites along the River Nene from Weedon as far downstream as Peterborough hints at such relationships. Earls Barton, where a celebrated pre-Conquest church still stands overlooking the Nene, bears a name element identifying it as an 'inland' component of a minster estate.* Might it have belonged to the minster at Northampton, or to the abbots of *Medeshamstede*? A second candidate for a home farm along the same river lies at Aldwincle, a very large and valuable estate at the time of the Domesday Survey that belonged partly to Peterborough – formerly *Medeshamstede* – Abbey and had, before the Conquest, been held 'for the supplies of the monks'.[49] It would be surprising if evidence for such relationships did not also emerge along the Trent, Severn and Thames, where so many monastic sites flourished in the eighth century. At Old Windsor, a hall complex and watermill, probably eighth-century in date, were discovered in the 1950s by Brian Hope-Taylor, the pioneering excavator of Yeavering.[50] Unlike the great palace site in Northumberland, the Old Windsor excavations have never been fully published. But the mill, naturally enough, suggests intensive production of grain and flour and the investment of significant capital. It was a complex affair, with three vertical wheels and a leat more than half a mile long:[51] a substantial engineering achievement. Less than a mile away on the opposite, east bank of the Thames, the village of Wraysbury has revealed excavated evidence for a contemporary complex lying next to St Andrew's church. A series of enclosure ditches yielded coins of kings Offa and Cœnwulf and John Blair suggests an inter-dependent relationship between the two sites.[52]

Much more extensive excavations at Yarnton on the tight, northernmost loop of the Thames above Oxford exposed a long-lived Anglo-Saxon settlement with a modest hall complex that, nevertheless, formed the core of an intensively managed settlement of fenced pens and paddocks, sunken-floored buildings, pits and

---

* Originally from *bere-tūn* – 'barley-tun'. Faith 1997, 36ff.

ancillary buildings across a site of some eight acres, with distinct functional zones: grain storage, crafts, a hen-house, perhaps. Increasingly widespread evidence of ditched livestock enclosures at this period is suggestive not just of controlling movement and security. Folding stock close to habitation at the centre of a farm implies the concentration of manuring, so that small fields close to the hall might be fertilised before ploughing to improve and maintain bread wheat and barley yields.[53] Yarnton's material finds included a small assemblage of Ipswich Ware pottery and a pattern-welded *seax* blade. Like burials, which are only very rarely found on sites of this period, the domestic detritus of such habitations seems to have been disposed of where archaeologists cannot find it – a source of ongoing frustration. On the other hand, analysis of soil samples allowed Yarnton's excavators to show that new crop types such as rye and legumes were part of an intensive farming régime here in the eighth century.[54] Flax was being grown and processed, indicative of linen textile production – itself a marker for élite consumers of fine cloth. Evidence of hay cropping on the Thames floodplain indicates a policy of over-wintering substantial breeding flocks or herds. Cattle drank from ponds that naturally formed in old, braided river channels on the site, which also allowed fish to be kept and were probably also used for the filthy business of retting flax. Yarnton seems to have been a dependent settlement of the wealthy* Thames-side minster complex at Eynsham, some two and a half miles to the south-west.[55]

It is still a moot point whether Mercia's powerful kings drove, promoted or merely profited from the abundance of the land's increasing wealth. In the ninth century their *de facto* capital at Tamworth was to have its own watermill; but whether as an initiative of Mercia's kings is impossible to say. They were not insensitive to the value of productivity in underpinning the power of the Mercian state and their own prestige; and they enjoyed

---

* Seemingly endowed with a huge 300-hide estate – the size of the smallest polities recorded in Tribal Hidage – possibly as early as the reign of Wulfhere (658–75). Kelly 2008.

its profits. We cannot say if royal permission was required to construct such marvels as watermills or if, as in later centuries, kings were able to extract a share of profits from them; but mills not only increased the speed and efficiency at which grain could be milled; they also centralised flour production – and mills would later become a focus for the taxation of arable surplus.

A king's dooms, which codified customary law governing crime and punishment, compensation, the resolution of land disputes and, to a more limited extent, the buying and selling of goods, monumentalised his reign as surely as his coins or great earthworks. Often, one detects a strong element of the reactive: fines for very specific offences like fighting – *gefeohte* – within the king's hall or for allowing a thief to escape.[56] More general customary laws included provisions for valuing property, livestock and human life while other clauses were intended to resolve disputes between neighbours. Truth was established by oath; and oath was guaranteed by status. Great swathes of customary law must, nevertheless, have remained unwritten.

It is a matter of regret that no law code of King Offa survives, and there is much to be said for the idea that his judgement was transmitted orally, through precedent and as opportunity arose, preserved in the memories of his household. It is possible that an Offan code was written but did not survive the devastations of the Viking Age, like so many other documents. A letter of Alcuin, written after Offa's death to an unnamed minister, perhaps the long-lived and trusted Brorda, does not quite answer the question.* He urges that the race of the Mercians…

… diligently observe the good, moderate and chaste customs (*mores bona et modestos et castos*) which Offa of blessed memory established (*instituit*) for them.

---

* Osberht was identified as the minister by William of Malmesbury; but several historians argue the likelihood that he was addressing Brorda (Ray and Bapty 2016; Wormald 1991). *Alcuin Letters* 46; *English Historical Documents* 202.

Much attention has been paid to the preamble of King Ælfred's code of a century later, in which the king claims to have mined the most just of the laws he had found among his predecessors: Ine of the West Saxons; Offa and Æðelberht of Kent.[57] There has been an understandable attempt to identify elements in Ælfred's dooms that might fossilise the Mercian legal roots – so far with little success. The distinguished Anglo-Saxon and legal historian Patrick Wormald doubted that they can be found.[58] Instead, and with Alcuin's letter in mind, he argued that both Alcuin and Ælfred were probably referring not to a lost set of dooms but to the canons of the Legatine Mission of 786. The provisions of these canons are understandably focused on liturgical and synodical matters, but they also touch on matters of state. A law code of Offa would give historians crucial insights into this key period in the emergence of the Mercian state and into the evolution of royal administration and social justice in the century before the Viking Age. Its absence or silence is history's loss.

*

Through the first half of the 790s the diplomatic impasse between Mercia and Francia rumbled on. Emissaries passed to and fro across the Channel. The death of Pope Hadrian in 795, after twenty-three years in office, seems to mark the end of one era and the beginning of another. His successor, Leo III, was consecrated the day after Christmas and he sent to Charlemagne, by envoy, Rome's standard and the keys to the tomb of St Peter as an act of submission. The Frankish king responded with a magnificent gift of treasure, part of an enormous hoard of booty captured that year from the Huns or Avars of Pannonia across the Danube, which Duke Frioli, the Frankish commander, had brought back to Aachen. In celebration of that famous victory Charlemagne distributed a part of the loot among his lords and bishops, as befitted a generous and all-powerful king.[59] Whether his new-found wealth and a spirit of magnanimity prompted the king to renew friendly relations with Mercia, or whether it was through the good offices of Alcuin and Abbot

Gervold – each with their own reasons for promoting entente – we cannot say. But at the beginning of 796 Alcuin drafted a letter to King Offa on behalf of his master, offering terms.[60] In it, Charlemagne acknowledged the receipt of letters brought by Mercian envoys, the contents of which had prompted the Frankish king to agree that genuine pilgrims from Mercia wishing to travel to the holy sites were to be granted leave to travel once more. Merchants, too, were welcome, so long as they conducted their business properly and paid the appropriate tolls – to men like Abbot Gervold – and so long as reciprocal arrangements applied across the Channel.

King Offa had, it seems, also requested from his Frankish counterpart some 'black stones' of a certain length and Charlemagne, seemingly puzzled by the request, asked Offa to send him a messenger who knew exactly what he meant so that Charlemagne could have them dispatched. One infers that some envoy had seen such marvels in the new palace at Aachen. The quid pro quo was that 'our people' – Frankish merchants, perhaps, or members of the court at Aachen – requested that the English cloaks they had been sent should be longer, like 'those we used to get in the past'.

Then there were gifts: dalmatics and palls for the English bishops, in the hope that they would pray for the soul of the blessed Pope Hadrian, while each of the three English metropolitans had been sent something from the Avar hoard. As for King Offa himself, Charlemagne sent him a belt and a Hunnish sword from the same treasure and two silk cloaks. The sword may be that which Edmund Ironside was bequeathed by his brother Æðelstan Æðeling in 1014...

> ... And ic geann Eadmunde minan breðer Þæs swyrdes
> Þe Offa cyng ahte...

> ... And I grant Edmund my brother the sword which
> King Offa owned.[61]

In a second letter from that year Alcuin assured Offa of his and Charlemagne's loving and trusting friendship – a diplomatic convention.[62] He informed Offa that Charlemagne was sending envoys to Rome 'for the jurisdiction of Archbishop Æðelheard' – a formal request that he be sent the pallium to which his office entitled him. The letter must have been written sometime in May or June 796, because Alcuin revealed in it that messengers sent to King Æðelred in Northumbria with gifts (yet more of the Avar treasure) had returned – having been delayed by extending their mission into *Scotia* – with the news that Æðelred had been killed at a place called The Cover.* A long and detailed entry for that year in the *Historia Regum* recorded portentously that an eclipse of the moon occurred between cockcrow and dawn on 28 March; and that the king's murder took place on 18 April.

King Æðelred was briefly replaced by one Osbald, but he was deposed and exiled after only twenty-seven days. In his place another Northumbrian exile,† Eardwulf, was made king: these were indeed tumultuous times. Charlemagne withdrew his gifts to the Northumbrian king, describing Alcuin's people as treacherous and perverse.

In the same letter Alcuin admitted to Offa that he had intended to bring the king his master's gifts in person and then return to Northumbria; but that it no longer seemed that any man was safe there: the earth of his homeland was stained with the blood of its princes, its holiest places ravaged by pagans, its altars desecrated. Such horrors were a lesson to all good kings. The same year, having retired to Tours, Alcuin wrote to Offa's daughter Æðelburh, abbess of Fladbury. Sending her a dress on behalf of Charlemagne's consort Liudgard, Alcuin hoped that she would enter the queen's name in her monastery's *liber vitae* – its list of patrons. He also sent her a plate and flask with which to make offerings – his own gifts, it seems. 'The infidelity of my country causes me such horror,' he confided.[63]

Northumbria's new king, Eardwulf, was consecrated on 26

---

* Likely *Coria*, the royal *vill* and former Roman town of Corbridge in the Tyne Valley.
† In Pictland? In Francia... In Mercia...? We do not know.

May of the same year, 796.[64] Since King Æðelred had been King Offa's son-in-law, a Mercian military expedition to avenge him, securing the position of his widow, Offa's daughter Ælfflæd, and potentially engineering *imperium* over Mercia's northern neighbour, would have been legitimate; expected even. But less than two months later, on 26 or 29 July, King Offa himself was dead, probably of old age. If he was buried in a suitably grand mausoleum in the royal church at Repton history does not record it. Instead, we have the late and uncorroborated testimony of Roger of Wendover, the thirteenth-century historian and another resident of St Albans, whose *Flores Historiarum* contains some credible material not found elsewhere. Some of what he says about King Offa is merely fanciful – a journey he was supposed to have made to Rome, for example, in pursuit of canonisation for Alban – but his information that Offa was buried in 'a certain chapel' outside the city of Bedford on the banks of the River Usk has some credibility.\* As it happens, the document drawn up at *Clofesho* in 798 recording the legal dispute over the monastery at Cookham notes that the church at *Bedeford* – probably Bedford – was to be the residual beneficiary of the Cookham estate after Queen Cyneðryð's death.[65] That church must, by then, have been a royal minster. Jeremy Haslam, who has studied many of the Early Medieval settlements defended or urbanised during the Viking Age, believes that Bedford was an Offan *burh*, constructed at a strategically important crossing of the Great Ouse, one of the five substantially navigable eastward-flowing rivers that drained into and gave access to the Wash.[66] He tentatively identifies St Paul's church as the site of the minster built by Offa to serve the *burh* at the centre of a royal estate.

If it is true, as I suggested in Chapter 7, that the *burh* network was conceived by Mercian kings as a means of accommodating an ever-larger peripatetic royal household, bloated with *ministri*, functionaries, *missi* and petitioners, then the urgency of threats

---

\* Giles 1849, 166. The Great Ouse; Usk may be an older Brittonic cognate.

by Scandinavian raiders may have given it a new military impetus during Offa's last decade. Unlike the Alfredan *burhs*, which were conceived from the start to function as fortified refuges for a large district, manned by defenders drawn from the same area, Mercian *burhs* show no signs of urban character. There are no planned or gridded street systems like those that survive at Wareham and Wallingford; instead, several (Northampton, Hereford, Bedford and Oxford, for example) show signs of having a single spinal street, physically and functionally linking bridge, defended enclosure and an extra-mural market place.[67] That is to say, they are camp sites that must often have been sparsely occupied until the royal household arrived and set up its tents: a market would materialise; commercial, judicial and governmental business would be transacted; feasting and song, poetry and the recital of pedigrees would exhaust them all; then the court would move on – vaporising, so to speak, leaving behind only the most basic infrastructure to be maintained until it reappeared.

At a site like Northampton, whose minster must have stood close to the large timber hall on the north side of the single thoroughfare, one must imagine the flurry of activity triggered by the arrival – or anticipated arrival – of the royal household. The encampments of the various parties would be located as a matter of negotiation and precedent: a *princeps* like Brorda; the *subreguli* of *Hwicce* and *Magonsaete*; bishops and archbishops, each acutely sensitive to rank and having secured a favourable site relative to that of the king's household, reflective of their status. Provisions would be drawn from storage – the great timber hall, perhaps – or arrive on the hoof or by wagon or boat from dependent home farms. Traders, merchants and opportunist petitioners would turn up; would congregate outside the walls and jostle for attention.

Where those *burh* defences also enclosed minsters, as at Hereford, Northampton, probably Bedford, Worcester, Winchcombe and elsewhere, there existed a continued focus for consumption and, perhaps, trade. It becomes increasingly difficult for archaeologists to separate secular and ecclesiastical urban characteristics; and maybe it doesn't matter, because ecclesiastical lords and entrepreneurs were also farmers and traders. *

In the later ninth century, in the days when Norse authority in Mercia was focused on the so-called Five Boroughs of Nottingham, Stamford, Derby, Lincoln and Leicester, those central places first identified and probably fortified under Æðelbald or Offa were developing market functions, driven as much by the processing of loot as by productive trade. Later, in the days of Edward the Elder, when West Saxon authority was extended into areas formerly held by Vikings, it was the *burgware* – 'dwellers in the *burh*' – who submitted to the king.[68] By the time of the Conquest, a hundred and more years later, the obligations through which regional *vills* provided labour to maintain the *burhs* had morphed into rights to build and maintain *hagae* – houses or halls – within the *burh*. They enjoyed rights to gather taxes and fines and sit in judgement on their peers, with royal interests represented by reeves of one sort or another.[69] Stephen Bassett argues that several of these *burhs*, specifically Hereford, Winchcombe and Tamworth, where timber-laced ramparts have either been proven or inferred, were systematically manned.[70] In King Offa's day I do not think that idea had been fully developed; but the structures were in place: a system waiting for a crisis.

King Offa was succeeded, as he and his queen had always intended, by their son Ecgfrið – aged perhaps twenty-five but unmarried, so far as one can tell. Four charters purporting to be genuine survive from his reign.[71] Two of these[†] are probably fabrications, possibly based on genuine late eighth-century material. Issued – if at all – at Chelsea, they involve modest

---

* Eleanor Rye and Tom Williamson have recently suggested that the name element *burh* in East Anglia may, in fact, point directly to the existence of a minster within its enclosing vallum or hedge. Rye and Williamson 2020.
† Electronic Sawyer S150; S151.

The line of the Dyke adapts to local conditions but drives
relentlessly north.

donations to St Albans Abbey. A third, issued at Bath, records a grant of three hides to Ecgfrið's 'faithful' *princeps* Æðelmund.*

The fourth, and most probably genuine of the group, is full of interest. It relates to an estate at Purton in Wiltshire, which King Offa was said to have taken from Malmesbury Abbey and which the new king now returned at the request of King Beorhtric of the West Saxons and Archbishop Æðelheard. Purton lies north-west of what is now Swindon and a little way south of Cricklade, close to the source of the Thames in an area of significant ongoing Mercian–West Saxon tensions. In order to smooth the deal, Abbot Cuthbert of Malmesbury and his brethren had managed to scrape together the huge sum of 2,000 shillings to pay the new king off. The text records that Ecgfrið took the advice of his bishops and *principes* in agreeing to the restoration of the thirty-hide estate, and their names are inscribed at the end. First to attest was King Beorhtric; second his queen, Ecgfrið's aunt Eadburh; then the archbishop; then the bishops; then Brorda, Offa's long-time friend and loyal *princeps*; then Æðelmund, the aforementioned ealdorman of the *Hwicce*; then another long-lived survivor of the Offan court, Esne.[72]

New political realities are in play here, which hint that the grant per se was probably of secondary importance in a set-piece of diplomatic theatre. The young King Ecgfrið needed the approval of his West Saxon neighbour; his aunt acted as intermediary; money – a lot of money – changed hands; old retainers and the Mercian-approved archbishop of Canterbury stood at the new king's shoulder, ensuring a smooth transition of power. The archbishop of Lichfield, whose authority did not extend south of the Thames, was nowhere to be seen.

King Ecgfrið may have been distracted, by such an important diplomatic obligation that set the seal on his relations with Wessex, from events stage right. In the south-east, the exiled Kentish 'priest' Eadberht returned dramatically from his Frankish exile and, with

---

* He was an ealdorman of the *Hwicce* and a charter witness as far back as the 760s (Electronic Sawyer S58). See below, Chapter 10.

Kentish support, threw off Mercian overlordship.[73] Archbishop Æðelheard fled into Ecgfrið's protection – 'deserting' his see, according to an admonitory letter to him from Alcuin.* In that same year, 796, the Welsh Annals record a battle fought at Rhuddlan close to what is now Rhyl in North Wales: Venedotian territory, west of the Dyke. Was Ecgfrið flexing his military muscles for the sake of it – to show his *comitates* and any would-be usurpers that he was made of the right stuff; or had the armies of Gwynedd taken the opportunity, after the old king's death, to test Ecgfrið's mettle? The outcome is unrecorded; but it did not matter: by December, after a reign of 141 days,[74] King Ecgfrið was dead – perhaps of battle wounds or illness; perhaps by the hand of an assassin. His reign is marked only by absence: no coins bearing his name have survived and the suspicion must be that he was unable to produce any – neither at Canterbury nor Ipswich nor, more remarkably, at London.

Alcuin, reflecting on another tumultuous year, wrote to a Mercian correspondent, who may be identified with the *princeps* Brorda, wondering if 'England's good fortune is nearly over'...

The noble youth [Ecgfrið] did not die through his own sins,
I believe; it was the vengeance of the father's blood that fell
upon the son.[75]

Whether his death is related to the battle at Rhuddlan or whether, as Alcuin's letter might suggest, he was the victim of a coup or a revenge attack by a kinsman of King Æðelberht of East Anglia, is unknowable. His burial place is unknown.

For all Offa's and Cyneðryð's carefully laid plans, their dynastic scheme had failed at the first hurdle. With the clarity of hindsight – provided by overwhelmingly West Saxon sources – it is tempting to calibrate their achievements against later Mercian kings' failure to counter the threat of Norse raids and then Mercia's conquest

---

* *Alcuin Letters* 49, 797. That Æðelheard fled into Mercia is shown by reference to a grant to the archbishop of a monastery at *Pectanege* – perhaps Patney in Wiltshire – by Ecgfrið (S1258) in 796 (Brooks 1984, 121).

by the Great Viking Army of the 860s. The reality was more complex.* Kings Æðelbald and Offa and their tight-knit councils had seen beyond the bellicose imperatives of earlier dynasts; had used their long reigns to construct tools of royal administration far more sophisticated than the expedient weapons of battlefield supremacy. They fought wars by other means. Their booking of land was calculated and refined; they subjected their neighbours and rivals by assimilating them into the Mercian orbit – rewarding them with resources from which they could profit and with the surety of a king's peace; with the reflected glory of their unassailable *imperium*. They led their *comitates* into battle when they had to but otherwise rarely. They accumulated political and material capital and controlled its distribution; patronised and sponsored an age of cultural exuberance typified by architectural grandeur; by the iconography of divine and imperial power manifest in coins and in earthworks, in metalwork, textiles, literature and sculpture.

No-one can say whose imagination and forethought drove the evolution of a system of *burhs* and trading places that both released the economic potential of the landlocked Mercian midlands and created the scaffolding for a new sort of royal power – the power of a state apparatus. The evidence for these *burhs* has been hard won and is far from complete. Archaeologists have been blindsided by the more famous, more tangibly evident and more militarily effective Alfredan *burh* system created in response to the threat of extermination a hundred years after Offa. The Offan *burhs* were less defensive; more administrative. The proof of the pudding is the success with which Mercians, adapting to violent new realities in the ninth century, were able to prosper so that the towns of the Danelaw became prime economic assets highly desirable to tenth-century West Saxon kings and to a later generation of conquerors. It does not much matter whether one credits Æðelbald, Offa, his queen, their bishops, abbots, ealdormen, merchants or foreign actors in this English revolution. It does matter that it is ascribed to Mercia: the crucible of the English state.

* And is explored in depth in *Ælfred's Britain*.

# Chronography III
## 798–918

Unless otherwise stated, narrative source entries are from the *Anglo-Saxon Chronicle*.

AC:   *Annales Cambriae*
ASB:  Annals of St Bertin
AU:   Annals of Ulster
EHD: English Historical Documents 500–1042
HR:   *Historia Regum*
NHF: Nithard's Histories of the Franks
RFA: *Royal Frankish Annals*
S:    Electronic Sawyer
WM: William of Malmesbury: *Gesta Regum Anglorum*

798      King Cœnwulf of Mercia harries Kent and Romney
Marsh and abducts King Eadberht *Praen* (HR);
appoints his brother Cuðred as king of Kent after
suppressing rebellion.
— *Lundenwic* suffers serious fire (HR).
— *Synod of Pincanheale*, with Archbishop Eanbald of
York presiding (HR).
— *Synod of Clofesho* – the fortunes of the monastery
at Cookham are recorded (S1258).
— Probable founding of a community of monks at
Winchcombe by King Cœnwulf.

800      Charlemagne crowned Emperor of the Romans at
Rome by Pope Leo III (RFA).
— On 24 December a great 'wind threw to the
ground by its immense force cities, houses and
villages'; severe flooding and tree loss; great
destruction of cattle (HR).

| | |
|---|---|
| 801 | King Eardwulf of Northumbria attacks King Cœnwulf of Mercia inconclusively after a long campaign. Peace brokered by church and 'nobles' (HR). |
| 802 | Death of King Beorhtric of Wessex. Ecgberht (grandfather of Ælfred) succeeds to the throne of Wessex. |
| 803 | *Synod of Clofesho*; Lichfield archbishopric is declared invalid after an exchange of letters between Pope Leo III and King Cœnwulf of Mercia; confirmation of the primacy of Canterbury under Archbishop Æðelheard; assertion of episcopal power against secular lordship (EHD 205, 210). |
| 804 | Alcuin of York dies as abbot at Tours. |
| | — Charter: Cœnwulf, king of Mercia, and Cuðred, king of Kent, to Seleðryð, abbess, and her familia at Lyminge; grant of land in Canterbury to serve as a refuge from the heathen (S160; EHD 82). |
| 805 | Archbishop of Canterbury Æðelheard dies; succeeded by Wulfred. |
| 806 | King Eardwulf driven from Northumbria into exile in Francia. |
| 807 | King Cuðred, brother of Cœnwulf of Mercia, dies; King Cœnwulf now sole ruler in Kent. |
| 808 | King Eardwulf restored in Northumbria (RFA). |
| 811 | Charter describes London as a 'royal town' (S168). |
| | — Archbishop Wulfred gathers small territories into one large contiguous estate by judicious charter acquisition at Eastry. |
| | — Death of Cynehelm, son of Cœnwulf of Mercia, legendarily at the hands of his sister (WM). Buried at Winchcombe Abbey. |
| 814 | Death of Charlemagne, aged 70 (RFA); succeeded by his son, Louis the Pious. |
| | — Archbishop Wulfred travels to Rome; returns 815. |
| 816 | Hostilities break out between King Cœnwulf and |

Archbishop Wulfred, who has been minting coinage independent of Mercian authority.
— *Synod at Chelsea* called by King Cœnwulf; limits powers of Archbishop Wulfred.
— Mercian attacks against Wales: between Clwyd and Elwy and as far as Snowdonia (Cotton Vespasian A xi).

818      Mercia attacks Dyfed (AC).

821      *Synod at London* restores lapsed power of Canterbury; Archbishop Wulfred has been restored to his see by this year (Brooks 1984, 181).
— West Saxon victory over Cœnwulf in 821 at *Cherrunhul* (Cherbury Camp, Oxfordshire). King Cœnwulf dies in Flintshire, buried at Winchcombe; succeeded by kinsman Ceolwulf.

822      Mercians under King Ceolwulf destroy fortress of Deganwy, assert control over Powys (AC).
— Ceolwulf I of Mercia grants land near Kemsing (near Sevenoaks) free from burdens except military service against pagans (S186).

823      Ceolwulf of Mercia deposed; succeeded by Ealdorman Beornwulf; Powys regains independence.
— 'Here the king of the East Angles and his subjects besought King Ecgberht to give them peace and protection against the terror of the Mercians': East Anglia submits to West Saxon overlordship.

824      (or 825) *Synod at Clofesho*: King Beornwulf attempts to resolve disputes with Wulfred. Charter records recovery of land by Wulfred acquired by Quœnðryð, who retains abbacies but gives up forty-seven *manentes* of land in Kent and Middlesex (S1434).

825      Battle of Ellendun (Wroughton, Wiltshire): defeat of Mercian King Beornwulf by Ecgberht, King of the West Saxons. Æðelwulf of Wessex annexes Sussex, Kent and Essex; rules Kent as *subregulus*. East Anglia asserts independence from Mercia, killing

King Beornwulf, who is succeeded by an ealdorman, Ludeca.

827      Ludeca, king of Mercia is slain by East Angles with five ealdormen; Wiglaf becomes king.

829      Conquest of Mercia by King Ecgberht of Wessex, who mints coins as ruler of *Lundenwic*. King Wiglaf exiled.

830      King Ecgberht of Wessex leads expedition to Wales to claim overlordship there. Wiglaf restored as king of Mercia – by rebellion.

835      Heathens 'devastate Sheppey' – the first large-scale Viking raid on an Anglo-Saxon kingdom.

839      Probable date of death of Wiglaf of Mercia; succeeded by Wigmund, then Beorhtwulf.

842      Slaughter at London, Rochester, *Quentovic*; *Hamwic* and *Northunnwig* (?Norwich) plundered.

843      Charles the Bald becomes king of West Francia (NHF).

849      King Wigstan (grandson of King Wiglaf of Mercia), killed by Beorhtfrið, son of Beorhtwulf; he is buried at Repton, where a cult of St Wystan develops.

851      350 Heathen ships arrive at the mouth of the Thames; they attack Canterbury and London, and put King Beorhtwulf of Mercia to flight (ASB).

853      Mercians under King Burghred receive West Saxon aid in pressing historic claims in Wales. Burghred given King Æðelwulf's daughter Æðelswið in marriage to seal the alliance.

857      Grant by King Burghred of Mercia to Alhwine, Bishop of Winchester, of a house and commercial rights at London (S208).

864      *Edict of Pîtres*: Charles the Bald reforms Frankish army and forms cavalry force; reforms coinage; orders construction of fortified bridges to block Viking incursions (ASB).

865      A great host comes to England and winters in East Anglia under its leader Ívarr. The East Anglians

submit; Kent submits; the Viking army moves inland and plunders the countryside.
— Death of King Æðelberht of Wessex; interred at Sherborne; succeeded by his brother, Æðelred.
— King Burghred of Mercia attacks Gwynedd and Môn (AU).

867 King Osberht of Northumbria expelled; succeeded by (? his brother) Ælle. Battle of York: city stormed by Northumbrian force, but Osberht and Ælle killed. Danish army under Hálfdan seizes York.

868 Burghred of Mercia asks King Æðelred of Wessex for support; he answers, with Prince Ælfred. They besiege Nottingham but the host cannot be dislodged. The Mercians make peace with the host, who construct a fort there. Ælfred marries Ealhswith, daughter of a Mercian ealdorman.

869 The host rides across Mercia to East Anglia and takes winter quarters at Thetford. Battle of Hoxne: the Danes under Ívarr (?) kill St Edmund, king of East Anglia (*Vita Edmundi*). He is succeeded by Oswald (of whom coins but nothing else is known). All East Anglian monasteries are 'destroyed'. *Medeshamstede* is attacked, its abbot and monks slain.

871 King Æðelred of Wessex dies; succeeded by his younger brother Ælfred.

873 The host moves to Repton; King Burghred is driven overseas with Queen Æðelswith (who dies in Pavia in 888). He dies in Rome and is buried in St Mary's in the *Scola Saxonum*. Ívarr dies; possibly interred at Repton. Mercia given into the hands of King Ceolwulf II, a 'foolish king's thegn'.

877 The southern host makes for Mercia: it is effectively partitioned by Danes, with King Ceolwulf II keeping the western part. Ælfred may have control of London.

878 Many forces submit to the host; King Ælfred and a small force flee into hiding in the Somerset marshes at

Athelney (Æðelweard's *Chronicle*). King Ælfred and King Ceolwulf decisively defeat Danes at Battle of Edington. Treaty with Danish King Guðrum.

879     The host relocates to East Anglia and 'shares out the land'; King Ælfred takes control of the London mint and raises the coin weight for pennies.
— Possible date for the death of Ceolwulf II.

881     Battle of Conwy: Defeat of Mercian army, under Ealdorman Æðelred, by Anarawd ap Rhodri of Gwynedd (AC). End of Mercian authority over Gwynedd and south-eastern Welsh kingdoms.

883     Possible date of a treaty establishing the Danelaw along line of the rivers Lea and Great Ouse as far as Watling Street at Stony Stratford.
— Æðelred is king/ealdorman of Mercia by this date: a charter records his grant of exemption of dues to Berkeley Minster endorsed by Ælfred (S218).

886     King Ælfred takes London from the Danes and gives it into the care of Ealdorman Æðelred. Possible date for a marriage alliance of King Ælfred's daughter, Æðelflæd, with Æðelred of Mercia.

899     King Ælfred dies; succeeded by his son, Edward (the Elder).

901     'Arrangements for the fortification of Worcester' by Ealdorman Æðelred and Æðelflæd (EHD 99 and S221).

907     Queen Æðelflæd 'restores' Chester.

910     King Oswald's Bardney remains are translated to Gloucester at the behest of Queen Æðelflæd. Battle of Tettenhall (Staffordshire): Mercia and Wessex engage with the returning Danish force; the Danes are defeated; kings called Hálfdan, Ásl and Ívarr are killed. King Edward takes London and Oxford. Æðelflæd builds a fortress at *Bremesburh* (?Bromsgrove, Worcestershire).

911     Ealdorman Æðelred of Mercia dies; succeeded by Queen Æðelflæd to 918.

912     Æðelflæd builds fortresses at *Scergeat* (unidentified) and Bridgnorth.
913     Æðelflæd builds fortresses at Tamworth and Stafford.
914     Æðelflæd builds fortresses at Eddisbury and Warwick.
916     Æðelflæd sends a force to Brycheiniog to punish its king for the murder of Abbot Egbert and companions; attacks the crannog on Llangorse lake, captures queen and thirty-three other hostages.
917     Æðelflæd uses Danish warlords' absence to capture Derby and its hinterland.
918     Æðelflæd receives the submission of the Men of York; gains control of Leicester.
        — 'Æðelflæd, a very famous queen of the Saxons' (AU), dies at Tamworth; King Edward takes Nottingham; then occupies Tamworth; succeeds to Mercian overlordship and Mercian hegemony in much of Wales. Ælfwyn, daughter of Æðelflæd, is removed from Mercia.

# Chapter 10

# Cœnwulf: High tide
# 796–821

❈

*A continuity candidate* • *Rebellion in Kent*
• *Clofesho again* • *The social contract*
• *Wessex tremors* • *Troublesome priests*
• *Matters arising*

Cœnwulf, father of St Cynehelm the martyr, [...] honourably received the crown of the kingdom of the Mercians, and held it with surpassing ability by the vigorous energy of his government.[1]

So reads the entry in the *Historia Regum* for the year 796. 'Times are dangerous', wrote Alcuin to a Northumbrian correspondent; 'the death of kings is a sign of misery...'.[2] He was very likely referring directly to the assassination of King Æðelred the same year; but the deaths of Offa, and of his son and immediate successor Ecgfrið, bore ill omens too. Offa had no other living sons; his son had no known heir. And yet the Mercian state, its institutions and infrastructure engineered across the previous century, was resilient. Those élite families of ealdormen, and the bishops and abbots who profited from royal patronage across almost the whole of the eighth century, had much to lose by indulging in a descent into civil war. Northumbria held a mirror up to the Mercian state to stare into; and it did not like what it saw. Wise counsel ensured the smooth transition of power into the hands of a rival dynasty that claimed descent, however spuriously, from a brother of Penda named Cœnwealh.

A Mercian genealogy drawn up in timely fashion at the beginning of the ninth century, and headed by Cœnwulf, claims his descent through six generations down to his father Cuðbriht – Cuthbert, who may be the ealdorman of the same name who appears as a witness to several charters under Offa and Ecgfrið.[3] Only one of these can be regarded as certainly genuine. It records the sale of a lease of ten hides at Swineshead in Lincolnshire, from

Abbot Beonna of *Medeshamstede* to the *princeps* Cuthbert for 1,000 shillings.[4] If Cuthbert's name had, in those other purported charters, been inserted as a retrospective witness, some suspicion must be cast on his importance as a Mercian royal councillor *before* Cœnwulf became king. On the other hand, the possibility that an otherwise unrecorded brother of Penda was indeed named Cœnwealh may be reinforced by the fact that Penda's son, Wulfhere, had chosen Cœnred (reigned 704–9) as the name of his eldest son, while his brother Æðelred had named his sons Ceolred (reigned 709–16) and Ceolwald.

On the other hand if, as this charter records, the new King Cœnred's father, Cuthbert, had been the *princeps* of Lindsey – in effect ruling there as Offa's governor – one would expect him to make more of a showing in a range of Offan charters as an attestor. If he was the recipient of large estates through grants – in Lindsey or Mercia proper – those charters do not survive. He is frustratingly obscure.

Nevertheless, in King Cœnwulf's early grants the names of men and women who had held high status in Offa's council are again present as councillors to the new régime. Queen Cyneðryð and her daughter Abbess Æðelburh appear alongside Cœnwulf's son Cynehelm in the first of the new king's charters to survive, recording his gift* of Glastonbury Abbey to the young Prince Cynehelm in 797.[5] And two years later, a document dealing with weightier matters in Kent is subscribed by, among others, the long-lived *princeps* Brorda,† Æðelmund (the *Hwiccian princeps*) and the long-serving Offan *dux*, Esne. Cœnwulf, describing himself in the Glastonbury charter as having been 'elected king by Almighty God', had inherited no mere kingdom but the full state apparatus of a Greater

---

* That is to say, the king confirmed the freedom of the abbey from all taxes, renders and obligations: a means of transferring the patronage – and, perhaps part of the income – to the prince.

† His death, most unusually for an ealdorman, is recorded in the *Historia Regum,* a year later in 799.

Mercia, his assumption of power endorsed by the presence in his entourage of the Mercian élite. Continuity was all; and the so-called Anglian Collection of royal genealogies shows how, at the beginning of the ninth century, Early Medieval kingdoms manipulated and refined their pedigrees to suit changing times.[6] It is quite conceivable that royal women played a significant role in curating and engineering such genealogies; Offa's dowager queen may have made herself indispensable to the new Mercian régime, endorsing the upstart king as a worthy successor to her famous husband.

King Cœnwulf established his credentials in short order. Gathering his forces, perhaps towards the end of 797 or in the early months of 798, the king...

> ... entering the province of the Kentish men with the whole force of his army, mightily devastated it in a lamentable pillage, almost to its utter destruction.[7]

King Eadberht *Praen*, the former priest who had been protected in exile by Charlemagne, had, in the aftermath of Offa's death, returned to Kent to reassert its independence. He was captured by Cœnwulf's men and taken into Mercia where he had his eyes torn out and his hands cut off 'because of the pride and deceit of those people' who had dared to shake off Mercian rule.* The devastation meted out by Mercian forces may have included a destructive attack on the cathedral church at Canterbury, because it is only after 798 that Christ Church's records and books survive in significant numbers.[8] An entry in the *Historia Regum* for the same year also tells of a destructive fire in *Lundenwic*, which killed a great number of men. Was this a Kentish raid that precipitated the punitive Mercian invasion, an accident or an attack by Norse pirates?

In the same year, more likely in 799, Cœnwulf installed his

* *Historia Regum* 798. See below in this chapter for evidence of his reprieve.

brother, Cuðred, as king in Kent.* Kent: the great prize for Mercian kings. Its many trading harbours, its moneyers and access to continental markets, its enviable relations with the Frankish court and with Rome and the great political influence of its archbishops, were key to Mercia's overlordship of all the southern kingdoms. Not even Offa had succeeded in imposing direct rule there. At Aachen the news must have been received with resignation – Eadberht *Praen* had been in exile at the Frankish court; he may have been sponsored on his return to Kent with Frankish money and possibly with fighting men; with gold, too. But in Aachen, as in Rome, other concerns may have been in play: Charlemagne was busy putting down a rebellion in Saxony.

The English church faced a crisis as serious as that which followed the plague of 664; and the church's troubles were also Mercia's troubles. The year of Offa's death had seen not just the exile from Kent of Archbishop Æðelheard, but also the death of Archbishop Eanbald of York and of the bishops of London and Lindsey; perhaps also of Hereford and Sherborne.[9] Æðelheard was safe at the Mercian court but unable to return to Kent in the face of local antipathy. Archbishop Hygeberht, at Lichfield, was in an uncomfortable position: his elevation – designed to facilitate the anointing of Prince Ecgfrið in the face of Archbishop Jænberht's opposition – was now much less relevant to Mercian royal interests.

Alcuin, in semi-retirement at Tours but politically pragmatic to the end, urged conciliation: Æðelheard should return to Kent; Kent should accept his authority; Canterbury should be restored to its position of primacy but Hygeberht should not be stripped of his *pallium*; the English must heal their disagreements.[10]

King Cœnwulf favoured direct, decisive action. He had already written to the pope in Rome the previous year, suggesting that Lichfield be stripped of its archiepiscopal status; that Canterbury

* Cuðred does not appear as a testator to the Oswulf grant of 798. See below.

should regain its primacy; that the now-mutilated upstart King Eadberht *Praen** be anathematised. But his envoy, Abbot Wada, travelling to Rome to deliver the letter, performed his diplomatic mission 'lazily, nay foolishly' according to Cœnwulf's own testimony; and so the following year he wrote again.[11] Now he sought to distance himself from his predecessor King Offa's policy, blaming it on a personal enmity towards Jænberht. Invoking the original Gregorian scheme of the early seventh century, he suggested tentatively that Britain's two metropolitans should be at York and London – rather neatly combining a Mercian and Kentish province into one at the former Roman capital. He enclosed a gift of 120 mancuses.

Pope Leo consulted his own councillors and the papal archives. Rejecting the idea of relocating Kent's metropolitan to London, he reminded the Mercian king that all the bishops of the English had requested the elevation of Lichfield but that Canterbury must, by historic precedent, retain its primacy. He agreed, however, to excommunicate the Kentish usurper. And regarding Offa, he reminded the new Mercian king that King Offa had promised to send to St Peter a sum of 365 mancuses annually for the support of the poor and for the provision of 'lights'. If Cœnwulf was to have his own way, he must go about it with the same subtlety as his illustrious predecessor; and be freer with the contents of his treasury.

King Cœnwulf's drive for executive action and direct governance nevertheless continued at pace. Among the momentous events recorded for 798 was the death of King Caradog of Gwynedd, killed 'by Saxons' according to the Welsh Annals – two years after the battle at Rhuddlan and likely in a punitive Mercian raid. Then, using his powers of patronage to cement Mercian authority in newly conquered Kent, Cœnwulf is to be found exchanging land south of the River Limen in south-east Kent with his *dux*, Oswulf. Oswulf, in turn, granted the land to the

---

* According to William of Malmesbury he was taken to Winchcombe and 'consoled with his liberty' – mercy, of a sort. Giles, 1847, 87.

abbey of Lyminge* on behalf of himself and his wife, Beornðryð – for the sake of their souls. Historian Michael Burghart's analysis of the growth of Mercian power reveals significant undercurrents here. Oswulf was no Mercian satrap, parachuted into conquered territory, but a Man of Kent who, elsewhere, is to be found witnessing the handful of known charters of King Cuðred as *comes* or *princeps*.[12] Burghart suggests that Oswulf was a senior Kentish ealdorman willing to throw in his lot with the new régime, legitimising Mercian *imperium* there while cementing his own regional power base.[13] Noting the significant location of the land in question, in Romney Marsh where Eadberht *Præn* seems to have been captured, Burghart offers the thought that Oswulf may have been granted the site of the victory; that he may even have been complicit in handing his anathematised king over to Mercian forces.

The occasion of the Oswulf grant is not recorded; but given the heavyweight witness list, which includes two archbishops and all the provincial bishops alongside the *princeps* Brorda, the *dux* Esne and many other familiar senior councillors, Burghart believes that the grant was promulgated at *Clofesho*, where even weightier matters were decided that year. No church council had met at *Clofesho* during King Offa's reign before the death of Archbishop Jænberht in 792. His successor, Archbishop Æðelheard, convened a council there in that year, confirming Kentish church privileges, and another in 794. Here also, in 742, King Æðelbald had first granted the Kentish church its rights and protections.

*Clofesho* was a suitable place for history to be wrought and, once again, it begs the question of its location. *Clofesho* must lie within Greater Mercia, since it was only used as a council site during the century and a half of Mercian *imperium* over the southern Anglo-Saxon kingdoms between 672 and about 830. All other church

* Founded in the seventh century by Queen Æðelburh, Lyminge is also the site of a very substantial royal palace complex, excavated over recent decades by Reading University archaeologists. https://www.lymingearchaeology.org/ retrieved 10.04.2023.

councils in this time took place within the diocese of London[14] and there is a strong logic in thinking that Archbishop Theodore's intentions at the Council of Hertford must place it within reach of Canterbury and of the major sees at Leicester, Dorchester-on-Thames, Lichfield and Lincoln.* Historians and archaeologists have pored over their maps of navigable rivers and Roman roads; have pinned tags on important royal *vills* and 'productive' sites, on the great minsters and the Offan *burhs*, in the hope of making a decisive identification – so far with no consensus. Since 1962 Brixworth, the most magnificent surviving church of the period, has been offered by some as a plausible location,[15] more or less on the basis of its grandeur; but it fulfils none of the accessibility criteria laid out by Catherine Cubitt, the historian of Anglo-Saxon church councils: that a site should be sought close to a navigable river and within two miles of a major Roman road.[16]

A plausible but unconvincing case has been made for a site in the vicinity of Hitchin;[17] a similar case might be made for St Albans, Northampton and any number of settlements where Ermine Street and Watling Street run close to important rivers. Oddly, none of the Thames minsters has drawn much support despite, or possibly because of, known council sites already existing at London, Brentford and Chelsea.

Keith Bailey has made the case that important Anglo-Saxon meeting places rarely coincided with great *vills* or minsters. The momentous council at Gumley in 749, at which Æðelbald imposed his 'common burdens', enjoys a dramatic topographic setting in a natural amphitheatre but there is no royal *vill* here; no minster church.[18] On grounds of accessibility Bailey offered Royston and Dunstable as possibilities, both of them sited where major Roman roads crossed the Icknield Way; and, as it happens, he was writing before both were highlighted as 'productive' sites,

---

* East Anglian bishops of Elmham and Dunwich, of Hereford, London, Worcester and Winchester, are less regular attendees. Archbishops of Canterbury, and those of the Mercian dioceses, are the most regular attendees. Keynes 1993, tables.

Brixworth, Northamptonshire: the finest standing Middle Saxon church
in England – but its foundation, around the year 800, is obscure.

where concentrations of coins and metalwork indicate commercial or diplomatic activity.* Neither should be ruled out. Nevertheless, applying a dispassionate logic to the problem, Bailey came to the conclusion that a site near Hertford, perhaps towards Ware in the parish of Great Amwell, where a twelfth-century field name, *Aldwic*, suggests the site of a Middle Saxon trading settlement, is the best candidate location for this most famous lost place in Anglo-Saxon England.[19] The topography is right: a concentration of *hoh*-names in the area; the navigable River Lea; the line of Ermine Street. The historical fact of Archbishop Theodore's presence here in 672 weighs in Hertford's favour, too, as does the

The ninth-century font at St Mary's, Deerhurst – perhaps the finest extant example of Mercian carving.

* See Chapter 6, p. 191.

proximity of the northern boundary of the diocese of London. For now, perhaps, it is the best bet.

Among other business, the 798 council at *Clofesho* recounted and resolved the very long-running dispute over the monastery and trading settlement at Cookham, on the River Thames in Berkshire.* The agreement reached at *Clofesho* between Archbishop Æðelheard and Abbess Cyneðryð, King Offa's widow, recounted how King Æðelbald had given Cookham to Canterbury; how, on the archbishop's death in 760, the deeds had been stolen and taken to King Cynewulf in Wessex; and how King Offa had seized the monastery for himself sometime in the late 760s or 770s. Cynewulf had ultimately returned the deeds to Canterbury but Offa had kept the monastery and its trading settlement and bequeathed them to his queen in her role as abbess of the royal monastery – perhaps also the site of King Offa's mausoleum– at *Bedeford*. Petty propriety aside, Cookham mattered because of its prime Thames-side location and its presumably profitable trading settlement, founded on extensive estates.

Now, the Mercian-appointed Archbishop Æðelheard and the elders of Kent agreed that Cyneðryð might keep Cookham and another minster at *Pectanege*,† in return for ceding 110 *manentes* of land in Kent to Christ Church. With one very neat land deal Æðelheard's stock at Canterbury was raised, Kentish resentments at Mercian *imperium* were soothed and, at the same time, the power and influence of King Offa's family there was reduced.

The suspicion that Cœnwulf sought actively to distance his rule from that of his predecessor is confirmed by another charter, dated to 799, in which he restored thirty *sulungs* of land in Kent to Canterbury – land that had once been granted (actually sold) by King Ecgberht to Christ Church but had been taken by King Offa and redistributed among his thegns. Offa's argument had been that, since Ecgberht was his 'thegn', he had no right to give the land away. This restoration is framed by the charter as a response

* See Chapter 7, p. 229.
† Possibly Patney in Wiltshire.

to a request from Archbishop Æðelheard, further reinforcing his credibility with the community at Christ Church and with the Kentish élite.[20] Nicholas Brooks notes that Æðelheard was asked to pay 100 mancuses to the king in compensation, far short of its market value,* and that the king or his brother must have paid off those thegns now dispossessed of their valuable Kentish holdings. They must have believed that it was a price worth paying.[21]

The dowager queen retained her family's prize possession of Cookham. With the untimely death of her son Ecgfrið and of her Northumbrian son-in-law Æðelred, Cyneðryð's once outstanding political status was now much diminished. She had been married to Offa since about 769, had acquired a very substantial portfolio of properties, had seen two of her daughters married to kings and another installed as the abbess of a wealthy minster at Fladbury.† She may have retained ties with family in Wessex and her daughter Eadburh was married to the West Saxon King Beorhtric. She had the rare distinction of having coins minted in her name and had been a prominent member of her husband's political councils. She may have played an active role, as other Early Medieval consorts did, in easing the transition of power to the new king; even of promoting his interests. If the deal brokered by Archbishop Æðelheard, which allowed her to retain Cookham, was a token of respect and reward, it also constituted something of a retirement gift: she is not heard of again.

Since the reigns of kings Æðelbald and Offa, the 'booking' of land as property to be held in perpetuity had been extended from grants to establish or endow religious houses to overtly secular transactions – sometimes as gift; often in exchange for other land and, increasingly, to raise cash. Property was commodified; minsters were secularised. It was increasingly the policy of Anglo-Saxon kings to raise funds from tolls, land sales, coin minting and

---

* Brooks shows that a single *sulung* of land was worth between 15 and 30 mancuses in this period.
† Later tradition had it that a fourth daughter, Æðelswið, witness to a charter of 787 (Electronic Sawyer S127), became abbess of Crowland.

their own private trading profits and fines as the basis of their income, instead of gathering booty and treasure from the riskier affair of battle and imposing punitive tributes on their defeated enemies. King Æðelbald's reservation of the 'common burdens' and the church's assertion of *its* privileges were maintained in an uneasy balance – a sort of social contract in which bishops, abbesses and abbots acted as territorial lords as much as – maybe even more than – they were ecclesiastical patrons, moral judges and men and women of prayer and teaching. Drawn from the same élites, often related by bonds of kinship and by common interests, they constantly reinforced, constantly renegotiated the integration of church and state. Kings – and their more astute ecclesiastical statesmen and women – never forgot Bede's warning that a king must be able to call on a caste of warriors, drawn from the lands of the *gens* and from tributary kingdoms, with which to defend his people and territory. But, as yet, no great Viking army had taken the field in an attempt to conquer any of the territories of the Insular kingdoms. Piratical Norse raids were infrequent and, in any case, impossible to defend against.

Mercian kings knew that their lands were surrounded: by openly hostile enemies – in Wales; by reluctant tribute states – in Kent and East Anglia; and connected with Wessex and Northumbria by uneasy alliances. These alliances were expedient: the success of diplomatic marriages depended on unknowable caprices, relationships and unforeseeable events. Over the not-so-distant horizon Charlemagne, crowned Emperor of the Romans by the pope on Christmas Day 800, hosted exiles of varying potency, maintaining his keen interest in the fortunes of the Anglo-Saxon kingdoms and interfering actively in pursuit of Frankish interests.

The fates remained contrary: civil war still raged in Northumbria; on Christmas Eve 800, a great storm threw down cities, houses and villages with immense force; there were severe floods, the loss of a great many trees and destruction of cattle.[22] Monasteries at Hartness and Tynemouth in Northumbria and at Portmahomack in northern Pictland were raided by Norse pirates. The following year a fire burned in London, destroying a great part of the town; and war came to Mercia. King Eardwulf of Northumbria attacked King Cœnwulf 'because he had given asylum to his enemies'.[23] Among those enemies was perhaps Alhmund, son of the former Northumbrian King Alhred (765–74), whom Eardwulf's men had slain the previous year and who was buried at Derby in the Mercian heartlands. Here he was venerated as a saint and martyr.[24] In response to this attack, the Mercian king raised an army and, gathering forces from across his tributary lands, made a 'long expedition' against the Northumbrians.[25] At length peace was brokered by the 'bishops and chiefs of the Angles'. * But, as James Campbell has pointed out, despite Mercia's apparent strengths and Northumbria's dynastic weaknesses, Mercian kings were never able to win and hold territory north of the Humber.[26] Their real power was political; economic; administrative; and circumscribed by geography.

When the record of armed conflict is so sparse, as it is for much of the eighth century, it is not at all clear how Mercian kings maintained their military capabilities except in the very general sense of imposing military service on all booked land and calling on their tributary ealdormen and *subreguli* to muster their own retinues.[27] King Ine of Wessex had, in his celebrated law code of the early eighth century, exacted a fine, *fyrdwite*, for 'neglecting' military duty: the *gesiðcund mon* – nobleman –who failed to answer the muster must pay the king 1,200 shillings and forfeit his land; a *ceorliscman* – a ceorl – was to pay sixty shillings.[28] The mechanics of the muster, the means of its enforcement and the numbers that might be involved are all uncertain.

An otherwise unremarkable charter sheds a little light. At a king's council held in 801 at *Caelichyth* – Chelsea, on the River Thames west of *Lundenwic* – a *comes* named Pilheard petitioned the king to confirm his inheritance of a grant of land made long before, by King Offa in 767.[29] Thirty *manentes* (hides) in Middlesex, between the heathen shrine of the *Gumeningas* (that

---

* Dorothy Whitelock renders an uncertain passage here as 'by the grace of the king of the *angels*'. Whitelock 1979, 276.

is, Harrow-on-the-Hill) and the *Lidding* (Wealdstone Brook) had been given to Abbot Stiðberht in exchange for an estate at Wycombe in the Chilterns. The confirmation made at Chelsea does not indicate which of these estates Pilheard now sought to have 'booked' in his favour; but he brought the title documents with him, and the explicit terms of the deal he now struck make this one of the most scrutinised documents of Cœnwulf's reign.

> I acquired from my most pious lord, the king of the Mercians, the freedom of those lands, that is, with 200 shillings, and afterwards each year in my lifetime and [in those] of my successors with 30 [shillings] and they should be free in perpetuity [...] from the payment of all royal taxes, works and burdens, and also the exactions of popular assemblies [*popularium conciliorum*], except only an account must be rendered for the three public burdens, that is, the building of bridges and fortresses, but also now only 5 men should be sent for the requirements of military service.[30]

In modern terms, Pilheard had acquired the freehold of this thirty-hide estate for 200 shillings and was to pay a ground rent of thirty shillings per year: a shilling a hide. This was a large tract of land, nominally supporting thirty family farms. The renders that Pilheard might receive from such a holding make the terms seem enviably easy.

The broader significance of this charter is that, for the first time, there is also an explicit statement of the relationship between the size of a holding and the number of warriors due for *fyrð-ware* – military service: thirty hides equals five men. Assuming, as the historian Eric John did in his 1960 analysis of land tenure, that Pilheard himself was also liable to answer for *fyrð-ware*, so making a total render of six warriors from the estate, the equation seems simple: a warrior for every five hides.[31] The so-called five-hide rule is well known from later sources: this is its earliest written record. Was this normal eighth- and ninth-century Mercian practice; or is it specified because it was novel?[32] The only other early Insular example of so explicit a statement is found in a Scottish source, the *Senchus Fer n'Alban*, whose amphibious Dál Riatan armies – a marine corps in effect – were mustered on the basis of an oarsman for every seven households. A Carolingian capitulary of 808 specified a man from every four *mansi*, or holdings.[33]

Now, if these were armed men, they must presumably supply their own arms. But it begs more questions than answers: were they mounted and, if so, must they provide their own mount? Who was to lead them – Pilheard, presumably; but, if so, did this *comes* bring *only* five warriors with him to fight for the king? Who fed these men while they attended? Were they compensated or paid with coin?

Very probably, a man of Pilheard's rank enjoyed proprietorial rights over many separate land holdings, most of which he will have enjoyed as hereditary *folc* land – effectively for a life interest but likely passed from one generation of a weapon-bearing family to the next and dependent on royal service. So Pilheard's retinue from his *folc*-lands may have amounted to a substantial warband. For such men 'booking' may still have been a comparative novelty in the early ninth century. The nominal five men who must answer the king's muster from Pilheard's newly acquired – or newly confirmed – thirty farms in Middlesex or the Chilterns implies that the other public works due from his estate were carried out by labour due from the remaining four-fifths of households, who were not liable to provide a warrior. These services may have included, besides periodic bridge- and fortress-work, the construction of halls at royal *vills*;[34] maintenance of the king's roads; messenger service and, perhaps, following the king's hunt. Such obligations are often only recorded very late in historical sources, sometimes as services commuted to rents or cash payment;[35] they can be useful in reconstructing late pre-Conquest estate relationships and they are fundamental, if tricky, tools for unpicking the political economy of Anglo-Saxon England. Pilheard's charter teases as much as it satisfies the historian.

In Tribal Hidage, that much-scrutinised tribute list from seventh-century Mercia or Northumbria, the 'original' lands of Mercia were assessed at 30,000 hides. Nominally, if a five-hide rule applied, Mercian kings would be able to raise a force of 6,000 warriors of combined *gesiðcund* men (like Pilheard) and *ceorlisc* men (his farmers) – twice the size of the Mercian force that I have suggested based on a ten-hide render. Including the tributary territories of Greater Mercia – the *Wreocansæte*, *Magonsæte*, *Hwicce* and Lindsey – would add another 4,000.* It seems unlikely that such forces took to the field of battle often, if ever. At the *Winwæd* in 655 Penda had been slain along with thirty *duces regii*, but that seems to have been a very large battle by Early Medieval standards. The sort of running campaign of battles, skirmishes and raids that seems to have characterised the war between Mercia and Northumbria in 801 is likely to have involved smaller forces over a longer period.

A letter written by Alcuin to Northumbrian correspondents in that same year reveals just how militarised northern society remained. He alludes to the 'troubles' in which Archbishop Eanbald (c.796–808) found himself, having, it seems, been accused of harbouring the king's enemies and of seizing property. Alcuin asks – metaphorically or literally is unclear – why the archbishop needs so many men in his *comitatus*; men who, it seemed, had more followers of the 'rank and file' below them than was proper. The suspicion is that the archbishop had drawn to himself a bunch of self-seeking thugs and criminals at the expense of the priests and monks whose legitimacy as *milites christi* was of greater spiritual and moral value.[36] He was playing at secular politics. It is an extraordinarily unguarded insight into the role of a great metropolitan of the English church – recalling Wilfrid's large and belligerent retinue of a previous century and the status of Bede's friend Archbishop Ecgberht, the brother of a king.

Looking for material evidence of the mustering system in operation under the Mercian kings, thoughts naturally turn to

---

* In *The First Kingdom* (Adams 2021a, 328–9) and above, Chapter 5, p. 156, I suggested that following King Ine's ten-hide render, a one man per ten hides rule might be realistic; even then, Mercia would have been able to field an impressively large army.

the provisions made for the construction and defence of the Dyke, with its *burh-weards, tot* hills and *burh-tūns.* A *fyrð* may have spent much of its time digging ditches and hauling materials, as modern soldiers do. And so inextricable are the provisions for bridge-work and defences that one must regard the Cromwell bridge as potentially just one of a number of such ambitious projects*. King Ælfred was to achieve a stunning victory against a Viking army in 895 by blocking the River Lea near Hertford with a fortified bridge.[37]

If the *burhs* of central Mercia seem like overkill in military terms, they nevertheless functioned as key nodes in the Mercian system of royal administration; and by the beginning of the ninth century very real threats were concentrating minds on defences against coastal raids by actual *paganos marinos.* In 804 kings Cœnred and Cuðred granted Abbess Seleðryð of Lyminge, and her *familia,* land of six acres' extent within the walls of Canterbury to serve as refuge from heathen attacks. Seleðryð was also abbess of the royal minster on Thanet and sister of a Kentish *comes,* Ealdberht.[38] How often, in reality, the obligations of the 'common burdens' were called upon is a moot point; the principle was paramount. But, King Cœnwulf, by inclination or necessity, called upon his *fyrð* more often than his predecessor.

In 802, a year after the conflict in the North, Offa's son-in-law, King Beorhtric of Wessex, died. According to Asser, writing at Ælfred's court in the 890s, he was accidentally poisoned by his Mercian wife, Eadburh. Asser tells the story in the same brief chapter of his *Vita Ælfredi regis* in which he ascribes the Dyke to King Offa.[39] The story of Eadburh's marriage to King Beorhtric is told as a conventional tale of the scheming royal wife who, greedy for power and jealous of rivals, wished to dispatch a young protégé of the king with poison. Instead, it was the king

---

* At Cromwell; perhaps at East Bridgford near Nottingham, on the Trent; possibly Wroxeter on the Severn, Hereford on the Wye, Godmanchester on the Great Ouse; Northampton; Irthlingborough; Oundle, perhaps, on the Nene; Hatton Rock on the Warwickshire Avon and along the Upper Thames.

who took the draught and was killed. Eadburh was said, by Asser, to have sought protection in Francia from Charlemagne and to have come to a sticky end.[40]

King Beorhtric was immediately succeeded by Ecgberht, the returning Frankish exile and son of a Kentish king. One can almost imagine his and Eadburh's ships passing each other in the English Channel – except that she may have been the messenger who brought the news to the Frankish court that propelled the exile to return and seize the throne. On the day of Ecgberht's accession, according to the 'E' recension of the *Anglo-Saxon Chronicle*, a great battle was fought at Kempsford in Wiltshire between the men of *Hwicce*, under Ealdorman Æðelmund, and the *Wilsætan* under Ealdorman Wiohstan. Both ealdormen were killed, but the Wessex men gained the victory. The death of both commanders is indicative of substantial casualties on either side; of a more than usually bloody fight in close combat. If the battle took place in Wiltshire the aggressors must, for reasons of geographical logic, have been the men of *Hwicce* – seeking to settle old tribal scores, or acting as the tool of official Mercian policy: testing the new king's military mettle. King Ecgberht's fledgling régime survived, ultimately to Mercia's cost. A suspicious charter dated to 799, but perhaps referring to the year of Ecgberht's accession, suggests a swift rapprochement: King Cœnwulf granted thirty hides of land to Balþun, abbot of Kempsey, in exchange for twelve hides at Harvington in Worcestershire; and...

... In the same year peace was made between the Mercians and the West Saxons, and a firm alliance was confirmed by oaths under the Kings Coenwulf and Ecgberht.[41]

Doubt has been cast on this clause, which Patrick Sims-Williams calls an 'antiquarian addition'; but the document may reflect Worcester traditions of such an event; and incentives for peace existed on both sides.[42] King Ecgberht cannot have been without internal rivals in his early reign, for Wessex had been riven by competing factions for decades. Cœnwulf's political priority in his early relations with Ecgberht must have been the security of the Mercian position in Kent, since Ecgberht was said to have been the son of a king of Kent and, therefore, eligible to challenge for its throne. Ecgberht is likely to have had Kentish supporters and to have been a potential focus for rebellion. A treaty was expedient.

Kent's political value is underlined by evidence from contemporary archaeology. *Lundenwic*, subject to a serious fire in 801, was already in decline by the end of King Offa's reign. The number of London moneyers declined quickly thereafter and from about 820 onwards coins issued there are rare finds. A decade later, it seems, no coins were being produced in London at all. Instead, the number of moneyers striking coins at Canterbury and Rochester doubled from six to twelve in the first half of the ninth century.[43] The blame for London's decline cannot be laid at the door of Norse pirates: the great Frisian trading settlement at Dorestad near the mouth of the Rhine was thriving during these same decades.[44] If Cœnwulf was putting all his commercial eggs in one Kentish basket, his successors would pay the price.

It may have been at King Cœnwulf's instigation, but more likely on his own initiative, that Archbishop Æðelheard, having presided over the canny deal that restored many lands to Christ Church, travelled through Francia to Rome, in 801, to consult with Pope Leo. Alcuin had urged him to stay in Kent after the rebellion of Eadberht *Præn*; instead Æðelheard had chosen Mercian exile and it had proved a canny move: he had been restored to Canterbury under Mercian patronage. But the matter of Lichfield was outstanding. Alcuin now gave the archbishop a letter of introduction to Charlemagne, suggesting that he refrain from any show of material pride – gold ornaments and silk garments on his entourage would not go down well with the emperor. Alcuin sent word to his people at St Judoc's, a minster near *Quentovic* retained by Alcuin for the hospitality of pilgrims, to receive Æðelheard and send him on his way. The letter is gracious and cordial; but the old monk, wise in years and steeped in the spiky sarcasm of Frankish court banter, could not resist a dig at a request of the archbishop's...

I have sent you, as you asked, the saddle I usually ride on, prepared in the way favoured by churchmen in this district, and also a horse to carry the saddle, with you sitting on it, if it please your grace...[45]

Alcuin would die at his monastery in Tours the following year. His enduring status as a scholar and intellectual bulwark of Carolingian kingship is equalled by his value, for historians, as a diplomat during the evolution of the Mercian state; as an insider opening a window onto royal and ecclesiastical politics; for fascinating asides on trade, arts and crafts, social mores and pen sketches of his immediate world. From the 'sooty roofs' of Tours to the mathematical curiosity of the 'wolf, the goat and the cabbage'* Alcuin, like Bede before him, bequeathed a unique view of the Early Medieval world; his death brings down the shutters on historians' intimacy with that world.

The successful outcome of Archbishop Æðelheard's mission to Rome was recorded in a papal letter to him dated to 802, and in the proceedings of a council held at *Clofesho* in October 803 after his return to England.[46] Containing a comprehensive rejection of King Offa's 'fraudulent' elevation of Lichfield as a metropolitan see, it confirmed the primacy of Canterbury. If Æðelheard's policy was to disassociate himself from Mercia and reinforce his stock in Kent, it also suited King Cœnred to implement a sort of *damnatio memoriae* of his predecessor. In truth, Lichfield had already lost its effective status as an archdiocese. Hygeberht had not been named as archbishop after about 799; his successor, Ealdwulf, was content to sign the proceedings at *Clofesho* as mere bishop.

Æðelheard did not enjoy the fruits of his diplomatic triumph for long; he died two years later in 805. He was succeeded by Wulfred, who must have been in place by the middle of the year

---

* A conundrum whose charm has survived the centuries, involving a boatman ferrying three items across a river but with room for only two of them at a time in his punt. If he leaves the goat alone with the cabbage or the wolf alone with the goat, he risks losing his payload.

because a grant issued by King Cuðred in July records a sale to him of a modest estate at Buckholt, on the northern, wooded scarp edge of Romney Marsh, for thirty mancuses.[47] It is one of only three surviving charters issued solely by King Cuðred during his reign in Kent – all from the same year or shortly thereafter. His other grants were to his *praefectus*, Æðelnoð – the sale of land at Eythorne in East Kent – and to his minister Ealdbeorht and Abbess Seleðryð of land, at Ruckinge.* Both of these were issued jointly with his brother, King Cœnwulf. The two brothers also had coins minted at Canterbury in the same few years, both copying the profile-with-diadem portrait pennies of King Offa.[48]

Archbishop Wulfred may, like his predecessor, have hailed from the lands north of the Thames; but he was no Mercian place man. Although his family enjoyed the fruits of holdings in Middlesex, Wulfred had been Æðelheard's own archdeacon in the *familia* of Canterbury. His loyalties were to the archdiocese and to his mother church. Newly enthroned archbishops, like secular lords, needed to cultivate relations with royal and regional patrons, as well as with their spiritual master in Rome and with their diocesan bishops. But Wulfred seems to have enjoyed a head start: he had money to spend, lots of it, and Michael Burghart hints at the possibility that he may have purchased his archdeaconship with the idea of being promoted to the highest office – a move, if true, reminiscent of a man like Wilfrid.[49] Nicholas Brooks has calculated that, during the first decade after his consecration, Wulfred was able to spend no less than 17,720 silver pence or 590 gold mancuses on the acquisition of estates and land parcels with which to enrich Canterbury's cathedral church.[50] No wonder that he began to mint coins, with his name on the obverse accompanying the bust of a tonsured prelate and with the king's name absent from the reverse. His

---

* These are the brother and sister recipients of those earlier, generous grants under King Offa that also included rights in the woods of Buckholt. See Chapter 9, p. 306. The abbess had also been the recipient of the grant in Canterbury for a refuge from heathen raiders.

coins are outstandingly well-crafted and attractive. Nicholas Brooks sees kings Cœnwulf and Cuðred as pursuing a policy of indulgence towards such a wealthy and self-assertive primate; another view might be that he was pushing his luck in rivalling the trappings of kingship.

Cuðred, the king's brother and proxy in Kent, died in 807 and from that time King Cœnwulf ruled the province directly. Mercian relations with Wulfred had already begun to sour a year later, judging by a letter written by Pope Leo to Charlemagne in which he alludes to the fact that 'Cœnwulf had not yet made peace with his archbishop'.[51] But the king seems to have been at pains to maintain cordial relations with Canterbury. A raft of charters traces the fascinating process by which Wulfred, like a latter-day press baron, consolidated a constellation of disparate holdings into an organised portfolio of resources clustered around Canterbury. Sometimes these were modest in size but high in value:

London, 1 August 811. Cœnwulf, king of Mercia, to Wulfred, archbishop; grant of 2 sulungs at *Appincg lond* in Rainham, 2 sulungs at *Suithhunincg lond* at Graveney near Faversham, and 2.5 *hagae** in Canterbury, all in Kent, in return for 126 mancuses [3,780 silver pence].[52]

Other transactions encoded more complex relations and legal histories:

Canterbury, 21 April 811. Wulfred, archbishop, to Christ Church; grant of 3 sulungs at *Folcwining lond* in the district of Eastry, 1 at *Liminum* and 1 at *Dunwaling lond*, Kent, in exchange for 4 sulungs at Bishopsbourne, Kent. The land at *Liminum* had been granted to Wulfred by King Coenwulf, king of Mercia, in exchange for land at Yarkhill,

---

* *Hagae* were halls with plots of land within city walls. They become a feature of late Anglo-Saxon towns, highly valued means of establishing commercial and judicial rights.

Herefordshire, which Wulfred had obtained from Queen Cyneðryð. The land at Bishopsbourne had been given to Christ Church by Aldhun, confiscated by King Offa, and then restored. *

It is difficult to know if Wulfred's self-aggrandising tendencies were at that point a matter of concern for the king, but the Canterbury man was still attending royal councils. The Kentish charters are prominent in the minds of historians partly because of the rich detail that they contain and partly because of the dearth of other contemporary sources. Alcuin was now dead. The chronicler of the northern Annals or *Historia Regum* had retired or died by 802 and the flurry of detailed entries in the *Anglo-Saxon Chronicle* that gives such a breathless immediacy to the last years of King Offa's reign slows to a trickle; then dries up entirely. Between 804 and 825 entries were made in only thirteen years; and these are often of one line only. Many of them relate solely to ecclesiastical matters.

Political events further north may have been of more pressing interest to the Mercian court. In 806 King Eardwulf of Northumbria was 'driven from his kingdom', according to the *Anglo-Saxon Chronicle*, after a decade-long reign. His army had invaded Mercia five years previously in response to Cœnwulf 'harbouring his enemies'.[53] Eardwulf's chequered career first comes to notice in 791, the year he was captured and brought to Ripon by King Æðelred's men and ordered to be killed. They attacked him and left him in a tent to die but by some miracle he was found alive and, five years later, he returned from exile and was crowned at York minster.[54] In the following years he saw off a number of attempts to replace him with rivals. His deposition in 806 is a reminder of just how volatile dynastic politics had become north of the Humber, in stark contrast to the rock-solid

---

* Yardhill (*Geardcylle*) is described in the very long charter text as being *on Magonsetum* – the first written occurrence of the name of the *Magonsæte*. Electronic Sawyer S1264.

stability of the Mercian state throughout the eighth century. How far King Cœnwulf's machinations contributed to Eardwulf's toppling cannot now be known but, exiled for a second time, he arrived at the Frankish court to plead his cause with an ailing Charlemagne, now in his late sixties.

The chronicler of the *Royal Frankish Annals* takes up his story. The emperor was at Nijmegen, where Eardwulf pleaded his case for support; then he travelled to Rome – perhaps to confer with Pope Leo, perhaps also as a pilgrim. It seems to have been a successful journey. After returning to the Frankish court...

> ... he was taken back to his own kingdom by the envoys of the Roman pontiff and the Lord Emperor. [...] As his envoy, the Deacon Aldulf, a Saxon from Britain, was sent.[55]

With such august sponsors Eardwulf was able to reassert his authority in Northumbria in 808 and one must infer that Archbishop Eanbald of York, himself accused by Alcuin of plotting with the king's enemies, had been persuaded to support the returning exile by Deacon Aldulf and by the letters (possibly also bribes) he brought with him. But then, the *Anglo-Saxon Chronicle* records that the archbishop died in the same year – and so perhaps fate was as effective a diplomat as the reassuring words of the pontiff's envoy and a chest full of cash.

It is unclear, in the absence of a contemporary chronicle, how long the restored King Eardwulf was able to maintain his power in Northumbria; but he was, in time, succeeded by his son, Eanred – who may still have been ruling in the 840s; then by a grandson, Æðelred. At any rate, the Frankish chronicle entry provides an intriguing Mercian footnote. It seems that when the envoys were returning to Francia, Deacon Aldulf was captured by 'pirates'.[56] King Cœnwulf magnanimously paid his ransom and returned him to Rome: a case, it seems, of diplomacy maintaining the international rule of law.

The quiescence of the English chroniclers during the decade after 804 is an illusion: an unexplained silence rather than an absence of events. A severe pestilence that ravaged Francia in 810 and reached Britain in the same year, killing thousands of animals, may have had devastating consequences for communities and trade.[57] One of its victims may have been the exotic war elephant Abul Abbas – a present to Charlemagne from Harun al Rashid, caliph of Baghdad – on its way to campaign with the Frankish army against Denmark's powerful King Godfrið. The presence of a huge Danish fleet, harrying the Frisian coast and provoking Charlemagne to raise a fleet of his own, feeds into a sense of dynamic forces in play, visible at the salty periphery of English events.

In Mercia the passing of the years is marked almost exclusively by the issuing of charters, often from King Cœnwulf to Archbishop Wulfred: in 809; in 811 – twice; in 812; in 814 – three times. Other favoured recipients include Bishop Deneberht of Worcester, the minsters at Thanet and Abingdon, and otherwise invisible members of the royal council: the *praefectus* Æðelnoð in Kent; a kinsman, Eanberht and the *comes* Swiðnoð.[58]

A charter of 811 recording the gift to Bishop Beornmod of three *sulungs* of land south of Rochester, with appurtenant swine pastures in the Weald, reveals something of the make-up of Cœnwulf's inner circle. Attesting it are the archbishop and the bishops of Worcester and Selsey; King Sigered of the East Saxons;* Queen Ælfðryð;† the *duces* Beornmod,‡ Eadberht, Eanberht and Cynehelm; Cyneberht *propinquo regis*, Cœnwald *propinquo regis* and Quœndryð *filia regis*. *Propinquus*, meaning 'near' in Latin, is tricky: someone close to the king, likely a kinsman.[59]

The king's daughter, Quœndryð, has her own history. It seems very likely that she was the daughter of Queen Ælfðryð, even though the name of another, perhaps previous royal wife, Cynegyð, is recorded on a suspect charter of 799.[60] Cyneberht

* Not for long; by 812 he was merely a *subregulus*. Electronic Sawyer S170.
† Who attests royal charters from 808.
‡ Apparently a coincidental namesake of the bishop.

and Cœnwald seem to have been cousins of Quœnðryð, probably sons of the now deceased *subregulus* Cuðred whose Kentish holdings they may have inherited.

More problematic is the figure of the *dux* Cynehelm, because a son of King Cœnwulf bearing that name is known from the Glastonbury charter of 798.* A strong later historical tradition has it that the royal prince was murdered by his sister, Quœnðryð, their father's heiress. The clearly retrospective entry in the *Historia Regum* recording King Cœnwulf's accession in 796 describes him as the father of the martyred St Cynehelm.† There is much to unpick here, including the origins of a royal abbey at Winchcombe, where King Cœnwulf was buried after his death.

The man described as a *dux* in the 811 Rochester grant and several others, and as *princeps* in two further charters, seems to be the same Cynehelm who appears at the bottom of the list of attestors to the 'Pilheard' charter at the turn of the century, which reserved to the king the military service of five men.‡ Cynehelm's charter-witnessing career spans eleven years and he becomes progressively more important as the decade matures. After 811 he disappears, presumed dead; so, it looks as though he enjoyed the orthodox career of a *gesið* in the king's household.

The vast majority of the grants that Cynehelm witnessed relate to land in Kent. The notable exception is what purports to be the foundation charter of Winchcombe Abbey, which only survives in a version edited by William of Malmesbury from an English translation of the original, dated to 811.[61] In the sort of aside that makes William of Malmesbury such an infuriating but occasionally enlightening filter, he mentions that the mutilated and anathematised usurper of Kent, Eadberht *Præn*, was 'consoled with his liberty' by King Cœnwulf at the ceremony of dedication.[62] If William is right that Eadberht *Præn's* liberty was

---

* Electronic Sawyer S152 – see above in this chapter.
† See above in this chapter.
‡ See above in this chapter.

granted 'soon after' his capture in 798, then the true foundation date for Winchcombe might be placed more than a decade earlier than 811; might, in fact, constitute an act of contrition by King Cœnwulf. At any rate, in the same passage William records the king's burial there after his death in 821; and then...

> His son Kenelm [Cynehelm], of tender age, and undeservedly murdered by his sister Quenðriða, gained the title and distinction of martyrdom, and rests in the same place [i.e. Winchcombe].

It is certain that Cœnwulf had a son named Cynehelm, because he is mentioned in a papal privilege of 798 confirming his ownership of the monastery at Glastonbury and alluding to his age, twelve years.[63] Patrick Sims-Williams argues that there must have been two distinct and separate Cynehelms – the name is not that uncommon in Anglo-Saxon records. The king's son and the charter witness are not the same man.[64]

To complicate matters, the twelfth-century *Winchcombe Annals* record that the original foundation there, of a convent of nuns, was made by King Offa in 787.[65] So there is no doubting strategic royal Mercian investment in the minster.[66] The royal prince supposedly murdered by his evil sister and interred at Winchcombe appears as a witness in no genuine charter* and, as history records, he did not survive to succeed his father as king. But the tradition of his murder led to the creation of a cult and pilgrimage site at the abbey and a popular hagiography, the eleventh-century *Vita Sancti Kenelmi*. He even enjoys a bit-part in the *Canterbury Tales*.† Sims-Williams may be right again in suggesting that, whatever the prince's true fate – and the deaths

---

* He is recorded as a witness to a charter, Electronic Sawyer S184, dated to 821, the final year of Cœnwulf's reign; but all scholars agree that this is a fabrication.

† In the *Nun's Priest's Tale*, lines 290–301. A summary of the traditional *Vita* of St Kenelm is found in Baring-Gould 1897, 427–8.

of young princes were a popular trope of medieval conspiracy theorists – the legend of the evil and jealous sister, Quœnðryð, may have emerged from a conflation of the real princess with Offa's queen Cyneðryð and *Beowulf's* Ðryð. It suited the abbey to exploit the fruits of a royal martyr cult: martyrs attracted pilgrims.

Winchcombe, on the northern edge of the Cotswold hills in prime wool-producing country, was or became a defended *burh* and possessed a mint in the late tenth century, later emerging as its own small shire in the eleventh. It was well placed: a day's ride north-west of the strategically important royal *burh* of Salmonsbury and the line of the Fosse Way while to the west, another day's ride away, lay the confluence of the rivers Severn and Avon, close to a number of important early minsters: Bredon, Fladbury and Deerhurst. The *Vita* of St Kenelm claimed that that King Cœnwulf had 'enclosed the town with a wall'; and a substantial length of Winchcombe's pre-Conquest earthwork defences survives into the present. As a secularised minster, invested with royal patronage and the remains of a king and a princely martyr, Winchcombe enjoyed a similar status, around the turn of the ninth century, to Hereford and Tamworth.*

King Cœnwulf's charters reflect an ongoing process by which powerful ecclesiastics – Wulfred in Kent and Bishop Deneberht in Worcester – established episcopal control over formerly independent minsters and their estates. Patrick Sims-Williams sees three key drivers for this trajectory: the wishes of their founders to become part of a greater *familia*; the 'intermittent piety' of Mercian rulers; and 'tenacious episcopal litigation'.[67] Nicholas Brooks suggests that a more pragmatic motivation may have been the efficient management of estates.[68] How far these changes affected the lives of ordinary people – the tenants, monks, nuns and semi-free farmers who lived out their lives on those lands – is difficult to judge.

---

* For a more detailed presentation and discussion of the excavated and documentary evidence, see Bassett 2008, 224.

Winchcombe in the Cotswold Hills: a royal Mercian minster foundation in the territory of the Hwicce; later a royal *burh* and shrine; later still the centre of its own shire.

Since the seventh century the tensions between, on the one hand, episcopal administration – which was both orthodox and aligned with the interests of tribal kingship – and the Benedictine loyalties of fiercely proprietorial abbots and abbesses on the other, had been played out at royal councils and synods: at Whitby in 664; at Hatfield in 679; at *Clofesho* in 747 and 798, and elsewhere.

Archbishop Theodore – a monk as much as he was an archbishop – had reinforced the independence of monastic *familiae* to choose their own abbots and abbesses at Hertford in 672. But Bede's warning about the abuse of proprietary minsters, the risk of their falling into the hands of lay people for their own venal ends, and so that they might 'book' lands for the benefit of their heirs, had come true. Eighth-century bishops attempted to resist such privatisation and secularisation by asserting their right to interfere in monastic dealings in order to defend canonical rectitude.[69] Understandably, the forces acting against episcopal 'interference' included those monastic leaders who believed that their independence from bishops' authority must be protected; and those members of the élite, including kings and queens, who were monastic proprietors themselves. Bishop Deneberht sought to circumvent open conflict by persuading lay minster proprietors to bequeath their family monasteries to the see of Worcester.[70]

Kings were caught up in these tensions. Demanding the right to intervene in such weighty judgements, and reserving the public burdens on religious holdings, they also benefitted increasingly from rights of hospitality at minsters where they themselves, or their collaterals, were the proprietors – hence the very blurred distinction at sites like Winchcombe, between the Benedictine religious community and the defended *burh* and royal mausoleum; and hence the overt commercialism of well-placed royal minsters like that on Thanet and the much-disputed Cookham. In 816 these tensions came to a head at a synod held at Chelsea. Here, Archbishop Wulfred and his bishops declared that ecclesiastical property must not be alienated – that is, sold or leased for a period longer than one lifetime. Every bishop was to maintain the right to appoint abbots and abbesses in his own diocese; and...

In monasteries where the regular life was endangered 'by penury on account of the rapacity of secular men', the bishop was authorised to 'defend the flock of Christ, rather than abandon it to the jaws of Wolves.'[71]

Wulfred had already laid the groundwork for this assertion of episcopal power; he had travelled to Rome in 814, the year in which Charlemagne died, to consult with Pope Leo, perhaps also to consult canon law precedents and muster support for his position. On his return to England there must, I think, have been a flurry of diplomacy between him and the bishops of the southern kingdoms, ensuring their support.

The king and his *duces* were present at the Chelsea synod and duly subscribed to its canons.[72] That year and the next Cœnwulf continued to grant privileges to Bishop Deneberht. As a monastic proprietor in his own right the king enjoyed the security of recent papal privileges – confirmations of the rights and ownership of his own monasteries and their possessions.[73] In any case, he may have been distracted: in 816 he was raiding in North Wales.

At the beginning of his reign Cœnwulf had intervened in the politics of Gwynedd, having King Caradog murdered, according to the Welsh Annals. In the days of Penda and his successors Mercian overlords and their Welsh counterparts – particularly in Gwynedd and Powys – had been in amicable alliance against Northumbria. As Mercian *imperium* spread to encompass almost all the southern kingdoms of the English, periodic hostilities and spells of Mercian hegemony in North Wales had been interspersed with Welsh counter-raids, up to the point when Powys threw off Mercian control sometime towards the end of King Æðelbald's reign, as recorded on the Pillar of Eliseg.* King Offa's concerted dyke-building efforts to create a defensive and offensive line of control, using the Dee–Severn–Wye line as his north–south

* For a comprehensive appraisal of Mercian–Welsh relations in this period, and to appreciate difficulties in their interpretation, see Charles-Edwards 2014, 424ff.

marker, showed how seriously Mercian kings took the threat from Wales and how far they wished to project their power over the Britons. But the relationship with the Welsh kingdoms was very different from that with Kent, which might be subdued and absorbed into a Greater Mercia by judicious political interference and régime change. Cœnwulf's policy seems to have been to assert tributary rights over Powys and Gwynedd and, when the need arose, to support one side against the other: to interfere, rather than conquer. The very sparse evidence allows no more than such unsubtle inferences.

From 813 to 816 the Welsh Annals contain references to civil wars in Powys and Gwynedd, in which Hywel ap Rhodri Molwynog captured *Mona* – Anglesey – and finally emerged victorious. In 816 a Mercian force, wading into these troubled waters, invaded the mountains of *Eryri* (Snowdonia) and Rhufoniog, a small Venedotian territory north-east of the Snowdonia massif. Two years later, in 818, Cœnwulf 'devastated' the Dyfed region of south-west Wales, suggestion of a broader, more aggressive Mercian policy against Wales. In 822, perhaps a year after Cœnwulf's death...

The fortress of Deganwy* is destroyed by the Saxons and they took the kingdom of Powys into their own hands.[74]

Among many questions that historians would like to resolve in this region of ancient cultural and political tensions are how far King Cœnwulf used, or was able to use, Offa's defensive line in his controlling or offensive operations against the Welsh kingdoms; and what role the earthwork known as Wat's Dyke, of which at least a part has been dated to the early ninth century,† played in his military and political planning.

Wat's Dyke runs for forty miles, some two or three miles east of, and roughly parallel to, its more monumental predecessor,

---

* At the mouth of the River Conwy; see below, Chapter 11, for another battle at Conwy.
† See Chapter 11.

reaching the Dee Estuary at Holywell. It connects the courses of north–south-flowing rivers and links royal centres at Maesbury and Bagillt, in a more settled arable landscape than its hilly counterpart to the west. If Wat's Dyke is in fact Cœnwulf's dyke, it projects a more defensive mind-set, protecting the Cheshire Plain and the lands of the *Wreocansæte*. Insubstantial as it often appears now, having been much interfered with, the original was no minor affair – in places the ditch was cut ten feet deep.[75] If Offa planned his dyke as a sort of doomsday statement of finality, Wat's – or Cœnwulf's – Dyke responded more to military expedience. If Offa's Dyke was still functional, which would require military manpower and devolved military administration of the marcher lands, then Wat's Dyke was designed to prevent active passage across it and may also have been used in part to absorb tributary labour – to keep idle hands busy. It may also have functioned as a line from which to launch offensive actions, like those recorded in the Welsh Annals; a funnel channelling military resources into North Wales. Precious mineral resources, such as North Welsh copper and the salt of the Cheshire brine springs, hint at intangible motivations that are now obscure. Resolution of these problems must wait for more detailed work and the definitive dating of both earthwork systems.*

If the king's attentions were focused, in the second decade of his reign, on control of the Marches, he was nevertheless unable to ignore the fallout from the synod at Chelsea. The battle ground was neither fortresses nor ramparts, but two Kentish minsters, at Reculver and Minster-in-Thanet; and the prize was neither booty nor cattle but the protection of lines of patronage and influence. Cœnwulf's battle with Archbishop Wulfred was fought with soft power; it was nevertheless executed with considerable prejudice on both sides.

It is not until nearly a decade after the event and after King Cœnwulf's death, at synods held at *Clofesho* in 824 and

---

* See Ray and Bapty 356ff for a discussion of Cœnwulf's possible motivations.

825, that the explosive tensions between king and archbishop were even partially resolved. In that moment of record – of legal retrospect – the bitterness with which the dispute was prosecuted comes to light; and then from an explicitly one-sided, Kentish perspective.[76] Sometime before 810 Seleðryð, abbess of Minster-in-Thanet and Lyminge, died. The island minster had been founded by Domne Eafa, a Kentish princess and wife of the *Magonsætan* King Merewalh, back in the late seventh century.* It was perfectly located, on the sheltered southern shore of Thanet, overlooking the navigable Wantsum Channel (now the course of the River Stour), for connections with Kent's many trading settlements and directly with Frisian and Frankish ports. It had been endowed generously with estates and trading privileges over the intervening decades. On Abbess Seleðryð's death it passed to Quœndryð, King Cœnwulf's daughter. She also came into possession of Winchcombe Abbey and of another Kentish foundation at Reculver – equally well placed in its ancient clifftop fortress site at the northern end of the Wantsum Channel on the Thames estuary.

As *de facto* king of Kent, Cœnwulf believed himself to have these appointments in his gift but Wulfred, the promulgator of the Canons of Chelsea in 816, claimed them for Canterbury. These two monasteries were valuable political and economic assets in their own right; but any precedent set on either side also had implications for other royal minster estates in Kent – at Hoo, Folkestone and Dover – and together these comprised a quarter of the kingdom's landed wealth.[77] It may be that for the king it was the principle that mattered – freedom from ecclesiastical interference in proprietary royal affairs. For Wulfred there was more at stake: these churches and their lands lay within

---

* According to the medieval Mildrið legend, possibly based on an eighth-century original, it was founded in expiation for the murder of two Kentish princes, in about 670. Domne Eafa was the mother of Abbess Mildburh of Much Wenlock and of Mildrið, who succeeded her mother as abbess in Thanet. Rollason 1982.

Canterbury's immediate diocesan orbit and, moreover, they were Kentish lands; not Mercian.

Between 816 and 824 no church councils are recorded, a reflection of open hostilities having broken out between the parties. The hugely detailed retrospective account contained in a restorative charter of about 827, running to nearly 2,500 words, records how, during the last year – or years – of the king's reign:

> ... Archbishop Wulfred, by the enmity, violence, and avarice of King Cenwulf, ... in witness of the whole people, had been deprived of all his rightful authority [in the two monasteries, at Minster-in-Thanet and at Reculver]. [...] Afterwards, also, the aforesaid King Cœnwulf, coming with his *Witan* to the royal residence at London, invited that archbishop thereto under his own guarantee, and with the surety of his chief men. Then [...] he ordered the bishop to be despoiled of every thing which belonged to his authority, and to be utterly exiled from this country, and never to return to it [...] until he had consented to this – namely, to return [as payment for the two monasteries] three hundred hides of land at *Logneshomme*, and to give up one hundred and twenty pounds of money. And the bishop, after long refusing this settlement, yet at last, under compulsion, unwillingly consented.[78]

In 825, under a new Mercian régime, King Beornwulf resolved at *Clofesho* that Abbess Quœnðryð should restore to Wulfred his rights to the deeds and the lost revenues of those minsters:

> She ought to restore to the bishop every thing which had been violently taken from him during all that time, and make good all injury, and add as much more over and above, and make compensation for the use during the same period.[79]

Only after another two years was a final deal thrashed out. Wulfred, under a new royal patron, regained title over the minsters and Quœnðryð, now unprotected by her father, was forced to concede possession. Not for the last time, the tensions that drove relations between English kings and their archbishops had threatened the political fabric that bound temporal and spiritual lordship in delicate equipoise.

King Cœnwulf's death in 821 marks the end of an extraordinary era of political dominance by Mercian kings across a century and a half. His brother and successor, Ceolwulf, was deposed after just two years and replaced by a usurper, Beornwulf. He was defeated and killed in a great battle in 825 at *Ellendun* (Wroughton, in Wiltshire) by the West Saxon forces of Ecgberht, who subsequently annexed Sussex, Kent and Essex. Ludeca, Beornwulf's immediate successor, was killed by 'the East Angles', who liberated their kingdom from Mercian rule.[80] King Wiglaf succeeded him in 827. By 829, eight years after Cœnwulf's death, King Ecgberht of Wessex was able to assert his overlordship over all the southern kingdoms.

In one sense at least, the West Saxon triumph over Mercia heralds the beginning of what historians, with hindsight, call the Viking Age.* Mercia seems to sink beneath the waves of a political and military inundation; the narrative thread is lost. But it is not the end of Mercia's story. The kingdom in which the foundations of an English medieval state were laid with almost revolutionary brilliance bequeathed a robust legacy.

---

\* Its story is taken up in *Ælfred's Britain*.

# Chapter 11

# Ceolwulf to Æðelflæd:
# All the king's horses
# 821–918

✤

*Tamworth • The Danelaw burhs • Repton*
*• A 'foolish king's thegn' • Annals of*
*Æðelflæd • Churches and charters • Legacies*

A memorial service, held at St Editha's church in Tamworth in June 2018 to mark the 1100th anniversary of the death of Æðelflæd, the 'Lady of Mercians', is packed out. The great and the good are there: minor royals; distinguished academics; historians and cast members from a television series called *The Last Kingdom*. St Editha's stands prominently at the centre of the modern town and, before the accretions of medieval castle, market buildings, eighteenth-century townhouses and modern shopping centres, commanded a spur of land overlooking the confluence of two rivers: the Anker and the Tame.

Æðelflæd, indomitable daughter of King Ælfred, consort of Ealdorman Æðelred and last independent ruler of Mercia, is commemorated by a mossy statue standing on a tall plinth beneath Tamworth's medieval castle. She holds a protective arm around the shoulder of her nephew, the future King Aðelstan. The year of her death, 918, was also her moment of triumph: she had received the submission of the Men of York and reasserted Mercian control over the town of Leicester – one of the Five Boroughs of the Danelaw. Had she survived there is a substantial chance that the old kingdom of Mercia might have been restored to something of its former importance; not, perhaps, to overlordship over all the southern kingdoms, but wielding military and economic power over much of central England and the Marches.

Tamworth was the '*worðig*' or 'enclosure' of the *Tomsæte*, dwellers of the lands between Watling Street and the River Trent in the extreme south-east corner of what is now Staffordshire.

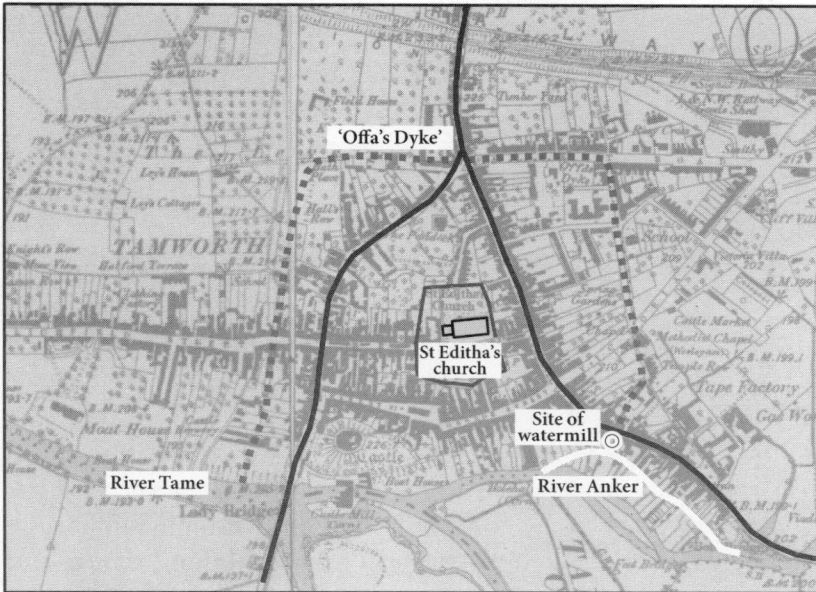

A nineteenth-century Ordnance Survey map showing the line of the 'Offan' defences, the early enclosure around St Editha's church, and the site of the Anglo-Saxon watermill.

Æðelflæd's forces had built a new *burh* there five years previously, part way through a brilliant, decade-long military and political campaign in partnership with her brother Edward* and following the revival of Mercian fortunes under her husband Ealdorman Æðelred.† A rash of *burh* constructions and military victories enabled allied West Saxon and Mercian forces to wrest control of the lands of the Danelaw from their Scandinavian lords. One after another, the strongholds of the Five Boroughs‡ submitted

---

* c.874–924, who reigned from his father's death in 899. King Ælfred had married a Mercian noblewoman, Ealswið, in about 868.

† Æðelred had allied with King Ælfred as far back as 886, when the Mercian leader was put in control of London; his marriage to Æðelflæd probably took place after the death of Ceolwulf II. He died in 911 and she succeeded him.

‡ Nottingham, Stamford, Lincoln, Leicester and Derby.

373

to Ælfred's children. In the immediate aftermath of Æðelflæd's death, twelve days before Midsummer,[1] her brother Edward 'the Elder' (899–924) occupied Tamworth and assumed lordship over all the former Mercian lands between the Thames and Humber. Two years later he seems to have been recognised as overlord of all the English and Welsh kingdoms. The days of an independent Mercian kingdom ended with Æðelflæd.

If Æðelflæd is still venerated at Tamworth, more tangible traces of Mercian royal patronage can just, even now, be detected in the town. On the earliest Ordnance Survey maps, dating to the middle of the nineteenth century, a label just below Albert Road reads 'Offa's Dyke', something of a surprise to those who think the Dyke should lie fifty or so miles west. This 'Dyke' is, in fact, the barely visible remnant of a rectilinear earthwork that once enclosed the whole of the spur at the confluence of the Anker and Tame* and, with it, St Editha's church and the *villa regia* of Mercia's eighth- and ninth-century kings. Excavations in the 1960s and 1970s confirmed its line and showed that there had been at least two construction phases of a ditched and banked rampart. One of these can be identified with Æðelflæd's recorded fortification of 913; the other probably belongs to the time of Offa, whose choice of Tamworth as a royal residence lies behind the ascription of the 'Dyke'.[2]

Tamworth, like Hereford and Winchcombe, was a favoured royal residence. King Offa issued charters here at Christmas in 781 *'in regali palatio in Tamouurðige'* and the site of the 'palace' in question has been identified as a rectangular inner enclosure surrounding the church.[3] Tamworth was a regular Christmas and Easter residence for Mercian kings thereafter, implying that it became something of a winter quarters for the Mercian court. Very substantial supplies and services from its hinterland must have been harnessed to provide for the household and Tamworth was perfectly situated to exploit the region's resources. Within a short distance lie the *burh* at Tutbury, perhaps superseded by Tamworth; Burton (on Trent); the substantial township at

* Some 55 acres.

Catholme; Repton, the site of a royal monastery and mausoleum; the intersection of Ryknield Street and Watling Street; and Lichfield, seat of Mercia's bishops and sometime archbishop. To the west lay the ample hunting grounds of Cannock Chase.

Part of the function of inner and outer precincts in a royal *burh* was to control access to the king's household; but in part these enclosures must also have been designed to store and protect the region's agricultural surplus – sufficient to provide for the royal household during the winter months – and to impose an idea of privacy and exclusivity. Within the enclosure, customs and laws specific to royal *vills* must have applied and gated entrances provided for the extraction of tolls as well as the checking of bona fides. The ramparts are likely to have been surmounted by an impressive wooden palisade and walkway although whether, as in King Ælfred's burghal scheme, there was provision for the walls to be garrisoned by local men as part of their military service is impossible to say.[4] But behind their ramparts, eighth- and ninth-century Mercian kings were more remote from their communities than their predecessors had been.

Two ancient river crossings and trackways, approaching from the south-west and south-east respectively, meet at Tamworth and head north, joining at precisely the point where the northern entrance to the *burh* was constructed.[5] The present street layout preserves their lines along Bolebridge Street, Gungate and Aldergate and the modern bridges that cross the rivers Anker and Tame use the same ancient crossing points. The course of the River Anker, flowing south-west then west below the town, has altered slightly over the centuries. It used to run a little further north, hard up against the south-east corner of the *burh* defences; and here, in the 1970s, excavations revealed a marvel of the Mercian age: a horizontal turbine water mill. Two successive mills, dating to the middle decades of the ninth century, were fed by a leat from the River Anker, some 500 yards upstream. The diverted river water was drawn through a chute from a millpond, directing a jet of water down onto wooden paddles – one of which was recovered during excavation – directly driving the millwheel in an upper chamber via an ungeared vertical shaft.

Such mills were not particularly powerful compared to later, more sophisticated overshot vertical-wheeled mills, which became such a common sight in Britain in later centuries – more than 6,000 of them at the time of the Domesday Survey. But they dramatically increased output compared to rotary hand querns. Mills were expensive and highly prestigious engineering projects, requiring the services of skilled craftsmen. Other Mercian mills have come to light, at Old Windsor on the Thames and at a site near Hereford. It would be surprising if, in future years, evidence of more mills associated with royal sites and minsters – like Cookham, perhaps, or along the River Nene – did not emerge.

Æðelflæd's career is not just of interest because she is the book-ending counterpart to King Penda's emergence in the seventh century as the first great Mercian overlord. As the Anglo-Danish towns of the Midland shires surrendered to West Saxon and Mercian forces in her later years, they come into focus for the first time since the reign of Cœnwulf. When Æðelflæd and her brother received the submissions of the Five Boroughs and of fortress-towns at Bedford, Huntingdon, Northampton, Towcester and Bakewell, they were 'liberating' settlements that had already been the focus of Mercian administrative investment in the seventh and eighth centuries. Their status under Scandinavian rule, as thriving settlements on a scale quite unknown in Wessex outside *Hamwic* – they might, just, be called towns by this time – allows historians to tentatively project their stories back through the deep obscurity of the intervening years.

The archaeologist Jeremy Haslam, looking for threads by which the important central places of King Offa's day might be tied to those of tenth-century Mercia, studied sixteen settlements that became the caputs of Early Medieval shires or regions.[*]

---

[*] Bedford, Cambridge, Godmanchester, Hereford, Leicester, Lincoln, London, Nottingham, Northampton, Oxford, Stamford, Tamworth, Winchcombe, Worcester, Canterbury and Norwich. Haslam 1987, 80.

Discounting Canterbury and Norwich, both of them only periodically subject to Mercian overlordship, he defined a key set of shared characteristics. Almost all became *burhs* under Ælfred or his half-Mercian children. Most were, at one time, Viking bases and many were certainly Norse trading settlements; and most were sited at defensible bridging points on rivers.* Each one had demonstrably or probably been an early ecclesiastical centre and nine of them show archaeological evidence for Early and/or Middle Saxon defences. Haslam draws the conclusion that Norse and, later, West Saxon rulers and administrators were inventing nothing new in the Mercian landscape; they took advantage of existing Mercian *burh* foundations – most of them, probably, founded under King Offa's rule after about 780, and surely emerging not from the ether, but as royal *vills*, the centres of regional lordship.[6]

Those towns prospered under Norse rule, as convenient and defensible craft centres, military bases and trading places. Most, if not all of them, were situated close to the heads of navigable rivers where these crossed Roman roads, so they connected the river-centric routeways so favoured by Scandinavian entrepreneurs. The topographies that had made them attractive sites for *burhs* and minsters, for investment through patronage and the extraction of tolls, also made them targets and assets for Viking leaders and their land-hungry followers. The majority of these sites lie along, and north-east of, the line of Watling Street, where nearly all Scandinavian place names are to be found. They are generally the sites that gave Mercian traders and ecclesiastics vital access to the North Sea and to the continent, and it is no wonder that they were exploited by seagoing, trading and raiding peoples who saw in them the same virtues.

When England was 'shired' in the tenth century these long-thriving central places became the county towns of the Midlands:

* Lincoln, Leicester, Tamworth and Winchcombe are his exceptions – although Tamworth seems to fulfil the criteria admirably and only Winchcombe lacks a potentially navigable river.

Northampton, Bedford, Leicester, Oxford, Lincoln, Nottingham and Derby – Stamford is the curious exception.* They are legacies of a palimpsest of ecclesiastical investment, Mercian royal patronage, trading enterprise and naturally wealth-giving locations. In turn, they acted as foci for the increasingly productive and specialised farms in their hinterlands. Many land holdings, as the charters reveal, were assembled into rational portfolios of resources, to be exploited under organised management, for profit. It is a moot point whether new Scandinavian lords, perhaps controlling those central places and hinterlands semi-independently, were more, or less, welcome to the indigenes than their Mercian counterparts. Perhaps they demanded fewer public burdens from their subject peoples.

*

Peering into the historical void before the end of Norse rule in northern Mercia, to span the gap between Cœnwulf and Æðelflæd, is no simple matter. The West Saxon narrative preserved in the *Chronicle* paints King Ecgberht's West Saxon triumph over Mercia in 829 and the exile of King Wiglaf in stark monotones. Subsequently the West Saxon king began to mint coins at London and then led an expedition into Wales to assert his overlordship there. But West Saxon rulers were no more popular in Mercia than Mercian kings had been in Kent. Mercia's lords rebelled and Wiglaf was restored. In 836 he was sufficiently independent of West Saxon interference to be able to grant privileges to the *Hwiccian* minster at Hanbury, lying on the saltway between Droitwich and Bidford-on-Avon: freeing it from royal dues on its woods, fields and meadows, salt pits and lead-furnaces; from entertaining the king or his ealdormen; from building royal

---

* As Paul Blinkhorn points out to me, Stamford's vibrant wheel-thrown pottery industry probably pre-dates the arrival of the Great Army in the 860s; so it was already a town of some importance, even if it was for obscure reasons not chosen as a shire centre.

residences and from the imposition of *fæstingmen*,* excepting only bridge-work and the repair of defences. Normal service had been resumed, it seems.[7]

Viking raids on the South Coast seem to have preoccupied Ecgberht from then on until his death in 839. In that year, also, King Wiglaf was very briefly succeeded by a son, Wigmund; then by Beorhtwulf (840–c.852). London, whose economic power had been weakened by competing Mercian and West Saxon interests and by increasingly destructive Norse raids on the continent, suffered its first recorded pirate attack in 842. King Beorhtwulf was issuing coins there in the same year, but using West Saxon dies from Kentish moneyers. London was raided again in 851 by a Norse army that had overwintered on English soil for the first time and, in defending what had once been a prime Mercian asset, King Beorhtwulf was put to flight. He may have died from wounds sustained in battle: a year later his son, Burghred, succeeded him and reigned until 873. Mercia was by now fully subordinate to West Saxon overlordship. In 853, the year of his accession, Burghred married Æðelswyð, daughter of the West Saxon King Æðelwulf, and received West Saxon support in a campaign against Mercia's old foes in Wales.

Mints in Kent and London now suspended coin production. King Burghred must, nevertheless, have been able to assert some level of control in London – either at the old *wīc* or in the ancient walled Roman city – when he granted Bishop Alhwine of Winchester commercial rights there in 857.[8] But the continental trading settlements, markets for the produce that flowed through England's *wīcs*, were also in decline. Dorestad's traders suffered from both an entirely natural shift in the course of the River Rhine and from Norse raids, and was sacked in 863; *Quentovic* had been raided in 842 and occasionally thereafter.[9] Even Paris, deep in the interior of Francia, was raided and burned in 861.

By small increments the Frankish king, Charles the Bald,

* An obligation to feed and lodge those going about the king's business. Whitelock 1979, 515.

evolved a new defensive tactic of constructing fortified bridges at critical navigational points, restricting the movement of pirate fleets: either they could not pass upriver or, if they did, could not return to the sea. In 864, in an edict issued at Pîtres in Normandy, he instituted reforms to the army and created responsive cavalry units, reinforcing defences along the River Seine: matching the raiders for speed and versatility. Francia became, finally, too risky a proposition for Norse armies.

Very likely as a result of Charles's success, in 865 a great Viking host, the *micel hæðen here* of the *Anglo-Saxon Chronicle*, landed from a vast fleet of ships in East Anglia, which fell under the rule of its commanders and ceased to be an independent kingdom thereafter. Its king, Edmund, was martyred. That heathen host, which had already subjected the Francia of Charlemagne's children to terrible depredations over the previous decade or so, was no mere raiding party but an immense force of warriors, slavers, traders, craftspeople and hangers-on: an entire piratical tribe on the move, so to speak. If historians were once inclined to downplay the size of the host, their doubts have been muted since the discovery by fieldwork of one of their camps at Torksey, on the east bank of the River Trent in Lincolnshire. More than 100 acres in extent and yielding thousands of objects belonging to traders, artisans, soldiers and their families, Torksey was a thriving encampment of thousands of people.[10]

Even under such grave external pressures, in that year a Mercian army went into Gwynedd and Môn and 'expelled the Britons from their lands'.[11] The Viking army swept north, taking York in 867 and killing Northumbria's kings. A year later one of its commanders split a force from the main body and laid siege to Nottingham. King Burghred appealed for his West Saxon neighbour's help; but together they could not dislodge the Viking army from the Trent, Mercia's symbolic artery, and Burghred was forced to sue for peace.[12] To the south, in 870 and through the following year, a desperate, cross-country skirmishing defence of Wessex led to the death of King Æðelwulf, then that of his fourth son and short-lived successor, Æðelred, leaving a last brother, Ælfred, to sue for peace. The *micel hæðen here* took possession of London, Mercia's jewel; and in 873 the army that had taken Nottingham moved to the Mercian royal minster at *Hreopadun*: Repton, fortifying it as a pirate base on the headwaters of the Trent.[*] The Mercian king was driven ignominiously into exile – whether by his own people or by the invaders is unclear. No Anglo-Saxon kingdom was equipped militarily or imaginatively to deal with such brilliant speed and improvisation, combining swift river-borne fleets with mounted raiding parties of hard-bitten veteran warriors bent on winning loot and land.

Excavations in the graveyard of St Wystan's church, Repton, by Martin Biddle and Birthe Kjølbye-Biddle[†] uncovered the D-shaped enclosure ditch constructed by the Viking force on the south bank of the Trent, which incorporated the minster church into its south side as a fortified gatehouse and military headquarters. A religious house had been established here some two centuries earlier as a daughter foundation of *Medeshamstede*. Guðlac had taken holy orders at Repton before his self-imposed exile in the Fens, and it was a favoured burial site of kings from the *Magonsætan* Merewalh to Æðelbald, to the eponymous Wystan – King Wigstan, who died in about 839.[‡] The royal mausoleum survives, miraculously, as a crypt, once visible above ground but now buried by earth, ashes and dust and accessible

---

[*] In those days Repton's church lay close to the south bank of the Trent – the course can still be traced in Old Repton Water; the main channel of the river now lies half a mile to the north.

[†] The Repton Stone was uncovered in 1979; the enclosure excavations took place up until 1988. Biddle and Kjølbye-Biddle 1985. The enclosure excavations are reported in Biddle and Kjølbye-Biddle 1992.

[‡] He was the son of King Wigmund and Ælfflæd, daughter of Ceolwulf I. A later historian, John of Worcester (early twelfth century), wrote that he had been killed by his cousin Beorhtfrið and thereby became a royal martyr. His body was buried alongside that of his grandfather, Wiglaf. Thorpe 1872, 72.

by a flight of steep, narrow stone steps. With its mouldy sense of suspended time, dimly lit recesses and marvellous, candy-twist columns, it may be the most atmospheric and evocative surviving monument of the age.

The Viking host stayed at Repton for a year; after its departure the church somehow resumed its minster functions, Christian and Viking burials intermingled around it. One burial, which caused considerable excitement when it was discovered, was of a man in his late thirties who had been dispatched with a powerful blow to the top of his femur. He was interred with a necklace of two beads and a plain Thor's hammer. By his left leg lay a sword and between his thighs lay the tusk of a wild boar.[13]

What later became the vicarage garden had, it seems, once been the site of a seventh- or eighth-century mortuary chapel that was repurposed, during the Viking hosts' occupation in 873–4, as a burial mound. First rediscovered in the seventeenth century, the mound was re-excavated by the Biddles and revealed the remains of 249 people, their bones stacked like the contents of a charnel house. What remained of a single, central burial – a Viking warlord – included finds of an iron axe, part of a double-edged sword, knives, keys, silver pennies and small items of jewellery.[14] Repton's half holy, half unholy remains are monuments to both the heyday of the Mercian state and to its demise.

When Burghred died in exile in Rome with his queen,* in the year that Repton was taken and fortified (873) or in the year after, he was, according to the *Chronicle*, succeeded by Ceolwulf II, a 'foolish king's thegn'. The *Chronicle* recorded that Ceolwulf swore an oath of submission to the *micel hæðen here* and gave them hostages, promising to fight with them. From

---

* A gold finger ring, inscribed EAÐELSVIÐ REGNA, was found in Aberford, West Yorkshire, in 1870. British Museum AF.458. It may have been hers, or a gift from her to her niece, Æðelflæd. She is said to have been buried in Pavia in 888, having died on her way to bring her brother's alms to Rome (*Anglo-Saxon Chronicle* recension 'E'). The king was buried in St Mary's church in the *Schola Saxonum*.

then on the story of the Viking Age is the story of a tributary Mercia and a resurgent Wessex; and of Ælfred. Improbably, this fifth son of King Æðelwulf and grandson of King Ecgberht, the former exile, had survived each of his brothers in turn. The death of his last brother, Æðelred, of wounds sustained in the frantic campaigning of 870–1 – the so-called Year of Nine Engagements – left Ælfred as the last man standing, aged twenty-one. His defensive tactic against seemingly overwhelming odds was, in the first instance, to engage them in running skirmishes; then to make peace by paying them off with treasure until, at midwinter 877, he was betrayed by a senior ealdorman. After the Viking army attempted his capture at Chippenham and his flight into the watery fastness of Athelney in the Somerset marshes, he enjoyed a brief respite – to lick his wounds and plot his counterattack.

In later, more reflective times Ælfred's Welsh courtier and bishop, Asser, wrote a flattering biography of the great warrior statesman and philosopher; and at his court the *Anglo-Saxon Chronicle* was first compiled in the form that has come down to the present day. So far as the *Chronicle* is concerned, the decades of warfare from 870 to 893, when Ælfred won his final military victories, were decades of West Saxon endeavour.

The 'foolish king's thegn', Ceolwulf II, last independent king of Mercia in the line of Penda, Æðelbald and Offa, was not quite the underwhelming, even cowardly figure sketched by the scribes of the *Chronicle*. To begin with, his name suggests that, far from being a mere king's companion, he was an Icling, perhaps a kinsman of King Ceolwulf I or, more likely perhaps, his nephew Cœnwald.* His royal authority was sufficient, in 875 – even after the loss of Repton and Nottingham – that he and King Ælfred jointly embarked on an impressive coinage reform that saw silver content raised fivefold.[15]

Then, two years later, in autumn 877, after signing an expedient peace treaty with Ælfred at Exeter...

---

* He attested a charter of 811. Electronic Sawyer S 165.

... the host went into the land of Mercia, and some of it they divided up, and some they granted to Ceolwulf.[16]

The 'some of it' granted to Ceolwulf must, looking with the benefit of hindsight, have meant Mercia south and west of Watling Street, the ancient watershed line through the Midlands. Ceolwulf II's Greater Mercia now consisted of the lands of the *Hwicce*, of the *Magonsæte* and *Wreocansæte*, and the territories of what are now Oxfordshire and Buckinghamshire. The valleys of the Trent, Ouse and Nene, lands prized and capitalised over so many decades, were lost; so too Lindsey and London. A rump of perhaps Tamworth, probably Lichfield, certainly Worcester, Hereford, Winchcombe, the headwaters of the Thames and the brine springs of Droitwich, remained under Mercian control. Mercia had been cut in half; Ceolwulf was king of its rump.

And yet... and yet... when metal detectorist James Mather discovered, and reported, a hoard of silver objects and coins from a field nestling beneath the north scarp of the wooded Chiltern hills near Watlington, Oxfordshire, in 2015, the Mercian king's historical stock experienced a sudden boost. Along with some hack-silver,* the hoard contained 203 coins issued by kings Ælfred and Ceolwulf II and also by Ceolnoð, the archbishop of Canterbury. The latest of these coins date the deposition of the hoard to the end of the 870s or the beginning of the following decade, after a great and decisive West Saxon military triumph at Edington in 878. Thirteen of them represent an extremely rare type known as the Two Emperors, which rather speaks for itself.

As I write, a high-profile court case has just concluded, resulting in the convictions of two men for conspiracy to sell just two coins of the Two Emperors series in a collection of over forty

---

* Chopped up bullion used as currency – very much a feature of the Viking period when metalwork was worth its scrap value: prized as portable wealth.

they had come by illegally – these are hot potatoes. * The coins are beautifully crisp, high-quality silver pennies bearing Roman-style busts on the obverse[†] and, on the reverse, double seated portraits of the Eastern and Western emperors, copied from a later fourth-century *solidus*, with the name of the moneyer.[17] The coins are thought to date to about 875, part of a much larger issue from ten, probably London-based moneyers, which indicates that their scarcity now is a matter of silence rather than absence. A second series of jointly-issued coins represented in the hoard, the Cross-and-Lozenge type, dates from later in the decade.

In *Ælfred's Britain*, I raised the possibility, first mooted by Jinty Nelson, that the Two Emperors coins were issued as a limited, commemorative series celebrating a joint West Saxon and Mercian victory at Edington.[18] The revisionist idea of a joint victory still holds water even if the coins are now known to date from a few years earlier, as part of a more general, collaborative currency reform. Mercia and Wessex were acting in alliance in those few years, with Ælfred as the senior partner. His otherwise improbable victory at Edington is more explicable if he was able to call on Mercian military support – from a Mercian king in the tradition of Æðelbald and Offa, not a mere foolish king's thegn.

Ceolwulf II seems to have died the year after Edington, perhaps as a result of wounds sustained in battle there.[‡] In his place one Æðelred was chosen as a probably unrelated Mercian ruler. Æðelflæd's husband – Ælfred's son-in-law – is a suitably enigmatic Mercian ruler, sometimes styled ealdorman, *dux* or *patricius* in his charters but in any case the effective ruler of Mercia under West Saxon overlordship. He first appears on the

---

* These were also found in 2015, in a field in Leominster, Herefordshire; but the finders had, illegally, not reported them. Source: www.theguardian. com/uk-news/2023/apr/27/two-men-guilty-of-trying-to-sell-history-changing-anglo-saxon-coins-illegally retrieved May 9, 2023.

† The Mercian version inscription reads CEOLWULF REX M or REX MER. Naylor and Standley 2022, 69.

‡ Thomas Charles-Edwards cites a Worcester king-list recording Ceolwulf's reign from 874 to 879. Charles-Edwards 2014, 490.

losing side in a battle at Conwy in 881 on the north coast of what is now Clwyd. An ancient fort at Deganwy, on the right bank of the Conwy on the Llandudno peninsula, was said to have been destroyed by the Mercians under King Ceolwulf I in 822, after which he took control of Powys.[19] At Conwy the Mercians were beaten by the sons of Rhodri Mawr in revenge for his death at the hands of Ceolwulf II in 878.[20] Their leader was 'Edryd Long-hair, king of Lloegr', according to a thirteenth-century collection of Welsh genealogies – that is, Æðelred.[21] Thomas Charles-Edwards, the historian of Early Medieval Wales, believes that Æðelred's catastrophic loss at Conwy was the trigger for Ælfred to assert his *imperium* over Mercia, rather than the death of Ceolwulf.[22] By 883 the first of Æðelred's seven surviving charters has him donating land to Berkeley Abbey in Gloucestershire, 'with the consent of King Ælfred and the whole Mercian *witan*'.*

By 886 at the latest King Ælfred had wrested control of London from his Danish rivals. The *Anglo-Saxon Chronicle* records that he entrusted the city to Ealdorman Æðelred. Either now, or some time previously, he had given his daughter, Æðelflæd, in marriage to Æðelred – the seal of a tributary and probably mutually beneficial West Saxon–Mercian alliance that held until her death in 918, more than thirty years later. When King Ælfred died in 899 at the age of fifty, among his many bequests was a sword, worth 100 mancuses, which he left to Æðelred.[23]

By 901 the Mercian royal couple were issuing joint grants – the most significant of these a record of major construction work, and a set of consequent privileges, at Worcester.† If historians, looking back, see the end of independent Mercian rule in these decades, the contemporary zeitgeist was more hopeful, looking to

---

* Electronic Sawyer S218. The Old English '*witan*' ( 'wise men'), for the royal council, is an early use for a term that becomes much more common in later Anglo-Saxon England. The charter is recorded in English in the collection known as Hemming's Cartulary. Hearne 1723, 103.

† Described more fully in *Ælfred's Britain*. Adams 2017, 185ff. Electronic Sawyer 223. *English Historical Documents* 99.

The ninth-century Mercian church at Deerhurst, Gloucestershire, on the banks of the River Severn: comparable to Brixworth in its architectural splendour, monument to a century and a half of Mercian supremacy.

a future of restored influence and power. The charter provides for the construction of fortifications 'for the protection of the people'. It grants to the church at Worcester half of the commercial – and probably judicial – rights belonging to the *burh*, 'in the market or in the street'. The rights are detailed: *land feoh* (land rent); *fiht-wite* (fines for fighting) – and for *stelan* (theft); for *woh ceapung* (dishonest trading); and for *burh wealler sceatinge* (damage to the *burh* wall). The exceptions, as so often, provide equally telling detail: the wagon-shilling taxes on the transport of salt go to the king 'as they have always done at Droitwich'. The solid pragmatism of 150 years of Mercian royal administration, jealous of its rights and generous with its patronage, stood intact in that part of Mercia where the royal writ still ran. As the events of the first two decades of the tenth century showed, Mercians had every right to believe that their writ would, one day, be restored over the whole of Mercian territory.

The almost fortuitous survival of something akin to a genuine Mercian annal – the so-called Mercian Register – is the faintly sounding echo of that hope. In Recensions B and C of the *Anglo-Saxon Chronicle* a fragmentary annal has been 'crudely' * inserted after the year 915, but incorporating entries dating back to 902. Modern editions of the *Chronicle* have rearranged them to come in chronological order so they make more sense: enough, in fact, to recognise the bones of a Mercian annal for the years 902–24, covering the period when Æðelflæd, later acting in consort with her brother Edward, was making striking territorial gains in the Midlands and the North. The great Victorian historian Charles Plummer went so far as to call these fragments the 'Annals of Æðelflæd'.[24] The integrity of the Mercian-centred entries has been convincingly shown by their accurate recording of a lunar eclipse and a comet in 904 and 905 respectively.[25] Further analysis over the years has allowed a much more co-ordinated view to be drawn of the military alliance between sister and brother. They were acting together to a plan, as the historian Frederick Wainwright was able to show...

> ... Æðelflæd's capture of Derby before Lammas (1 August) 917 coincided with the confused fighting around Towcester, Bedford, *Wigingamere*, and Tempsford, a strategic combination of operations which both caused and took advantage of the dispersal of Danish forces. Her occupation of Leicester in 918 was, in all probability, also part of a concerted action which carried Edward northwards to Stamford.[26]

Just as significant is the Mercian Register entry for 919, referring back to the end of the previous year, which allows

---

* According to Wainwright (1945).

historians to nuance the supposedly seamless transfer of power from Æðelflæd on her death bed to her brother King Edward, and his unification of the two kingdoms...

> *Her eac wearð Æðeredes dohtar Myrcna hlafordes ælces onwealdes on Myrcum benumen 7 on Westsexe alæded þrim wucan aer middum wintra. Seo wæs haten Ælfwyn...*

> Here also, the daughter of Æðelred, Lord of the Mercians, was deprived of all authority in Mercia and was led into Wessex three weeks before Midwinter. Her name was Ælfwynn. *

The Mercian core lands are hidden from view throughout the period of Mercian *imperium*; and that is, ironically, largely a function of the treaty that King Ælfred and King Guðrum agreed after the West Saxon/Mercian victory at Edington in 878. The first clause in that treaty defines a border between the English – the *Angelcynn* – and what was later called the Danelaw, after which those lands north and east of the line – East Anglia, Essex, Lindsey, North Mercia and Middle Anglia – were lost to Mercian, or West Saxon, rule. That line of treaty ran...

> up the Thames, and then up the Lea, and along the Lea to its source, then in a straight line to Bedford, then up the Ouse to the Watling Street[27]

It seems certain that those charters that must once have existed, kept at Leicester, *Medeshamstede*, Repton, Breedon-on-the-Hill and other Mercian *scriptoria* on the 'wrong' side of that line, and which would have floodlit the dark corners of Mercian

---

* However sinister the entry sounds, the chances are that Ælfwynn was placed in a monastic house. She may be the Ælfwynn named as beneficiary in a charter of 948 (Electronic Sawyer S 535): 'A.D. 948. King Eadred to Ælfwyn, a religious woman; grant of 6 mansae [...] at Wickhambreux, Kent, in return for 2 pounds of purest gold.' *Anglo-Saxon Chronicle*. Swanton 1996, 105.

history, were destroyed during the period when minsters were being systematically looted, their communities dispersed. But it is to the surviving charters, those recording Mercia's peripheral interests, that historians turn once again to peer behind the curtain of West Saxon spin. And it is worth emphasising the richness of that resource: there are far more surviving Mercian charters than from any other of the early Anglo-Saxon kingdoms. Nowhere else can historians read between the lines of church and state politics, of local and regional power, of economy and the rural landscape, as well as they can in western and southern Mercia, along the Thames Valley and in Kent, where Mercian kings deployed the political capital of *imperium* to invest their power in land and territorial lordship.

From the 160 years between King Cœnwulf I's death in 821 and the emergence of Ealdorman Æðelred in the 880s, a mere thirty-one Mercian royal charters survive, compared to more than a hundred during the previous century. Most date from the reign of King Beorhtwulf (840–?52), many of them issued at Tamworth. They follow a familiar pattern of small grants of land and concessions to bishops, ministers and monastic houses, showing active royal patronage in spite of the diminished *imperium* of Mercia's kings. The geography of these charters is illuminating: Evesham, Worcester and Gloucester – popular beneficiaries – lay in the *Hwiccian* heartlands; so too did Hanbury and Berkeley.

A grant of privileges made at Repton to the *familia* of the monastery at Breedon-on-the-Hill, in return for 180 mancuses and fifteen hides in Derbyshire and Northamptonshire, may reflect the straitened circumstances of diminished royal coffers and London's declining trading capacity.* Mercian kings needed cash and they raised it from land sales. But more extreme measures were sometimes necessary to enrich the treasury. King Beorhtwulf, in the year of his accession, took five estates away from the church at Worcester and gave them to his own

---

* Electronic Sawyer S197: dubious, but probably based on an authentic record.

men – supporters, one imagines, in his bid for power. Bishop Heahberht went to Tamworth to petition the king for their restoration, brandishing the original charters as evidence. The king and his council agreed to their restoration – for a price: four very choice horses, a ring worth thirty mancuses, a skilfully wrought dish, two silver horns and, for the queen, two good horses, two goblets and one gilt cup.[28]

A charter of 844 or 845 attested by, among others, Queen Sæðryð, contracts with a thegn as follows...

> I, King Beorhtwulf, give to my thegn Forðred an estate of nine hides in Wootton* for him to have in perpetual inheritance, and to give after his day to whoever has earned it from him by humble obedience: Cisseðebeorg [Cisseða's barrow]; *Feowertreowehyl* [Four-tree hill]; *Eanburgemere*; *Tihhanhyl* [Tihha's hill], and land of two hides out along the River Ray in perpetual inheritance. And he gave as the estate-price thirty mancuses and nine hundred shillings for the land in perpetual inheritance.[29]

This is small beer for a once-mighty monarchy. King Beorhtwulf's son and successor, Burghred, raised 300 shillings in 857 by freeing a monastery, belonging to Bishop Ealhun of Worcester at Blockley in Gloucestershire, from the obligation to feed and maintain hawks and falcons and the king's huntsmen; from feeding and maintaining those men called in Old English *Walhfereld*, and from lodging them and other mounted men. Dorothy Whitelock speculated in her transcription of the grant that *Walhfereld* – literally 'Welsh expedition' – might mean a troupe of king's messengers, like those *Cyninges horsewealh* mentioned in King Ine's laws.[30] But one might alternatively speculate that these troops had formerly been tasked with patrolling the great Dyke; and that it was, in fact, no longer being maintained as assiduously as it had once been.

* Wotton Underwood, Buckinghamshire.

Seven years later Burghred and his West Saxon Queen Æðelswyð – sister of the future King Ælfred – granted to Alhhun, bishop of Worcester, five hides in return for 'precious' objects worth 400 shillings, with an annual render of thirty shillings to be paid to the church at Eynsham.[31] The charter was written at Bath in 864, the year before the arrival of the *micel hæðen here*, when Mercian kings and their consorts still ruled the vast bulk of the English Midlands. Their last grant is dated 869.

A charter of King Ceolwulf II dating from 875, in which he leased six hides from the bishop of Worcester at Daylesford, Worcestershire, for the term of four lives, again addresses the issue of mounted men:

> That the whole diocese of the *Hwicce* is to remain absolved and secure from feeding the king's horses and those who lead them.[32]

This was an avowed act of pious generosity from the king: remission of a burden on the ecclesiastical purse for the sake of the king's soul. But one wonders what his horses were to be fed on after that. It reinforces the suspicion that, in those difficult times, the king had fewer horses to feed.

A number of charters belonging to this period, which purport to grant lands and privileges to churches but are patent forgeries, is almost as illuminating in measuring the shrinkage of Mercian political capital. King Wiglaf's supposed 833 grant to Siward, the abbot of Crowland minster, founded more than a century before at the shrine of St Guðlac, is a post-Conquest fabrication. An 851 charter of King Beorhtwulf to the same abbot, Siward, is also a later forgery, as is another of 868.[33] Those monasteries, like Crowland, whose communities were overrun or enslaved during the Viking Wars, lost their original charters as their estates were appropriated or broken up. When they were refounded during the tenth century or later, traditions that maintained knowledge (real or wishful) were retrospectively written, or entirely invented, in efforts to restore those lands and privileges. That many of these traditions seem to have credible roots is remarkable in itself. That many of the forgeries are clumsy or implausibly written is only evident thanks to the dedication of generations of scholars who have pored over, debated and edited them for all students and researchers to use.

Mercia's most celebrated religious foundations enjoyed, or suffered, contrasting fortunes after the middle of the eighth century. The key to institutional survival lay in retaining possession of those large estates accumulated under royal or noble patronage across the seventh and eighth centuries. When monastic records were lost or collections of documents were broken up or fragmented, estates became easy prey, both to cash-strapped kings and to regional warlords. Occasionally, as with the land books of Ely or of the community of Cuthbert at Lindisfarne, records of 'thefts' were kept as assiduously as those of gifts and in later centuries, during a monastic reform movement in the tenth century, retrospective claims were sometimes tested by legal judgement or indulged by royal prerogative. A charter of Ealdorman Æðelred, dating to the last decade of the ninth century or the first decade of the tenth, records the renewal of a grant made to his minister Cuðulf under King Burghred, 'the ancient land book having been carried off by pagans'.[34]

The first phase of Viking attacks was one of raiding for booty; for books that might be ransomed back to their owners; church treasures to be sold for scrap or recycled into more attractive forms for their Viking owners. Serfs, monks, nuns and prisoners of war were traded as slaves; some were perhaps sold back to their families. In Yorkshire it was remembered that the Danish leader Hálfdan shared out the land among his followers; and numerous Scandinavian place names are witnesses to incoming settlers.

But that is not the whole story. The destruction of churches as an act of anti-Christian, heathen animosity is less well documented than the rapid adoption by Scandinavian settlers of English social mores. Fresh generations of Norse landlords assimilated themselves to the language and customs of their new lands and became active, sometimes enthusiastic patrons of the church. Norse names including the element *Kirby* or *Kirkby* abound in gazetteers of northern English villages and towns and a vibrant, hybrid sculptural tradition that blends common elements of Norse and Christian world views – animal and plant imagery; sacrifice; judgement and apocalypse – can still be found in churches, churchyards and museums from Cumbria and Yorkshire to the Midlands.

Few, if any functioning monasteries survived unscathed under Scandinavian rule in East and Middle Anglia or in Essex.[35] No East Anglian bishops are known between the 860s and 940s. But some known church sites, like that of the episcopal see at North Elmham in Norfolk, have been shown by excavation to have continued in use and the cult of St Edmund, martyred by Vikings in 869, was acknowledged and patronised by later Danish rulers. *Medeshamstede's* charter records at least partially survived those turbulent times and Peterborough became one of the wealthiest and most powerful of medieval cathedrals. Local saints' dedications were preserved at Northampton, Bedford, Oundle and at St Guðlac's shrine at Crowland, where later tradition remembered its refounding in 948. Abbots of Crowland are recorded from at least the late tenth century onwards, even if later incumbents felt it necessary to forge charters relating to its pre-Viking rights.[36]

The royal minster at Bardney in Lindsey, founded by King Æðelred and his Northumbrian Queen Osðryð in 675, was the subject of a remarkable coup in about 909, when the *Anglo-Saxon Chronicle* records that St Oswald's relics were transferred from their resting place there to the new royal Mercian monastery at Gloucester. Whether Æðelflæd paid a ransom for them or led a raiding party to seize them is unclear; but the minster and its shrine must still have been intact.

South-west of the line of Watling Street and the River Lea, the border demarcated by treaty between Ælfred and Guðrum after the decisive Battle of Edington in 878, minster survival is far better attested. The existence of so many *Hwiccian* charters is testament in itself to Worcester's continuing importance to

Mercian royal interests, as is the record of its fortification, market place and functioning judiciary in 901. The bishops of Worcester not only survived but, in general, thrived, partly by bringing more and more minster foundations and estates under their control.[37] Worcester's bishops were nothing if not pragmatists. A grant of 872 conferring a two-hide estate in Warwickshire from Wærferð, bishop of Worcester, to a thegn called Eanwulf for twenty gold mancuses was agreed because of the 'very pressing affliction and immense tribute' that he had been forced to give up 'in that same year when the pagans stayed in London'.[38] This may relate to a certain 'profitable little estate' that the bishop's predecessor had been granted by King Burghred in 857.[39] Such extortions became a fact of life. Elsewhere many early minsters retained their status as mother parishes into later centuries and excavations at minster sites like Eynsham show apparently uninterrupted occupation through those centuries.[40]

Mercia's finest standing Anglo-Saxon building, All Saints', Brixworth, is its own witness and monument to changing fortunes. It survives, with a substantially intact core of original eighth- and early ninth-century work. In the ninth century a spiral staircase turret was added at the west end and the eastern apse was refurbished. The side aisles were removed in later centuries; the splendid south door was modified around the time of the Conquest.[41] Even so, this great Mercian achievement, so redolent of those years of *imperium* and royal patronage, comes with no early history; its builder unknown.

The same applies to the once-stupendous stone hall at Northampton, just a few miles to the south at the head of the densely settled Nene Valley. Northampton stood on the front line of tensions between the Danelaw settlements and what one might call 'free' Mercia: the base for a Scandinavian force ruled by an earl, probably from the 880s until the campaigns of Æðelflæd and Edward; and as late as 940 it suffered an attack from a conquering army under Óláfr Guðrøðsson.[42] It was attacked and burned again in 1040.[43] But its value as a trading and production settlement seems to have been sustained up to the point where it became the shire town of a new royal administrative reform movement. It was a centre of pottery production from the late ninth century onwards, although the great stone hall seems to have fallen into disuse at the beginning of the tenth century. King Edward appears to have constructed *burh* fortifications here, with the minster and stone hall at their centre, in about 917 and even if, as a recent review of Early Medieval Northampton suggests, true urbanism with street frontages, merchants and production workshops was slow to develop in the town, by the time of the *Domesday Survey* it supported 300 houses and 87 burgesses: a substantial settlement.[44]

*

As an independent kingdom – or as a confederacy under its dynastic overlords – Mercia wielded supreme power over the English kingdoms, sometimes also the Welsh, from the 680s until the 820s: 140 years of successful administration, royal patronage and more or less consistent political unity. Mercia's kings treated with the great Charlemagne and with popes, as serious European players. Its topographic disadvantages, landlocked and surrounded by potentially hostile rivals, were turned to advantage. Control of London gave Mercia access to continental markets and courts and its traders, producers and ecclesiastical entrepreneurs found ways to exploit the east-facing rivers that emptied into the Wash and the North Sea. Its administrators learned the business of coinage; used it to project royal power and create a functioning economy. They found ways to exact profits from salt and wool, from lead and from the sale of estates. A gradual understanding of the Roman communications network – the Fosse Way, Watling Street, Ryknield Street and Ermine Street – promoted the idea of central places – the *burhs* – where resources could be concentrated and converted to economic and political assets and from which military and judicial control could be exercised. The blueprint they invented and refined was eventually appropriated by Mercia's ancient rival Wessex.

Mercian kings manipulated ecclesiastical and secular lords in subject kingdoms and provinces, demoting former kings to willing vassals; extending their networks of influence and loyalty into the far corners of Kent, Essex, East Anglia and Sussex. They expended much energy on framing relations with archbishops in Canterbury, not always successfully. They were rarely passive players; always active dynasts and diplomats, warlords and ecclesiastical patrons – as generous in their benefactions as their legendary heroes and moral compasses dictated they should be; as brutally cynical as the realpolitik of the eighth and ninth centuries demanded. Mercian sculpture and the few surviving products of their metallic arts and literature are treasures worthy of a great state; their architecture, what little we have of it, was suitably grand; suitably formidable and confident.

Mercian kings deployed their kin as proprietorial assets in establishing a dense network of minsters through which family wealth was invested, secured and used to project regional influence and make a commercial profit. Many of their 'plays' come from the standard Machiavellian rule book of Old Testament kingship, refined by Bede for the times. Others seem novel: the clusters of functional settlements that orbited the *burhs*; the extraordinary dykes constructed to control their western marcher lands; their canny use of land grants to concentrate and profit from the huge range of landed resources that they could tap.

Mercian queens and other royal women enjoyed a unique status in Mercia – more so even than the exceptional consorts and kinswomen of seventh-century Northumbria. From Osðryð, the wife of Æðelred who promoted minster foundations and royal cults and found herself on the wrong end of a blood feud, to Offa's queen Cyneðryð, who issued her own coins, assembled an enviable portfolio of properties and disputed with archbishops; from Cœnwulf's equally steadfast and acquisitive daughter Quœnðryð, to Queen Sæðryð, consort of King Beorhtwulf, who witnessed every one of his charters; from Ælfred's sister Æðelswið – another consistent charter witness, who followed her husband Burghred into exile – to his own daughter Æðelflæd, 'Lady of the

Mercians', Mercian royal women were public figures, patrons, entrepreneurs and politicians in their own right; sponsors and consumers of literature, art and craft; correspondents of Alcuin and Boniface; curators too, one suspects, of genealogy and tradition. If, as the historian Pauline Stafford argues, ninth-century Mercian kings' sons almost never succeeded their fathers, royal women nevertheless helped these dynasts of less successful generations to maintain power.

Mercia's cultural failure either to produce – or preserve – a body of historical narrative or secular literature beyond the literary confines of the charter and, perhaps, the outstanding poetic masterpiece *Beowulf*, left a gaping hole for a triumphant West Saxon narrative to fill at their erstwhile rival's expense. It has encouraged historians to probe tentatively at the edges of the darkness with their torches and sticks without, so to speak, turning the floodlights on. To attempt something approaching a chronicle of Mercia's heyday is to throw no more than another lighted candle into the shadows. There are many dim recesses and side passages still to be illuminated.

# Acknowledgements

Among the indispensable resources now available to the Early Medievalist the online Prosopography of Anglo-Saxon England or PASE and the Electronic Sawyer are outstanding. The Corpus of Early Medieval Coin Finds, supplemented by the ongoing project to map all the finds from the Portable Antiquities Scheme, likewise keep the modern scholar abreast of the latest material.

I have benefitted from insightful conversations with my friends and colleagues, especially Colm O'Brien and Brian Roberts, Paul Blinkhorn, Duncan Wright and Andy Chapman. Thanks to Jo Story I had the lucky chance to explore the tower of All Saints', Brixworth. I would like to express my thanks to the generations of scholars who have produced fine English editions of all the major contemporary source material; and to all the excavators who have brought their work to publication – no mean feat.

Final thanks go to Sarah (for patience and encouragement), to Richard Milbank, my editor, and to the creatives at Head of Zeus who make such fine books.

# Bibliography

Abels, R.P. 1988 *Lordship and Military Obligation in Anglo-Saxon England*. London: British Museum Publications

Adams, H. 1905 *Essays in Anglo-Saxon Law*. Boston: Little, Brown and Co.

Adams, M. 2013 *The King in the North: The Life and Times of Oswald of Northumbria*. London: Head of Zeus

Adams, M. 2017 *Ælfred's Britain: War and peace in the Viking Age*. London: Head of Zeus

Adams, M. 2018 *Unquiet Women: from the Dusk of the Roman Empire to the Dawn of the Enlightenment*. London: Head of Zeus

Adams, M. 2021a *The First Kingdom: Britain in the Age of Arthur*. London: Head of Zeus

Adams, M. 2021b St Columba as a Territorial Lord. *Donegal Annual* 73: 85–92

Adams, M. and O'Brien, C. 2021 A sparrow in the temple? The ephemeral and the eternal in Bede's Northumbria. In Hüglin, S., Gramsch, A. and Seppänen, L. (eds) *Petrification Processes in Matter and Society*, 155–66. European Association of Archaeologists. Cham: Springer

Alexander, M. (Trans & Ed) 1973 *Beowulf*. London: Penguin

Allott, S. 1974 *Alcuin of York, c. AD 732 to 804. His life and letters*. York: William Sessions, 1974

Anderton, M. (ed) 1999 *Anglo-Saxon Trading Centres: Beyond the Emporia*. Glasgow: Cruithne Press

Arnold, T. 1879 *The History of the English by Henry, Archdeacon of Huntingdon*. London: Longman

Atherton, M. 2005 Mentions of Offa in the Anglo-Saxon Chronicle, *Beowulf* and *Widsith*. In Hill, D. and Worthington, M. (eds) *Æthelbald and Offa: Two Eighth-Century Kings of Mercia*: 65–74

Attenborough, F.L. 1922 *The Laws of the Earliest English Kings*. Cambridge: Cambridge University Press

Bailey, K. 1989 The Middle Saxons. In Bassett, S. (ed) *The Origins of Anglo-Saxon kingdoms*: 108–22

Bailey, K. 2000 Clofesho revisited. In Griffiths, D. (ed) *Anglo-Saxon Studies in Archaeology and History* 11: 119–32

Bailey, R. 1980 *The Early Christian Church in Leicester and its Region. Vaughan Paper 25*. Leicester: University of Leicester

Bailey, R. 1988 *The Meaning of Mercian Sculpture*. Sixth Brixworth Lecture. Leicester: Friends of All Saints' Church Brixworth

Banham, D. and Faith, R. 2014 *Anglo-Saxon Farms and Farming. Medieval History and Archaeology*. Oxford: Oxford University Press

Bannerman, J. 1974 *Studies in the History of Dalriada*. Edinburgh: Scottish Academic Press

Baring-Gould, S. 1897 *The Lives of the Saints*. London: John C. Nimmo

Bassett, S. (ed) 1989 *The Origins of Anglo-Saxon Kingdoms*. London: Leicester University Press

Bassett, S. 2007 Divide and rule? The military infrastructure of eighth- and ninth-century Mercia. *Early Medieval Europe* 15: 1, 53–85

Bassett, S. 2008 The Middle and Late Anglo-Saxon Defences of Western Mercian Towns. In Crawford, S. and Hamerow, H. *Anglo-Saxon Studies in Archaeology and History* 15: 180–239. Oxford: Oxbow Books

Bassett, S. 2020 The Role of Mercian Kings in the Founding of Minsters in the Kingdom of the Hwicce. In Langlands, A. and Lavelle, R. (eds) *The Land of the English Kin: Studies in Wessex and Anglo-Saxon England in Honour of Professor Barbara Yorke*: 373–86

Biddle, M. and Kjølbye-Biddle, B. 1985 The Repton Stone. *Anglo-Saxon England* 14: 233–92

Biddle, M. and Kjølbye-Biddle, B. 1992 Repton and the Vikings. *Antiquity* 66, 36–51

Birch, W de G. (ed) 1887 *Cartularium Saxonicum : A Collection of Charters Relating to Anglo-Saxon History*. Volume 2, AD 840–947. Cambridge: Cambridge University Press edition 2012

Blackburn, M. 1995 Money and Coinage. Chapter 20 in McKitterick (ed): *The New Cambridge Medieval History Volume II*: 538–59

Blair, J. 1989 Frithuwold's kingdom and the origins of Surrey. In Bassett, S. (ed) *The Origins of Anglo-Saxon kingdoms*: 96–107

Blair, J. 1996 The Minsters of the Thames. In Blair, J. and Golding, B. (eds) *The Cloister and the World: Essays in Medieval History in Honour of Barbara Harvey*: 5–28. Oxford: Clarendon Press

Blair, J. 2005 *The Church in Anglo-Saxon Society*. Oxford: Oxford University Press

Blair, J. 2018 *Building Anglo-Saxon England*. Oxford: Princeton University Press

Blair, J. 2020 Beyond the Billingas: From Lay Wealth to Monastic Wealth on the Lincolnshire Fen-Edge. In Langlands, A. and Lavelle, R. (eds) *The Land of the English Kin: Studies in Wessex and Anglo-Saxon England in Honour of Professor Barbara Yorke*: 387–406

Blinkhorn, P. 1999 Of cabbages and kings: production, trade and consumption in Middle Saxon England. In Anderton, M. (ed) *Anglo-Saxon trading centres: beyond the emporia*: 4–23

Boutflower, D.S. 1912 *The Life of Ceolfrid, Abbot of the Monastery at Wearmouth and Jarrow*. Sunderland: Hills & Co.

Bradley, S.A.J. 1982 *Anglo-Saxon Poetry*. London: J.M. Dent

Brady, L. 2017 The Welsh Borderlands in the Lives of St Guthlac. In Brady, L. *Writing the Welsh Borderlands in Anglo-Saxon England*. Manchester: Manchester University Press

Brady, L. 2018 Crowland Abbey as Anglo-Saxon Sanctuary in the Pseudo-Ingulf Chronicle. *Traditio*, 73, 19–42. https://doi:10.1017/tdo.2018.1 retrieved 16.05.2023

Brooks, N. 1971 The development of Military Obligations in Eighth- and Ninth-century England. In Clemoes, P. and Hughes, K. (eds) *England Before the Conquest: Studies in Primary Sources presented to Dorothy Whitelock*: 69–84

Brooks, N. 1984 *The Early History of the Church of Canterbury. Studies in the Early History of Britain*. Leicester: Leicester University Press

Brooks, N. 1989a The Creation and Early Structure of the Kingdom of Kent. In Bassett, S. (ed) *The Origins of Anglo-Saxon kingdoms*: 55–74

Brooks, N. 1989b The Formation of the Mercian kingdom. In Bassett, S. (ed): 159–70

Brown, M.P. and Farr, C.A. (eds) 2001 *Mercia: An Anglo-Saxon Kingdom in Europe. Studies in the Early History of Europe*. London: Leicester University Press

Burghart, M.A. 2007 *The Mercian Polity 716–918*. Unpublished PhD Thesis, Kings College, London

Buteaux, S. and Chapman, H. 2009 *Where Rivers Meet: The Archaeology of Catholme and the Trent-Tame Confluence. CBA Research Report 161*. York: Council for British Archaeology

Campbell, A. 1967 *Æthelwulf: De Abbatibus*. Oxford: Clarendon Press

Campbell, J. 1986 *Essays in Anglo-Saxon History*. London: Hambledon Press

Campbell, J. 2000 *The Anglo-Saxon State*. London: Hambledon and London

Campbell, J. 2003 Production and distribution in Early and Middle Anglo-Saxon England. In Pestell, T. and Ulmschneider, K. (eds) *Markets in Early Medieval Europe: Trading and 'Productive' Sites 650–850*: 12–19

Carr, R.D., Tester, A. and Murphy, P. 1988 The Middle-Saxon settlement at Staunch Meadow, Brandon. *Antiquity* 62: 235, 371–7

Cavill, P. 2015 The naming of Guthlac. *Nottingham Medieval Studies* 59: 25–47

Chadwick, N.K. 1963 The Celtic background of Early Anglo-Saxon England. In Chadwick, N.K. (ed) *Celt and Saxon: Studies in the Early British Border*. Cambridge: Cambridge University Press

Chambers, R.W. 1963 *Beowulf: An Introduction to the Study of the Poem with a Discussion of the Stories of Offa and Finn*. Cambridge: Cambridge University Press

Chapman, A. (ed) 2021 The Archaeology of Medieval *Northampton*. *Northamptonshire Archaeology* 41. Northampton: Northamptonshire Archaeological Society

Charles-Edwards, T.M. 2014 *Wales and the Britons 350–1064*. Oxford: Oxford University Press

Chick, D. 2005 The Coinage of Offa in the Light of Recent Discoveries. In Hill, D. and Worthington, M. (eds) *Æthelbald and Offa: Two Eighth-Century Kings of Mercia*: 111–22

Cole, A. 2013 The Place-name Evidence for a Routeway Network in Early Medieval England. *BAR British Series* 589. Oxford: Archaeopress

Colgrave, B. 1927 *The life of Bishop Wilfrid by Eddius Stephanus*. Cambridge University Press

Colgrave, B. 1940 *Two Lives of Saint Cuthbert*. Cambridge University Press

Colgrave, B. 1956 *Felix's Life of Saint Guthlac*. Cambridge University Press

Coupland, S. 2002 Trading places: Quentovic and Dorestad reassessed. *Early Medieval Europe II*: 3, 209–32

Coupland, S. 2023 A Coin of Queen Fastrada and Charlemagne. Early Medieval Europe doi: https://doi.org/10.1111/emed.12640

Cowie, R. and Blackmore, L. (eds) 2012 *Lundenwic: Excavations in Middle Saxon London, 1987–2000*. Museum of London Archaeology Monograph 83

Cramp. R. 1977 Schools of Mercian Sculpture. In Dornier, A. (ed) *Mercian Studies*: 191–233

Crossley-Holland, K. 1999 *The Anglo-Saxon World: an Anthology*. Oxford World's Classics. Oxford University Press

Cubitt, C. 1995 Anglo-Saxon Church Councils 650–850. *Studies in the Early History of Britain*. London: Leicester University Press

Cubitt, C. 2013 Appendix 2: The Chronology of Stephen's Life of Wilfrid. In Higham, N.J. (ed) 2013 *Wilfrid: Abbot, Bishop, Saint: Papers from the 1300th Anniversary Conferences*: 334–47

Davies, W. and Vierck, H. 1974 The Contexts of Tribal Hidage: Social Aggregates and Settlement Patterns. *Frühmittelalterliche Studien* 8, 223–93

Davis, R. 2000 *The Book of Pontiffs (Liber Pontificalis): The Ancient Biographies of the First Ninety Roman Bishops to A.D. 715*.Translated Texts for Historians 6. Liverpool: Liverpool University Press

Dornier, A. (ed) 1977 *Mercian Studies*. Leicester: Leicester University Press

Dugdale, W. 1693 *Monasticon anglicanum, or, The history of the ancient abbies, and other monasteries, hospitals, cathedral and collegiate churches, in England and Wales with divers French, Irish, and Scotch monasteries formerly relating to England*. London: Sam Keble

Dumville, D.N. 1976 The Anglian Collection of Royal Genealogies. *Anglo-Saxon England* 5: 23–50

Edwards, J.F. 1987 The transport system of medieval England and Wales: a geographical synthesis. Unpublished PhD thesis, University of Salford. Retrieved from: file:///C:/Users/Admin/Desktop/travel%20in%20med%20England%20PhD.pdf November 2015

Edwards, N. (2009) Rethinking the pillar of Eliseg. *Antiquaries Journal* 89: 143–77

Ekwall, E. 1960 *The Concise Oxford Dictionary of English place-names*. 4th edition. Oxford: Clarendon Press

Everson, P. and Stocker, D. 2023 Guthlac at Medeshamstede? *Early Medieval Europe* 31 (20, 194–219)

Fairweather, J. 2005 *Liber Eliensis: A history of the Isle of Ely from the seventh century to the twelfth*. Woodbridge: Boydell Press

Faith, R. 1997 The English peasantry and the Growth of Lordship. *Studies in the Early History of Britain*. London: Leicester University Press

Finberg, H.P.R. 1961 *The Early Charters of the West Midlands*. Leicester: Leicester University Press

Foot, S. 2013 Wilfrid's monastic empire. *Wilfrid Abbot, Bishop, Saint: Papers from 1300th Anniversary Conferences*. In Higham, N.J. (ed) 2013: 27–39

Forte, A., Oram, R. and Pedersen, F. 2005 *Viking Empires*. Cambridge University Press

Fouracre, P. 1995 Frankish Gaul to 814. In McKitterick, R. (ed) *The New Cambridge Medieval History Volume II*: 85–109

Fox, Sir C. 1955 *Offa's Dyke: A Field Survey of the Western Frontier-works of Mercia in the Seventh and Eight Centuries A.D.* London: The British Academy

Fraser, J.E. 2009 Caledonia to Pictland: Scotland to 795. *The New Edinburgh History of Scotland*. Edinburgh University Press

Ganz, D. (trans and ed) 2008 *Einhard and Notker the Stammerer: Two Lives of Charlemagne*. Harmondsworth: Penguin Classics

Garmonsway, G.N. (ed) 1972 *The Anglo-Saxon Chronicle*. London: J.M. Dent

Gelling, M. 1984 *Place-names in the Landscape*. London: John Dent

Gelling, M. 1992 The West Midlands in the Early Middle Ages. *Studies in the Early History of Britain*. London: Leicester University Press

Gelling, M. and Cole, A. 2014 *The Landscape of Place-names*. 2nd Edition. Donington: Shaun Tyas

Giles, J.A. 1847 *William of Malmesbury's Chronicle of the Kings of England*. London: Henry G. Bohn

Giles, J.A. 1849 *Roger of Wendover's Flowers of History*. London: Henry G. Bohn

Green, C., Green, I. and Dallas, C. 1987 Excavations at Castor, Cambridgeshire in 1957–8 and 1973 *Northamptonshire Archaeology* 21: 109–48

Hadley, D.M. and Richards, J.D. 2016 The winter camp of the Viking Great Army, AD 872–3, Torksey, Lincolnshire. *Antiquaries Journal* 96, 23–67

Halsall, G. 2003 *Warfare and Society in the Barbarian West 450–900*. London: Routledge

Hamerow, H. 2012 Rural Settlements and Society in Anglo-Saxon England. *Medieval History and Archaeology*. Oxford: Oxford University Press

Hardy, A., Charles, B.M. and Williams, R.J. 2007 *Death and Taxes: The Archaeology of a Middle Saxon Estate Centre at Higham Ferrers, Northamptonshire*. Oxford: Oxford Archaeology 2007

Harrison, M. and Embleton, G. 1993 *Anglo-Saxon Thegn AD 449–1066*. Oxford: Osprey

Haslam, J. 1986 The ecclesiastical topography of Early Medieval Bedford. *Bedfordshire Archaeology* 17, 41–50

Haslam, J. 1987 Market and fortress in the reign of Offa. *World Archaeology* 19, 1, 76–93

Hawkes, J. 2001 Constructing Iconographies: Questions of Identity in Mercian Sculpture. In Brown, M.P. and Farr, C.A. (eds) *Mercia: An Anglo-Saxon Kingdom in Europe*: 230–45

Hearne, T. (ed) 1723 *Hemingi Chartularium Ecclesiae Wigorniensis*, 2 vols. Oxford

Hey, G. 2004 Yarnton: Saxon and Medieval Settlement and Landscape : Results of Excavations 1990-96. *Thames Valley landscapes monograph* 20

Higham, N.J. 2005 Guthlac's Vita, Mercia and East Anglia in the first half of the Eighth century. In Hill, D. and Worthington, M. (eds) *Æthelbald and Offa: Two Eighth-Century Kings of Mercia*: 85–90

Higham, N.J. (ed) 2007 Britons in Anglo-Saxon England. *Publications of the Manchester Centre for Anglo-Saxon Studies* 7 Woodbridge: Boydell Press

Higham, N.J. (ed) 2013 *Wilfrid: Abbot, Bishop, Saint: Papers from the 1300th Anniversary Conferences*. Donington; Shaun Tyas

Higham, N.J. and Ryan, M.J. (eds) 2010 *The Landscape Archaeology of Anglo-Saxon England*. Woodbridge: Boydell Press

Hill, D. and Cowie, R. (eds) 2001 Wics: The Early Medieval trading centres of Northern Europe. *Sheffield Archaeological Monographs* 14. Sheffield Academic Press

Hill, D. and Worthington, M. (eds) 2005 *Æthelbald and Offa: Two Eighth-Century Kings of Mercia. BAR British Series* 383

Hinde, J. H. (ed) 1868 *Symeonis Dunelmensis Opera et Collectanea.* Publications of the Surtees Society volume 51. Durham: Surtees Society/ Andrews and Co.

Hodges, R. 2000 *Towns and trade in the Age of Charlemagne*. London: Duckworth

Hooke, D. 1985 *The Anglo-Saxon Landscape: the Kingdom of the Hwicce*. Manchester: Manchester University Press

Hopkins, K. 1980 Taxes and Trade in the Roman Empire 200 B.C.–A.D. 400. *Journal of Roman Studies* 70, 101–25

Jackson, D.A., Harding, D.W. and Myres, J.N.L. 1969 The Iron Age and Anglo-Saxon Site at Upton, Northants. *Antiquaries Journal* 49: 202–21

James, P. 1996 The Lichfield Gospels: the question of provenance. *Parergon* 13, 2: 51–61

Jewell, R. 2001 Classicism of Southumbrian Sculpture. In Brown, M.P. and Farr, C.A. (eds) *Mercia: An Anglo-Saxon Kingdom in Europe*: 246–62

John, E. 1960 Land Tenure in Early England. *Studies in Early English History,* Volume 1. Leicester University Press

Johnson, J. 1850 *A Collection of the Laws and Canons of the Church of England*. Volume 1. Oxford: John Henry Parker

Jones, G.R.J. 1979 Multiple Estates and Early Settlement. In Sawyer, P.H. (ed) *English Medieval Settlement*: 9–34

Jones, R. and Page, M. 2006 *Medieval Villages in an English Landscape: Beginnings and Ends*. Macclesfield: Windgather Press

Kelly, S.E. 1992 Eighth-Century Trading Privileges from Anglo-Saxon England. *Early Medieval Europe* 1, 3–28

Kelly, S.E. 2008 An Early Minster at Eynsham, Oxfordshire. In Padel,

O.J. and Parsons, D.N. (eds) *A Commodity of Good Names: Essays in Honour of Margaret Gelling*, 79–85

Keynes, S. 1993 The Eleventh Brixworth Lecture: The Councils of *Clofesho*. *Vaughan Paper* 38. Leicester: University of Leicester Press

Keynes, S. 2005 The Kingdom of the Mercians in the Eighth century. In Hill, D. and Worthington, M. (eds) *Æthelbald and Offa: Two Eighth-Century Kings of Mercia*: 1–26

Keynes, S. and Lapidge, M. 1983 *Alfred the Great: Asser's Life of King Alfred and other contemporary sources*. Harmondsworth: Penguin

Kirby, D.P. 1977 Welsh Bards and the Border. In Dornier, A. (ed) *Mercian Studies*: 31–9

Kirby, D.P. 2000 *The Earliest English Kings*. London: Routledge

Krapp, G. P. and Dobbie, E.V.K. (eds) 1936 The Exeter Book. *The Anglo-Saxon Poetic Records* Vol. III. New York: Columbia University Press

Langlands, A. and Lavelle, R. (eds) 2020 The Land of the English Kin: Studies in Wessex and Anglo-Saxon England in Honour of Professor Barbara Yorke. *Brill's Series on the Early Middle Ages* 26. Leiden: Brill

Lapidge, M., Blair, J., Keynes, S. and Scragg, D. (eds) 2014 *The Wiley Blackwell Encyclopaedia of Anglo-Saxon England*. Chichester: John Wiley

Lapidge, M. and Herren, M. 2009 *Aldhelm: The Prose Works*. Woodbridge: D.S. Brewer

Leahy, K. 2003 Middle Anglo-Saxon Lincolnshire: An Emerging Picture. In Pestell, T. and Ulmschneider, K. (eds) *Markets in Early Medieval Europe: Trading and 'Productive' Sites 650–850*: 138–54

Levison, W. 1946 *England and the Continent in the Eighth Century: the Ford Lectures Delivered in the University of Oxford in the Hilary Term, 1943*. Oxford: Oxford University Press

Lockett, N. Undated Worcestershire in the Roman period. *West Midlands Regional Research Framework for Archaeology*, Seminar 3. Retrieved from https://www.birmingham.ac.uk/Documents/college-artslaw/caha/wmrrfa/3/NeilLockett.doc, 24 April 2023

Looijenga, T. and Vennermann, T. 2000 The Runic Inscription on the Gandersheim Casket. In *Das Gandersheimer Runenkästchen*. Internationales Kolloquium, Braunschweig, 24–6 März 1999: 111–20. Herzog Anton Ulrich-Museum Braunschweig

Losco-Bradley, S. and Kinsley, G. 2002 Catholme: An Anglo-Saxon Settlement on the Trent Gravels in Staffordshire. *Nottingham Studies in Archaeology* 3. Nottingham: University of Nottingham

Loveluck, C. 2007 Rural settlement, lifestyles and social change in the later first millennium AD: Anglo-Saxon Flixborough in its wider context. *Excavations at Flixborough* Volume 4. Oxford: Oxbow

Loyn, H.R. and Percival, J. 1975 *The Reign of Charlemagne: Documents of Medieval History 2 London*. Edward Arnold

McClure, J. and Collins, R. (eds) 1994 *Bede: The Ecclesiastical History of the English People*. Oxford World Classics

McKerracher. M. 2018 *Farming Transformed in Anglo-Saxon England: Agriculture in the Long Eighth Century*. Oxford: Windgather Press

McKitterick, R. (ed) 1995 *The New Cambridge Medieval History Volume II*. Cambridge: Cambridge University Press

Maddicott, J.R. 1997 Plague in Seventh-Century England. *Past and Present 156*, 1: 7–54

Maddicott, J.R. 2005 London and Droitwich, c. 650–750: Trade, industry and the rise of Mercia. *Anglo-Saxon England 34*: 7–58

Maitland, F.W. 1897 *Domesday Book and Beyond: Three Essays in the Early History of England*. Cambridge: Cambridge University Press. Fontana Library paperback edition 1960

Martin, R. 2005 The Lives of the Offas: the Posthumous Reputation of Offa, King of the Mercians. In Hill, D. and Worthington, M. (eds) *Æthelbald and Offa: Two Eighth-Century Kings of Mercia*: 49–54

Meadows, I. (ed) 2019 *The Pioneer Burial: A high-status Anglian warrior burial from Wollaston, Northamptonshire*. Oxford: Archaeopress

Meaney, A. 2005 Felix's *Life of Guthlac*: History or hagiography. In Hill, D. and Worthington, M. (eds) *Æthelbald and Offa: Two Eighth-century Kings of Mercia*: 75–84

Mellows, W.T. (ed. & trans.) 1941 *The Peterborough Chronicle of Hugh Candidus*. Peterborough: Peterborough Natural History, Scientific and Archæological Society

Metcalf, D.M. 1977 Monetary affairs in Mercia in the time of Æthelabld (716–57). In Dornier, A. (ed) *Mercian Studies*: 87–106

Metcalf, D.M. 2003 Variations in the Composition of the Currency at Different Places in England. In Pestell, T. and Ulmschneider, K. (eds) *Markets in Early Medieval Europe: Trading and 'Productive' Sites 650–850*: 37–47

Morris, J. (ed) 1980 Nennius. British History and the Welsh Annals. *Arthurian Period Sources* Volume 8. London: Phillimore

Naismith, R. 2012 Money and power in Anglo-Saxon England: the southern English kingdoms 757–865. *Cambridge Studies in medieval life and thought*. Cambridge: Cambridge University Press

Naylor, J. and Richards, J.D. (2010) A 'Productive Site' at Bidford-on-Avon, Warwickshire: salt, communication and trade in Anglo-Saxon

England. In Worrell, S., Egan, G., Naylor, N., Leahy, K. and Lewis, M. (eds.) A Decade of Discovery: Proceedings of the Portable Antiquities Scheme Conference 2007. *British Archaeological Reports* British Series 520: 193–200

Naylor, J. and Standley, E. 2022 *The Watlington Hoard: Coinage, kings and the Viking Great Army in Oxfordshire, AD875–880.* Oxford: Archaeopress

Nelson, J.L. 1986 'A king across the sea': Alfred in Continental perspective. *Transactions of the Royal Historical Society* 36, 45–68

Nelson, J.L. 1991 The Annals of St Bertin. Ninth Century Histories, Volume 1. Manchester medieval Sources

Nelson, J.L. 2017 Losing the Plot? 'Filthy Assertions' and 'Unheard-of Deceit' in Codex Carolinus 92. In Naismith, R. and Woodman, D. (eds.) *Writing, Kingship and Power in Anglo-Saxon England,* 122–35

Niblett, R. 2010 Offa's St. Albans. In Henig, M. and Ramsay, N. (eds) *Intersections: The Archaeology and History of Christianity in England, 400–1200: Papers in Honour of Martin Biddle and Birthe Kjølbye-Biddle,* 129–34

Noble, F. 1983 Offa's dyke reviewed. *British Archaeological Reports, British Series* 114, ed. Gelling, M. Oxford: Archaeopress

O'Brien, C. 2002 The Early Medieval Shires of Yeavering, Breamish and Bamburgh. *Archaeologia Aeliana* Series 5, 30: 53–73

O'Brien, C. and Adams, M. 2016 The identification of Early Medieval monastic estates in Northumbria. *Medieval Settlement Research* 31: 15–27

Offer, C. 2002 *In Search of Clofesho: The Case for Hitchin.* Norwich: Tessa Publications

Oosthuizen, S. 2017 *The Anglo-Saxon Fenland.* Oxford: Oxbow books

Owen, H.W. and Morgan, R. 2007 *Dictionary of the Place Names of Wales.* Llandysul: Gomer

Palmer, B. 2003 The Hinterlands of Three Southern English *Emporia*: Some Common Themes. In Pestell, T. and Ulmschneider, K. (eds) *Markets in Early Medieval Europe: Trading and 'Productive' Sites 650–850*: 48–61

Parsons, D. 1977 Brixworth and its monastery church. In Dornier, A. (ed) *Mercian Studies*: 173–90

Parsons, D. 1988 St. Boniface – Clofesho – Brixworth. In Much, F.J. (ed) *Baukunst des Mittelalters in Europa.* Stuttgarter gesellschaft für Kunst und Denkmalpflege

Pelteret, D.A.E 1995 Slavery in Early Medieval England: from the reign oof Ælfred until the twelfth century. *Studies in Anglo-Saxon History* VII. Woodbridge: Boydell Press

Pestell, T. and Ulmschneider, K. 2003 *Markets in Early Medieval Europe: Trading and 'Productive' Sites 650–850*. Oxford: Windgather Press

Potts, W.T.W. 1974 The Pre-Danish Estate of Peterborough Abbey. *Proceedings of the Cambridge Antiquarian Society* 65, 13–27

Pretty, K. 1989 Defining the Magonsaete. In Bassett (ed) *The Origins of Anglo-Saxon Kingdoms*: 171–83

Rahtz, P.A. 1977 The archaeology of West Mercian towns. In Dornier, A. (ed) *Mercian Studies*: 107–30

Rahtz, P.A. and Meeson. R. 1992 An Anglo-Saxon watermill at Tamworth. *Council for British Archaeology Research Report* 83

Ray, K. 2015 *The Archaeology of Herefordshire: An Exploration*. Wooton Almeley: Logaston Press

Ray, K. and Bapty, I. 2016 *Offa's Dyke: Landscape and Hegemony in Eighth-Century Britain*. Oxford: Oxbow Books

Roberts, B.K. 2010 Northumbrian Origins and Post-Roman Continuity: an Exploration. In Collins, R. and Allason-Jones, L. (eds) *2010 Finds from the Frontier: Material Culture in the 4th–5th Centuries*. Council for British Archaeology Research report 162

Rollason, D.W. 1982 *The Mildrith Legend: A Study in Early Medieval Hagiography in England*. Leicester: Leicester University Press

Rumble, A.R 2001 Notes on the Linguistics and Onomastic Characteristics of Old English *Wīc*. In Hill, D. and Cowie, R. (eds) *Wics: The Early Medieval Trading Centres of Northern Europe*: 1–2

Rye. E and Williamson, T. 2020 Naming Early Monasteries: the Significance of *Burh* in East Anglia. *Medieval Archaeology* 64, 2: 226–43

Salisbury, C. 1995 An 8th-century Mercian bridge over the Trent at Cromwell, Nottinghamshire, England. *Antiquity* 69: 1015–18

Sawyer, P.H. 2013 *The Wealth of Anglo-Saxon England*. Oxford: Oxford University Press

Scholtz, B.W. (trans) 1972 *Caroloingian Chronicles: Royal Frankish Annals and Nithard's Histories*. University of Michigan Press

Sharp, S. 2005 Æðelberht, King and Martyr: the development of a legend. In Hill, D. and Worthington, M. (eds) *Æthelbald and Offa: Two Eighth-Century Kings of Mercia*: 59–63

Sharpe, R. (trans and ed) 1995 *Adomnán of Iona: Life of Saint Columba*. London: Penguin

Sims-Williams, P. 1990 Religion and Literature in Western England, 600–800. *Cambridge Studies in Anglo-Saxon England* 3. Cambridge: Cambridge University Press

Squatriti, P. 2002 Digging ditches in Early Medieval Europe. *Past and Present* 176: 11–65

Stafford, P. 2001 Political women in Mercia, Eighth to early Tenth Centuries. In Brown, M.P. and Farr, C.A. (eds) *Mercia: An Anglo-Saxon Kingdom in Europe*: 35–49

Stenton, D.M. (ed) 1970 *Preparatory to Anglo-Saxon England: being the collected papers of Frank Merry Stenton*. Oxford: Clarendon Press

Stevenson, J. 1855 (ed. and trans.) *The Historical Works of Simeon of Durham, Volume III, Part II*. London: Seeleys

Stevenson, W.H. 1914 Trinoda Necessitas. *English Historical Review* 29: 689–703

Stiff, M. 2001 Typology and Trade: a Study of the Vessel Glass from Wics and Emporia in North-west Europe. In Hill, D. and Cowie, R. (eds) *Wics: The Early Medieval trading centres of Northern Europe*, 43–9

Story, J. 2003 *Carolingian Connections: Anglo-Saxon England and Carolingian Francia, c. 750–870 Studies in Early Medieval Britain 2*. Aldershot: Ashgate

Sutherland, D.S. 2014 *The Building of Brixworth Church*. Brixworth: The Friends of All Saints' Church Brixworth

Swanton, M. (trans and ed) 1996 *The Anglo-Saxon Chronicle*. London: J.M. Dent

Talbot, C.H. (Trans. and ed.) 1954 *The Anglo-Saxon Missionaries in Germany*. London: Sheed and Ward

Taylor, H.M. 1987 St Wystan's Church, Repton, Derbyshire: A reconstruction essay. *Archaeological Journal* 144, 1: 205–45

Thacker, A. 2020 St Wærburh: the multiple identities of a regional saint. In Langlands, A.J. and Lavelle, R. (eds) *The Land of the English Kin*: 443–66

Thomas, C. 1981 *Christianity in Roman Britain to AD 500*. London: Batsford

Thorn, F. and Thorn, C. (eds) 1979 *Domesday Book 21: Northamptonshire*. Chichester: Phillimore

Thorpe, B. (ed) 1872 *Florentii Wigorniensis monachi chronicon ex chronicis*. 2 vols. London, 1848–49

Thorpe, L. 1974 *Gregory of Tours: The History of the Franks*. Harmondsworth: Penguin

Tolkien, J.R.R. 2014 *Beowulf: a translation and commentary*. Edited by Christopher Tolkien. London: Harper Collins

Tyler, D.J. 2005 Orchestrated Violence and the 'Supremacy of the Mercian Kings. In Hill, D. and Worthington, M. (eds) *Æthelbald and Offa: Two Eighth-century Kings of Mercia*: 27–34

Wainwright, F.T. 1945 The Chronology of the 'Mercian Register'. *The English Historical Review* 60: 238, 385–92

Wallace-Hadrill, J.M. 1971 *Early Germanic Kingship in England and on the Continent. The Ford Lectures delivered in the University of Oxford in Hilary Term 1970.* Oxford: Clarendon Press

Watts, V. 2004 *English place-names.* Cambridge: Cambridge University Press

Webb, J.F. and Farmer, D.H. (trans and eds) 1983 *The Age of Bede.* Harmondsworth: Penguin

Webster, L. 2012 *Anglo-Saxon Art.* New York: Cornell University Press

Webster, L. and Brown, M. (eds) 1997 *The Transformation of the Roman World AD 400–900.* London: British Museum Press

Wheeler, H. 1977 Aspects of Mercian art: the Book of Cerne. In Dornier, A. (ed), *Mercian Studies*, 235–44

Whitelock, D. 1958 *The Audience of Beowulf.* Oxford: Clarendon Press

Williams, G. 2021 Mercian Coinage and Authority. In Brown, M.P. and Farr, C.A. (eds) *Mercia: An Anglo-Saxon Kingdom in Europe*: 210–29

Williams, J.H., Shaw, M. and Chapman, A. 2021 Anglo-Saxon Northampton Revisited. In Chapman, A. (ed) *The Archaeology of Medieval Northampton*: 25–77

Williamson, T. 2000 *The Origins of Hertfordshire.* Manchester: Manchester University Press

Wilson, D.M. 1984 *Anglo-Saxon Art.* London: Thames and Hudson

Wood, I. 2008 Monasteries and the geography of power in the age of Bede. *Northern History* XLV.1: 11–25

Woolf, A. 2005 Onuist son of Uurguist: tyrannus carnifex or a David for the Picts? In Hill, D. and Worthington, M. (eds) *Æthelbald and Offa: Two Eighth-century Kings of Mercia*: 35–42

Wormald, P. 1991 In Search of Offa's Law Code. In Wood, I. and Lund, N. (eds) *People and Places in Northern Europe 500–1600: Essays in Honour of Peter Hayes Sawyer*: 25–46

Wright, D.W. 2015 Early Medieval settlement and social power: the Middle Anglo-Saxon 'Home Farm'. *Medieval Archaeology* 59, 24–46

Wright, D.W. and Wilmott, H. 2024 Sacred Landscapes and Deep Time: Mobility, Memory, and Monasticism on Crowland. *Journal of Field Archaeology.* DOI: 10.1080/00934690.2024.2332853

Yorke, B. 2005 Æthelbald, Offa and the Patronage of Nunneries. In Hill, D. and Worthington, M. (eds) *Æthelbald and Offa: Two Eighth-century Kings of Mercia*: 43–8

Yorke, B. 2013 *Kings and Kingdoms of Early Anglo-Saxon England.* Abingdon: Routledge

Zaluckyj, S. 2001 *Mercia: The Anglo-Saxon Kingdom of Central England.* Logaston: Logaston Press

# Endnotes

## Primary sources

The Electronic Sawyer: Anglo-Saxon charters online at https://esawyer.lib.
cam.ac.uk/about/index.htm
Adomnán of Iona: *Vita Colombae*. Sharpe 1995
*Alcuin Letters*. Allott 1974
*Anglo-Saxon Chronicle*. Garmonsway 1972; Swanton 1996
*Annals of St Bertin*. Nelson 1991
*Annals of Ulster*. Corpus of Electronic Texts Edition online at https://celt.
ucc.ie/published/T100001A/index.html
Anonymous: *Vita Ceolfridi*. Boutflower 1912
Anonymous, *The earliest life of Gregory the Great*. Colgrave 1968
*Asser: Life of King Ælfred*. Keynes and Lapidge 1983
Bede: *Historia Abbatum*. Webb and Farmer 1983
Bede: *Historia Ecclesiastica Gentis Anglorum*. McClure and Collins 1994
Bede: *Lives of the abbots of Wearmouth and Jarrow*. Webb and Farmer 1983
Bede: *Prose Life of St Cuthbert* and *Anonymous Life of St Cuthbert*.
Colgrave 1940
*Cartularium Saxonicum*. Birch 1887
*Chronicon Scottorum*. https://celt.ucc.ie/published/T100016
Eddius Stephanus: *Vita Wilfridi*. Colgrave 1927
Einhard: *Life of Charlemagne*. Ganz 2008, 33
*English Historical Documents*. Whitelock 1979
Felix: *Vita Sancti Guðlaci*. Colgrave 1956
*Gregory, History of the Franks*. Thorpe 1974
*Historia Regum*. Stevenson 1855; *English Historical Documents* 3
'Nennius': *Historia Brittonum*. Morris 1980
Prosopography of Anglo-Saxon England online at: https://pase.ac.uk
*Royal Frankish Annals*. Scholtz 1972
*St Boniface Correspondence*. Talbot 1954, 65–152
*St Boniface Letters*. Emerton 2000
Welsh Annals: *Annales Cambriae*. Morris 1980

## Introduction

1. *Annales Cambriae sub anno* 630.
2. Bede, *Historia Ecclesiastica* II.14 Morris 1980, 46

## Chapter 1

1. Bede, *Historia Ecclesiastica* Book II, Chapter 20
2. Bradley 1982, 512
3. Bede, *Historia Ecclesiastica* Book II, Chapter 14
4. Felix, *Vita Sancti Guðlaci* I
5. Felix, *Vita Sancti Guðlaci* X
6. Garmonsway 1972, 24; 25
7. Brooks 1989b, 164
8. Bede, *Historia Ecclesiastica* Book III, Chapter 24
9. Hooke 1985; Bassett 1989
10. Tyler 2007, 93
11. Bede, *Historia Ecclesiastica* Book II, Chapter 2
12. Bede, *Historia Ecclesiastica* Book III, Chapter 21
13. Bede, *Historia Ecclesiastica* Book II, Chapter 5
14. Bede, *Historia Ecclesiastica* Book III, Chapter 18; 'Nennius', *Historia Brittonum* 65
15. Bede, *Historia Ecclesiastica* Book II, Chapter 16
16. Laws of King Ine Cap. 33. Attenborough 1922, 47
17. Bede, *Historia Ecclesiastica* Book II, Chapter 24
18. Meadows 2019. The helmet became known as the Pioneer helmet, after the construction company, Pioneer Aggregates.
19. Harrison and Embleton 1993
20. Bede, *Historia Ecclesiastica* Book IV, Chapter 22
21. Bede, *Historia Ecclesiastica* Book III, Chapter 1
22. Bede, *Historia Ecclesiastica* Book III, Chapter 9; Sims-Williams

1990, 29; www.ancienttexts.org/library/celtic/ctexts/cynddylan.html retrieved 14.04.2023
23. 'Nennius', *Historia Brittonum* 65
24. *Anglo-Saxon Chronicle*
25. Bede, *Historia Ecclesiastica* Book III, Chapter 7
26. Yorke 1990, 62–3
27. Bede, *Historia Ecclesiastica* Book III, Chapter 18
28. Bede, *Historia Ecclesiastica* Book III, Chapter 16
29. Bede, *Historia Ecclesiastica* Book III, Chapter 17
30. Bede, *Historia Ecclesiastica* Book III, Chapter 21. McClure and Collins 1994, 144
31. Bede, *Historia Ecclesiastica* Book III, Chapter 21
32. 'Nennius', *Historia Brittonum* 65
33. Bede, *Historia Ecclesiastica* Book III, Chapter 24
34. Bede, *Historia Ecclesiastica* Book III, Chapter 24
35. *Anglo-Saxon Chronicle sub anno* 653/4.
36. Bede, *Historia Ecclesiastica* Book III, Chapter 24
37. Webster 2012, 121ff

## Chapter 2

1. Bede, *Historia Ecclesiastica* Book III, Chapter 24
2. Bede, *Historia Ecclesiastica* Book III, Chapter 24
3. Mellows 1941, 2
4. Bede, *Historia Ecclesiastica* Book III, Chapter 21
5. Garmonsway 1972, 29ff
6. Stenton 1933; Potts 1974
7. Electronic Sawyer S68; S72
8. Stenton 1933
9. Thomas 1981, 113–21
10. Oosthuizen 2017
11. Bede, *Historia Ecclesiastica* Book III, Chapter 21

12. Watts 2004
13. Chadwick 1963, 335ff
14. Eddius Stephanus, *Vita Wilfridi* 18
15. Adams 2021b
16. Bede, *Historia Ecclesiastica* Book II, Chapter 14
17. Bede, *Historia Ecclesiastica* Book III, Chapter 25
18. Bede, *Historia Ecclesiastica* Book IV, Chapter 3
19. Fairweather 2005
20. Bede, *Historia Ecclesiastica* Book III, Chapter 25
21. Bede, *Historia Ecclesiastica* Book III, Chapter 27
22. *Annals of Ulster* 664–5
23. Anonymous, *Vita Ceolfridi* 14
24. Bede, *Historia Ecclesiastica* Book IV, Chapter 1
25. Bede, *Historia Ecclesiastica* Book III, Chapter 27; *Anglo-Saxon Chronicle*
26. Bede, *Historia Ecclesiastica* Book III, Chapter 29
27. Bede, *Historia Ecclesiastica* Book III, Chapter 28
28. Bede, *Historia Ecclesiastica* Book III, Chapter 30
29. Bede, *Historia Ecclesiastica* Book II, Chapter 3
30. Bede, *Historia Ecclesiastica* Book III, Chapter 7
31. Bede, *Historia Ecclesiastica* Book IV, Chapter 22
32. Hopkins 1980, 104
33. Maddicott 2005, 11; Blair 1989, 98ff; Electronic Sawyer S1165
34. Kirby 2000, 96
35. Maddicott 2005, 28; Electronic Sawyer S1822
36. Bede, *Historia Ecclesiastica* Book III, Chapter 29
37. Bede, *Historia Ecclesiastica* Book III, Chapter 28; Eddius Stephanus, *Vita Wilfridi* 14
38. Cubitt 2013, 342
39. Bede, *Historia Ecclesiastica* Book III, Chapter 14
40. Eddius Stephanus, *Vita Wilfridi* 14
41. Eddius Stephanus, *Vita Wilfridi* 15
42. Bede, *Historia Ecclesiastica* Book IV, Chapter 3
43. Eddius Stephanus, *Vita Wilfridi* 16
44. Adams and O'Brien 2021
45. Bede, *Historia Ecclesiastica* Book IV, Chapter 3
46. Bede, *Historia Ecclesiastica* Book IV, Chapter 3
47. Bede, *Historia Ecclesiastica* Book IV, Chapter 1
48. Bede, *Historia Ecclesiastica* Book IV, Chapter 5
49. Kirby 2000, 96
50. Eddius Stephanus, *Vita Wilfridi* 20
51. Bede, *Historia Ecclesiastica* Book IV, Chapter 12
52. Arnold 1879, 62

## Chapter 3

1. Keynes and Lapidge 1983, 138–9
2. The concept of the 'cultural corelands' was developed by the historical geographer Brian Roberts, to whom I am most grateful for many discussions on the subject. Roberts 2010
3. Losco-Bradley and Kinsley 2002, 3
4. Ekwall 1960, 516
5. Gelling and Cole 2014, 200
6. Losco-Bradley and Kinsley 2002; Buteaux and Chapman 2009
7. Laws of King Ine Cap.40. Attenborough 1922, 40
8. Caps 55; 58; 6.1; 23.3; 39; 43; 25; 20
9. Blair 2018, 196
10. Bede, *Historia Ecclesiastica* Book IV, Chapter 12

11. Electronic Sawyer S1803, S1804, S1805
12. Thacker 2020
13. Barking: Electronic Sawyer S1171; S1246; Tetbury: S71; Gloucester: S70; Fladbury: S76. Faith 1997
14. Attenborough 1922, 24–5
15. *English Historical Documents* 60. Whitelock 1979, 488
16. Burghart 2007
17. Bede, *Historia Ecclesiastica* Book IV, Chapter 21
18. Bede, *Historia Ecclesiastica* Book IV, Chapter 21. Littleborough, the former Roman fort close to a crossing of the Trent on the road between Lincoln and Doncaster, is a likely candidate for the battle site.
19. Bede, *Historia Ecclesiastica* Book IV, Chapter 17; *Anglo-Saxon Chronicle* 675; Eddius Stephanus, *Vita Wilfridi* 33–4
20. Wood 2008, 23; Anonymous, *The earliest life of Gregory the Great*, 18
21. Bede, *Historia Ecclesiastica* Book III, Chapter 11
22. Bede, *Historia Ecclesiastica* Book III, Chapter 11
23. British Museum *Add. MS* 34,633. Finberg 1961, 200
24. Finberg 1961, 217
25. Pretty 1989, 176
26. Sims-Williams 1990, 60–1, from the *Vita Sancti Mildburgae*
27. Pretty 1989, 178
28. Finberg 1961, 204
29. Finberg 1961, 204–5
30. Finberg 1961, 206
31. Adams 2021
32. Bede, *Historia Ecclesiastica* Book IV, Chapter 6
33. See Bailey 1980 for problems with the dating of the Middle Anglian see and its location at Leicester.
34. See Bailey 1980, 7 for discussion of the primary source for many early bishops: the episcopal lists of the early ninth century.
35. Anonymous, *Vita Ceolfridi* 33
36. Lockett undated
37. See also Faith 1997, 26
38. Finberg 1961: 86. Electronic Sawyer S197
39. Hill and Cowie 2001, 112
40. Cowie and Blackmore (eds) 2012, xxiii
41. The preamble to his laws mentions *his* Bishop Earconwald of London. Attenborough 1922, 36
42. Bede, *Historia Ecclesiastica* Book IV, Chapter 22
43. Eddius Stephanus, *Vita Wilfridi* 25
44. Lapidge and Herren 2009, 169–70; Foot 2013, 27
45. Eddius Stephanus, *Vita Wilfridi* 24
46. By inference from his charter to Aldhelm dated 685. Electronic Sawyer S1169. Eddius Stephanus, *Vita Wilfridi* 40; Yorke 2013, 108
47. Bede, *Historia Ecclesiastica* Book V, Chapter 19; Eddius Stephanus, *Vita Wilfridi* 51
48. Bede, *Historia Ecclesiastica* Book IV, Chapter 26
49. Eddius Stephanus, *Vita Wilfridi* 43
50. Potts 1974: 13
51. Eddius Stephanus, *Vita Wilfridi* 45
52. Bailey 1980
53. Electronic Sawyer S53
54. Electronic Sawyer S1803
55. Electronic Sawyer S1804; S1805
56. Dugdale 1693 Vol 1, p.377.

https://wellcomecollection.
org/works/nq4qcmzu/
items?canvas=589&manifest=2
retrieved 26.04.2023

57. Electronic Sawyer S1806; Keynes
1993
58. *Anglo-Saxon Chronicle* E; Bede,
*Historia Ecclesiastica* Book V,
Chapter 24
59. Eddius Stephanus, *Vita Wilfridi* 47
60. Eddius Stephanus, *Vita Wilfridi*
48

## Chapter 4

1. Felix, *Vita Sancti Guðlaci* I
2. Colgrave 1956, 174; Cavill 2015
3. Felix, *Vita Sancti Guðlaci* XVI–
XVII
4. Abels 1988, 30
5. Cavill 2015
6. Electronic Sawyer S68; S72
7. Sims-Williams 1990, 60–1;
Biddle and Biddle 1985, 235
8. Taylor 1987
9. See above: Primary sources
10. Felix, *Vita Sancti Guðlaci* XXIV
11. Oosthuizen 2017
12. Meaney 2005, 78; Everson and
Stocker 2023
13. Felix, *Vita Sancti Guðlaci*
XXVIII
14. Wright and Wilmott 2024. I
am indebted to Duncan Wright
for drawing my attention to his
report.
15. Felix, *Vita Sancti Guðlaci* XXXI
16. Meaney 2005; Adams 2017, 144
17. Lindy Brady (2017) overstates
the case for a career in the Welsh
borderlands.
18. Oosthuizen 2017; Higham 2005,
88
19. Meaney 2005, 77
20. Felix, *Vita Sancti Guðlaci*
XXXVIII–XL; XLII; XLV, for
example

21. Meaney 2005
22. Felix, *Vita Sancti Guðlaci* XXXV
23. Bede, *Historia Ecclesiastica* Book
II, Chapter 12
24. Bede, *Prose Life of St Cuthbert*
24
25. Adomnán of Iona, *Vita
Colombae* Book I, Chapter 1
26. Felix, *Vita Sancti Guðlaci* XLIX
27. *Liber Exoniensis*: Exeter Book,
fols. 87a–88b: a tenth-century or
earlier collection of Anglo-Saxon
poetry. Krapp and Dobbie 1936,
154–6. Translation by Bradley
1995, 342
28. Also from the *Liber Exoniensis*.
Krapp and Dobbie, 1936 134–7;
translation by Crossley-Holland
1999, 52
29. Bede, *Prose Life of St Cuthbert*
17; Colgrave 1956, 180
30. Bede, *Lives of the abbots of
Wearmouth and Jarrow* I
31. Bede, *Historia Ecclesiastica* Book
II, Chapter 2
32. Electronic Sawyer S65
33. Eddius Stephanus, *Vita Wilfridi*
LIV
34. *English Historical Documents*
164; Stenton 1971, 142–3
35. Bede, *Historia Ecclesiastica* Book
V, Chapter 13
36. Bede, *Historia Ecclesiastica* Book
V, Chapter 19
37. Davis 2000, 94
38. Bede, *Historia Ecclesiastica* Book
V, Chapter 19; Yorke 1990, 50;
53–4
39. Bede, *Historia Ecclesiastica* Book
V, Chapter 19; Eddius Stephanus,
*Vita Wilfridi* 55
40. Eddius Stephanus, *Vita Wilfridi*
LXVII
41. *Anglo-Saxon Chronicle* 1070
42. Laws of King Ine Cap. 13.1
Attenborough 1922, 41
43. Halsall 2003

44. *Maxims* I.3 Crossley-Holland 1999, 349
45. Eddius Stephanus, *Vita Wilfridi* LXVII
46. Bede, *Historia Ecclesiastica* Book V, Chapter 19; Watts 2004, 455
47. Kirby 2000, 108 fn. 64 from the *Chronicle of Evesham Abbey*; cf Finberg 1961, 170
48. Finberg 1961, 89; 205
49. *St Boniface Letters* II
50. *St Boniface Letters* LVII
51. *Anglo-Saxon Chronicle* 715
52. *St Boniface Letters* LVII
53. Yorke 2013, 111
54. *Anglo-Saxon Chronicle* 'A' recension, 716
55. Felix, *Vita Sancti Guðlaci* XLVI
56. Felix, *Vita Sancti Guðlaci* XLVII
57. Felix, *Vita Sancti Guðlaci* L
58. Colgrave 1956, 192–3
59. Felix, *Vita Sancti Guðlaci* LI
60. Colgrave 1956, Introduction
61. Electronic Sawyer S82; Keynes 2005, 2
62. Colgrave 1956, 165
63. Felix, *Vita Sancti Guðlaci* LII
64. Colgrave 1956, 15

## Chapter 5

1. Biddle and Kjølbye-Biddle 1985, 251ff
2. Prosopography of Anglo-Saxon England https://pase.ac.uk/jsp/pdb?dosp=VIEW_RECORDS&st=PERSON_NAME&value=3033&level=1&lbl=Ofa Electronic Sawyer S90: A.D. 742 (Clofesho). Æthelbald, king of Mercia, to the Kentish churches; confirmation of privileges.
3. Campbell 1986
4. Campbell 2000, 238
5. For example, the princeps Tondberht of the Gywre. Bede,

*Historia Ecclesiastica* Book IV, Chapter 19; footnote on p. 406
6. Maddicott 2005, 8
7. Electronic Sawyer S53. A.D. 693
8. Electronic Sawyer S146; *English Historical Documents* 78
9. *Anglo-Saxon Chronicle*
10. Attenborough 1922, 37
11. *Anglo-Saxon Chronicle* 725; Kirby 2000, 106
12. *Anglo-Saxon Chronicle* 728; 730, when it records Oswald's death
13. Kirby 2000, 106
14. Maddicott 2005, 12
15. Charles-Edwards 2014, 162
16. Felix, *Vita Sancti Guðlaci* Li
17. O'Brien 2002; O'Brien and Adams 2016
18. Electronic Sawyer S94
19. Bassett 1989, 18–19
20. Electronic Sawyer S89; S1257
21. The Rodings: Bassett 1989, 21–3
22. Jones 1979
23. Blair 2018, 118; 126–7
24. Campbell 2003
25. Electronic Sawyer S146; *English Historical Documents* 78, p. 507–8
26. Hill and Cowie 2001, 106
27. Bede, *Historia Ecclesiastica* Book II, Chapter 3
28. Cowie and Blackmore 2012, 100
29. Cowie and Blackmore 2012, 108
30. Cowie and Blackmore 2012, 116
31. Sawyer 2013, 2–3
32. Blackburn 1995
33. Metcalf 2003
34. Metcalf 1977, 88
35. Metcalf 1977, 97
36. Metcalf 1977, 99; Metcalf 2003, 42–4
37. Hill and Cowie 2001, 84
38. Electronic Sawyer S91
39. *St Boniface Correspondence* Letter 44
40. Wheeler 1977; Brown 2001
41. Metcalf 2003, 47

42. *St Boniface Letters* VII, dated 720
43. *St Boniface Letters* VII, dated 735–6
44. *St Boniface Letters* XXVI
45. Stiff 2001
46. Cowie and Blackmore 2012, 176
47. Cowie and Blackmore 2012, 180
48. Campbell 2003, 242
49. Electronic Sawyer S1412
50. Electronic Sawyer S1804: Campbell 2000, 232–3
51. Electronic Sawyer S102: *English Historical Documents* 64
52. McClure and Collins 1994, 343ff
53. Maddicott 2005, 25
54. Hooke 1985, 125 for the map
55. Electronic Sawyer S1822; Finberg 1961, no. 198; Hooke 1985; Maddicott 2005, 28–9
56. Electronic Sawyer S 102: *English Historical Documents* 64, 489
57. Maddicott 2005, 32
58. S97; Maddicott 2005, 40. See also Faith 1997, 26–7 on Evesham's British origins.
59. Maddicott 2005, 43
60. Electronic Sawyer S98; Maddicott 2005, 45; Blair 2018, 196
61. Kelly 1992, 12
62. Metcalf 2003, 45
63. Naylor and Richards 2010; Maddicott 2005, 46
64. Hooke 1985, 125
65. Cole 2013
66. Blair 1996, 15
67. Electronic Sawyer S1258 (Cookham), S1165 (Chertsey), S1171(Barking)
68. Bede, *Historia Ecclesiastica* Book V, Chapter 13
69. Electronic Sawyer S103a; S103b; S1788
70. Electronic Sawyer S94; S96 and S89 – the so-called Ismere diploma

## Chapter 6

1. Maddicott 1997, 20
2. Bede, *Historia Ecclesiastica* Book III, Chapter 27
3. McKerracher 2018, 30
4. *Anglo-Saxon Chronicle*
5. Bede *Continuatio*. *English Historical Documents* 5
6. *Anglo-Saxon Chronicle*; *Historia Regum*
7. Sims-Williams 1990, 52
8. Blackburn 1995, 90
9. Chapter XVII. Keynes and Lapidge 1983, 132
10. Bede, *Historia Abbatum*, Chapter XVII
11. McClure and Collins 1994, 347
12. McClure and Collins 1994, 351
13. www.reading.ac.uk/news/2022/ Research-News/Anglo-Saxon-monastery-was-important-trade-hub retrieved 13.07.2022; *Current Archaeology* 380, November 2021: 26–31
14. https://research.reading.ac.uk/ middle-thames-archaeology/ wp-content/uploads/ sites/180/2022/03/Cookham-2021-Interim-final-v4.pdf retrieved 13.07.2022
15. Electronic Sawyer S1258; *English Historic Documents* 79
16. Palmer 2003, 53–4
17. Hamerow 2012, 98
18. McKerracher 2018, 76–7
19. Blair 2005, 255
20. Green et al. 1987, 135
21. Carr et al. 1988; Loveluck 2007
22. Blair 2018, 248
23. McKerracher 2018, Chapter 6
24. Metcalf 1977, 91
25. Leahy 2003
26. Blair 2020, 392
27. Webster 2012, 140
28. Hamerow 2012
29. McKerracher 2018, 82

30. Blinkhorn 1999
31. *St Boniface Letters* LXV
32. Electronic Sawyer 90
33. *St Boniface Letters* L
34. *St Boniface Letters* LVII
35. *St Boniface Letters* LXII
36. Keynes 1993, 25
37. Keynes 1993, 6
38. Johnson 1850, 245
39. Johnson 1850, 238
40. Halsall 2003, 75
41. Blair 2018, 183
42. Blair 2018, 184; Electronic Sawyer S92
43. Salisbury 1995
44. Blair 2018, 189
45. Blair 2018, 236ff
46. Adams 2017, 272ff
47. Gelling 1984; Gelling and Cole 2014; Cole 2013
48. Blair 2018, 193
49. Electronic Sawyer S1272; *Cartularium Saxonicum* 455
50. Blair 2018, 204
51. Blair 2018, 206
52. Bassett 2020, 378, allows for the likelihood; see also Sims-Williams 1990, 147ff
53. Bassett 2020, 377
54. Electronic Sawyer S90
55. Electronic Sawyer S92
56. Electronic Sawyer S96
57. Yorke 2013, 112
58. *Historia Regum*
59. Woolf 2005, 38
60. *Historia Regum*
61. *Anglo-Saxon Chronicle* 755 for 757
62. *Historia Regum*
63. *St Boniface Letters* XCII

## Chapter 7

1. Bradley 1982, 336ff; Alexander 1973; *Anglo-Saxon Chronicle* 'A' recension, *sub anno* 755
2. Atherton 2005, 69
3. Chambers 1963, 540ff
4. Electronic Sawyer S117; S146; *English Historical Documents* 78
5. Levison 1946, 257
6. *Bede Continuato. English Historical Documents* 5. Translation online at www.1066.co.nz/Mosaic%20DVD/library/bede/Continuation.htm retrieved 22.05.2023
7. Charles-Edwards 2014, 417
8. *Historia Regum.* Stevenson 1855, 449
9. *Annals of Ulster.* https://celt.ucc.ie/published/T100001A/ sub anno 764, retrieved 06.01.2023
10. *The Wanderer*, line 113. Crossley-Holland 1999, 52; Bede, *Historia Ecclesiastica* Book II, Chapter 13
11. *Historia Regum*: Stevenson 1855, 449
12. Sawyer 2013, 67
13. *Gregory, History of the Franks* Book 10, XIX
14. Williamson 2000, 64
15. Haslam 1987, 80–81
16. Electronic Sawyer S33
17. Electronic Sawyer S29
18. Brooks 1989b
19. Electronic Sawyer S105: see Brooks 1984, 112 for a detailed account
20. Electronic Sawyer S34
21. *Cartularium Saxonicum* 194
22. Brooks 1984, 114–5; *English Historical Documents* 80
23. Electronic Sawyer S110
24. Electronic Sawyer S155; *English Historical Documents* 80, and see below in Chapter 10
25. Naismith 2012, 219
26. *Anglo-Saxon Chronicle*
27. *Historia Regum*
28. Electronic Sawyer 108
29. *Anglo-Saxon Chronicle* recension 'E' 755 for 757

30. Yorke 1990, 113
31. *English Historical Documents* 79; Electronic Sawyer S1258 and see below in Chapter 10
32. *English Historical Documents* 79
33. Naismith 2012, 63
34. Stafford 2001, 40
35. Electronic Sawyer S59, S110, S111, for example
36. Stafford 2001, 38
37. Lines 1945 onwards in the Penguin edition. Alexander 1973; 1633 in the Tolkien (2014)
38. Lines 1957–60 in Chambers 1963, 540
39. Adams 2018, Chapter Four
40. Stafford 2001, 40
41. Levison 1946, 29–31; Stafford 2001, 40
42. Electronic Sawyer S125. See below Chapter 9 for more detail from this charter.
43. See, for example, Brooks, N. 1989a
44. www.heritagegateway.org.uk/ Gateway/Results_Single.aspx?u id=322257&resourceID=19191 retrieved 18.1.2023
45. Blair 2018, 204–5; www. nottshistory.org.uk/articles/ tts/tts1927/margidunum1.htm retrieved 21.01.2023
46. Blair 2018, 204–5
47. Blair 2018, 204
48. Bede, *Historia Ecclesiastica* Book II, Chapter 16; Blair 2018, 206
49. Electronic Sawyer S114
50. Blair 2018, 212
51. Blair 2018, 214–6
52. *Cartularium Saxonicum* 232
53. Blair 2018, 218–9
54. Bassett 2008, 186ff
55. Blair 2018, 262
56. Edwards 1987
57. Ray and Bapty 2016, 319; Ray 2015, 221
58. Bassett 2008, 226ff
59. Blair 2018, 259–60
60. Jackson et al. 1969. I am grateful to Roger Miket for drawing my attention to this site.
61. Blair 2018, 210–11
62. Electronic Sawyer S1184
63. Jones and Page 2006, 65–6, following a suggestion by Glenn Foard.
64. Chapman 2021
65. Williams et al 2021
66. Carr et al. 1988
67. Loveluck 2007
68. Sutherland 2014
69. Sutherland 2014, 31
70. Ray and Bapty 2016, 324 n. 97
71. Blair 2018, 238
72. Blair 2018, 237ff
73. Hardy et al. 2007
74. Hardy et al. 2007, 48ff
75. Hardy et al. 2007, 94
76. Hardy et al. 2017, 133
77. Electronic Sawyer S 1412

## Chapter 8

1. Naismith 2012, 56–7
2. Williams 2001, 216
3. Adams and O'Brien 2021
4. Naismith 2012, 97
5. Chick 2005, 111
6. Electronic Sawyer S155; *English Historical Documents* 80; and see Williams 2001, 218
7. Asser, *Life of King Ælfred* 14
8. *Henry VI* Part Three, Act 5, Scene 2
9. Fox 1955, xxiii
10. Fox 1955, 277
11. Owen and Morgan 2007
12. Ray and Bapty 2016, 79
13. Ray 2015, 219–22
14. Ray and Bapty 2016, 80
15. Fox 1955, 71; Ray and Bapty 2016, 167
16. Ray and Bapty 2016
17. Ray and Bapty 2016, 24, 125ff

18. Ray and Bapty 2016, 162
19. Gelling 1992, 112; Ray and Bapty 2016, 82.
20. Squatriti 2002, 40; Ray and Bapty 2016, 275
21. Gelling 1992, 119 and 121, Figure 49
22. Blair 2018, 207
23. Gelling 1992, 121
24. Ray and Bapty 2016, 294–5
25. Ray and Bapty 2016, 266
26. Noble 1983, 105; original transcription by Felix Liebermann, *Die Gesetze der Angelsachsen*, 3 volumes, Volume 1, 374–8. Halle 1903–16
27. Ray and Bapty 2016, 25ff
28. Ray and Bapty 2016, 19
29. Ray and Bapty 2016, 24ff
30. Ray and Bapty 2016, 20
31. Blair 2018, 229
32. Blair 2018, 118–9
33. Blair 2005, 126–7
34. Bailey 1988, 2
35. See Wilson 1984 for images and a general survey; and Cramp 1977 for a review. For illustrations and examples, Leslie Webster's survey *Anglo-Saxon Art* (2012) is the outstanding modern work.
36. Bailey 1988, 12–13
37. Looijenga and Vennermann 2000
38. Wilson 1984, 84; Lapidge et al. 2014, 204; Webster 2012, 142
39. Webster in Lapidge et al. 2014, 204
40. Webster 2012, 112
41. James 1996, 55–6
42. James 1996, 56
43. Bede, *Historia Ecclesiastica* Book IV, Chapter 3
44. Webster 2012, 26
45. James 1996; and see also a summary of the Lichfield Angel discovery: file:///C:/Users/Admin/Desktop/lichfield-angel.pdf retrieved 20.03.2023
46. Webster 2012, 26
47. James 1996, 51
48. James 1996, 52
49. Adams 2017, 220–3
50. Brooks 1984, 114 and see Chapter 10
51. Electronic Sawyer S105
52. Electronic Sawyer S155; *English Historical Documents* 80 and see above in Chapter 7
53. Story 2003, 61
54. Story 2003, 62
55. *English Historical Documents* 191
56. *English Historical Documents* 205; Levison 1946, 31; Story 2003, 60
57. Capitulary X. Wormald 1991, 30
58. Capitulary XVII. Wormald 1991, 31
59. Capitulary XIX. Wormald 1991, 31
60. Capitulary XIII. Wormald 1991, 31
61. Wormald 1991, 32
62. Story 2003, 83
63. Garmonsway 1972, 53–4. Entry in E recension sat 785
64. Brooks 1984, 119
65. *English Historical Documents* 210

## Chapter 9

1. Wallace-Hadrill 1971, 98
2. *Historia Regum*
3. *Historia regum* 792
4. *Alcuin Letters* 41
5. Wallace-Hadrill 1971, 116; Asser, *Life of King Ælfred* 15
6. *Alcuin Letters* 10, to Adelhard, 790
7. *Alcuin Letters* 50, 797
8. Bede, *Letter to Ecgberht*. McClure and Collins 1994, 350

9. *Historia Regum; Anglo-Saxon Chronicle* 793
10. *Alcuin Letters* 36
11. *Alcuin Letters* 40
12. Coupland 2023
13. *English Historical Documents* 20
14. Wallace-Hadrill 1971, 115
15. Story 2003, 186
16. Einhard, *Life of Charlemagne* XX
17. Loyn and Percival 1975: 39, 132–4
18. Loyn and Percival 1975,: 39, 132–4; Wallace-Hadrill 1971, 114; Story 2003, 198
19. Nelson 2017, 127
20. Brooks 1984, 117
21. *Historia Regum* 791
22. Brooks 1996
23. *Alcuin Letters* 46
24. Sharp 2005, 60
25. Story 2003, 139
26. *English Historical Documents* 196
27. *Anglo-Saxon Chronicle* for 839
28. *Anglo-Saxon Chronicle* for 855
29. Story 2003, 139ff
30. *Alcuin Letters* 40
31. Brooks 1984, 114
32. *Alcuin Letters* 31
33. Naismith 2012, 171
34. *English Historical Documents* 20
35. *Alcuin Letters* 40
36. *Alcuin Letters* 9, 790
37. *Alcuin Letters* 19
38. Electronic Sawyer S125
39. Electronic Sawyer S128
40. Electronic Sawyer S134
41. Maitland 1897. Quotation from the 1960 paperback reprint, p. 204
42. *English Historical Documents* 77; Electronic Sawyer S1257
43. Electronic Sawyer S137
44. Levison 1946, 29–30 for the papal privileges and see above, Chapter 7 and below, Chapter 10; 249 for Winchcombe
45. Electronic Sawyer S916; Martin 2005, 50
46. Bede, *Historia Ecclesiastica* Book I, Chapter 7; Niblett 2010
47. Niblett 2010, 132
48. Wright 2015, 35, 41
49. Domesday Northamptonshire 6a,27. Thorn and Thorn 1979, 222a
50. Blair 2018, 292
51. Hamerow 2012, 153
52. Blair 2018, 292
53. Hey 2004, Chapter 2; Blair 2018, 298–300
54. Hey 2004; Blair 2018, 292; Hamerow 2012, 95ff
55. Blair 2018, 292; Kelly 2008
56. Laws of King Ine Cap. 6; Laws of King Ine Cap. 36. Attenborough 1922, 38–9; 48–9
57. Attenborough 1922, 63
58. Wormald 1991
59. *Royal Frankish Annals* 796
60. *Alcuin Letters* 40
61. Atherton 2005, 65
62. *Alcuin Letters* 41
63. *Alcuin Letters* 42. Allott 1974, 55
64. *Historia Regum*
65. *English Historical Documents* 79
66. Haslam 1986
67. Haslam 1987, 87–8
68. *Anglo-Saxon Chronicle* for 915
69. Maitland 1897
70. Bassett 2007
71. Electronic Sawyer S148; S149; S150; S151
72. Electronic Sawyer S149
73. *Anglo-Saxon Chronicle* 794 for 796
74. *Anglo-Saxon Chronicle* 755
75. *Alcuin Letters* 46, 797; *English Historical Documents* 202

## Chapter 10

1. *Historia Regum.* Stevenson 1855, 459
2. *Alcuin Letters* 7, to Archbishop Eanbald
3. Dumville 1976, 37; Electronic Sawyer S132; S136, S1412; S150; S151
4. Electronic Sawyer S1412
5. Electronic Sawyer S152
6. Dumville 1976
7. *Historia Regum* 798. Stevenson 1855, 460
8. Brooks 1984, 121
9. Brooks 1984, 122
10. *Alcuin Letters* 49 and 50
11. *English Historical Documents* 204: Cœnwulf, king of Mercia, to Pope Leo III, 798
12. Electronic Sawyer S39
13. Burghart 2007, 32–3
14. Brooks 1984, 14; 124
15. Parsons 1988, 377; Keynes 1993
16. Cubitt 1995
17. Offer 2002
18. Bailey 2000
19. Bailey 2000, 128–9
20. Electronic Sawyer S155. EHD 80
21. Brooks 1984, 131
22. *Historia Regum*
23. *Historia Regum* 801
24. Kirby 2000, 132
25. *Historia Regum.* Stevenson 1855, 463
26. Campbell 2000, 89
27. Brooks 1971; Abels 1988
28. Attenborough 1922, 52–3
29. Electronic Sawyer S106; EHD 73
30. Electronic Sawyer S106. https://esawyer.lib.cam.ac.uk/charter/106.html retrieved 11.04.2023
31. John 1960, 80ff
32. Brooks 1971, 70
33. Bannerman 1974; Brooks 1971, 70
34. Brooks 1971, 71
35. O'Brien 2002, for example
36. *Alcuin Letters* 21. Allott 1974, 30
37. *Anglo-Saxon Chronicle* 896 for 895
38. Electronic Sawyer S160; S4134; *English Historical Documents* 82
39. Asser, *Life of King Ælfred* 14
40. Asser, *Life of King Ælfred* 15
41. Electronic Sawyer S154
42. Sims-Williams 1990, 171; Kirby 2000, 149
43. Sawyer 2013, 80–81
44. Coupland 2002
45. *Alcuin Letters* 51
46. Electronic Sawyer S1431a; *English Historical Documents* 210
47. Electronic Sawyer S40
48. Naismith 2012, 124–5
49. Burghart 2007, 51
50. Brooks 1984, 132
51. Brooks 1984, 133
52. Electronic Sawyer S168
53. *Historia Regum* 801
54. *Historia Regum* 796
55. *Royal Frankish Annals* 808
56. *Royal Frankish Annals* 809
57. *Royal Frankish Annals* 810; Welsh Annals: *Annales Cambriae* 810
58. Electronic Sawyer S171; S185; S172; S179; S180 Bishop Deneberht; S141(Æðelnoð); S159 (Eanberht); S173 (Swiðnoð)
59. Electronic Sawyer S165
60. Stafford 2001, 42 n.5
61. Electronic Sawyer S167; Levison 1946, 249
62. Giles 1847, 87
63. Levison 1946, 32
64. Sims-Williams 1990, 167 n. 111. The Prosopography of Anglo-Saxon England names 14 Cynehelms across two centuries.
65. Sims-Williams 1990, 165

66. Bassett 2008, 213ff; Blair 2018, 243–4
67. Sims-Williams 1990, 169
68. Brooks 1984, 138
69. Brooks 1984, 177
70. Brooks 1984, 179; Sims-Williams 1990, 173ff
71. Brooks 1984, 176, quoting from Canon 8 of the Synod
72. Burghart 2007, 53
73. Levison 1946, 255
74. Welsh Annals: *Annales Cambriae* 818
75. Ray and Bapty 2016, p. 183
76. Brooks 1984, 180ff; Electronic Sawyer S1434; S1436
77. Burghart 2007, 55
78. Electronic Sawyer S1436; S1434. Burghart 2007, 273–4; translation from Adams 1905, 324–7
79. Brooks 1984, 102; Adams 1905, 326
80. *Anglo-Saxon Chronicle*

## Chapter 11

1. *Anglo-Saxon Chronicle* recension 'E' s.a. 918; recension 'A' s.a. 922
2. Rahtz and Meeson 1992; Bassett 2007, 65ff
3. Electronic Sawyer S120; S121 – spurious but with a possibly original basis; Rahtz and Meeson 1992, 4
4. Bassett 2007, 71
5. Rahtz and Meeson 1992, 5
6. Haslam 1987, 90
7. *English Historical Documents* 85; Electronic Sawyer S190
8. Electronic Sawyer S208
9. *Annals of St Bertin*
10. Hadley and Richards 2016
11. *Annals of Ulster* 864
12. Asser, *Life of King Ælfred* 9
13. Biddle and Biddle 1992, 40–1
14. Biddle and Biddle 1992, 42–5

15. Sawyer 2013, 82
16. *Anglo-Saxon Chronicle* Swanton 1996
17. Naylor and Standley 2022, 67; 100
18. Nelson 1986, 60
19. Welsh Annals: *Annales Cambriae*
20. Welsh Annals: *Annales Cambriae* 877
21. Charles-Edwards 2014, 490
22. Charles-Edwards 2014, 492
23. *English Historical Documents* 96
24. Garmonsway 1972, 93
25. Wainwright 1945, 386–7
26. Wainwright 1945, 387
27. *English Historical Documents* 34, Whitelock 1979, 416–17
28. *English Historical Documents* 86. Electronic Sawyer S192
29. Electronic Sawyer S204
30. Laws of King Ine Cap 33. Attenborough 1922, 46–7; *English Historical Documents* 91; Electronic Sawyer S315
31. Electronic Sawyer S210
32. *English Historical Documents* 95; Electronic Sawyer S215
33. Electronic Sawyer S189; S200; S213
34. Electronic Sawyer S222
35. For a detailed account see Blair 2005, 298ff
36. Brady 2018
37. Sims-Williams 1990, 169
38. *English Historical Documents* 94; Electronic Sawyer S 1278
39. *English Historical Documents* 92; Electronic Sawyer S 208
40. Blair 2005, 307
41. Sutherland 2014
42. *Historia Regum*
43. Chapman 2021, 44
44. Chapman 2021, 55

# Image credits

Author photos except:

# Index

Illustrations are indicated by numbers in *italics*.

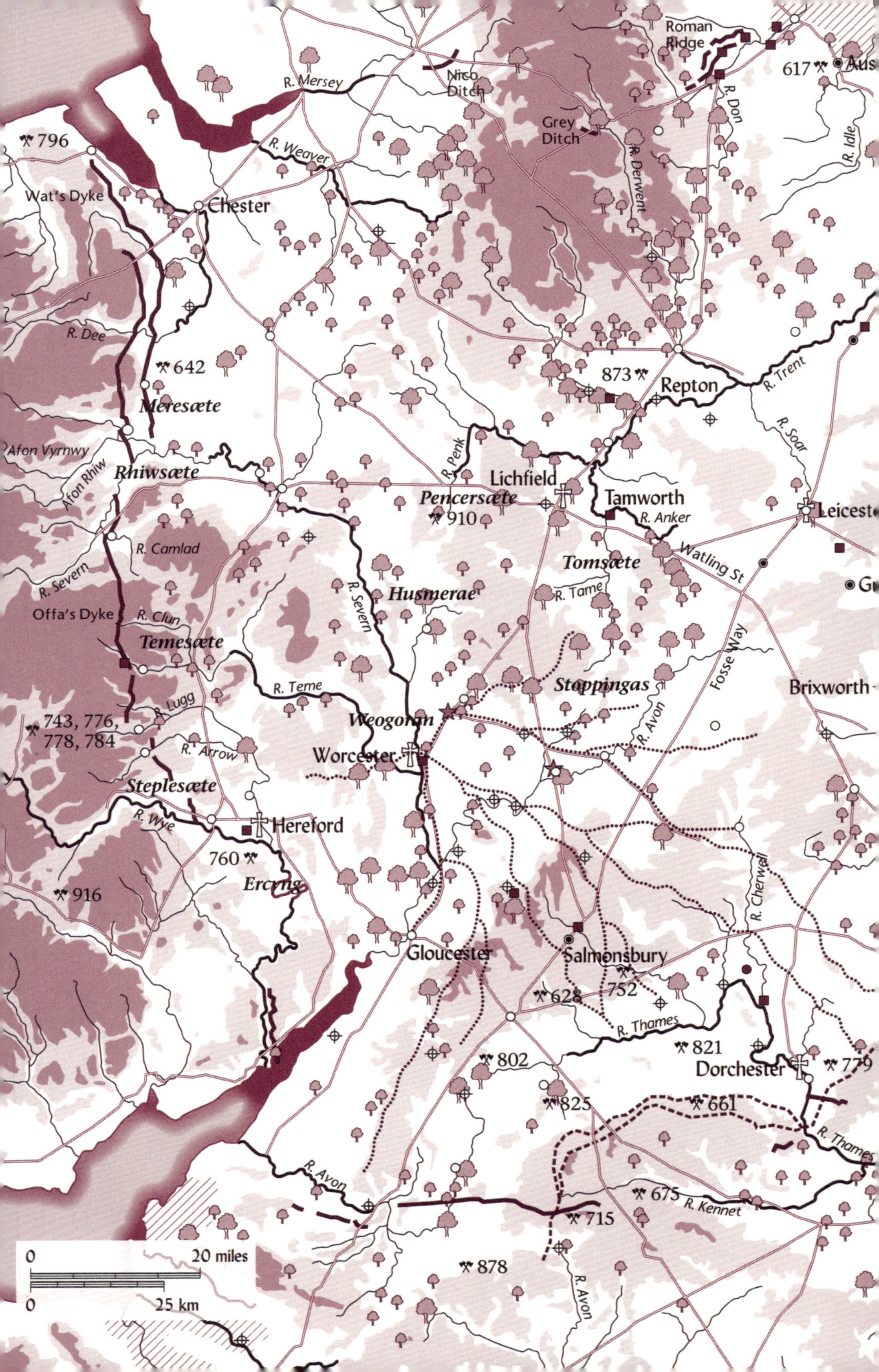

Roman Ridge
R. Mersey
Nico Ditch
617
R. Don
Grey Ditch
796
R. Weaver
R. Derwent
R. Idle
Wat's Dyke
Chester
R. Dee
642
Meresæte
Afon Vyrnwy
Afon Rhiw
Rhiwsæte
R. Camlad
R. Severn
Offa's Dyke
R. Clun
Temesæte
873
Repton
R. Trent
R. Soar
R. Penk
Lichfield
Pencersæte
910
Tamworth
R. Anker
Watling St
Tomsæte
R. Tame
Husmerae
R. Severn
Fosse Way
Stoppingas
Brixworth
R. Teme
R. Lugg
743, 776, 778, 784
R. Arrow
Weogoran
Worcester
R. Avon
Steplesæte
R. Wye
Hereford
760
Ercyng
916
R. Cherwell
Gloucester
Salmonsbury
752
628
R. Thames
821
802
Dorchester
779
825
661
R. Thames
R. Avon
675
R. Kennet
715
878
R. Avon
0
20 miles
0
25 km